Springer Series in Materials Science

Volume 275

The Springer Series in Materials Science covers the complete spectrum of materials physics, including fundamental principles, physical properties, materials theory and design. Recognizing the increasing importance of materials science in future device technologies, the book titles in this series reflect the state-of-the-art in understanding and controlling the structure and properties of all important classes of materials.

More information about this series at http://www.springer.com/series/856

Turab Lookman • Xiaobing Ren
Editors

Frustrated Materials and Ferroic Glasses

Springer

Editors
Turab Lookman
Theoretical Division
Los Alamos National Laboratory
Los Alamos, NM, USA

Xiaobing Ren
Frontier Institute of Science and
Technology and State Key Laboratory for
Mechanical Behaviour of Materials
Xi'an Jiaotong University
Xi'an, China

Center for Functional Materials, National
Institute for Materials Science
Tsukuba, Ibaraki, Japan

ISSN 0933-033X ISSN 2196-2812 (electronic)
Springer Series in Materials Science
ISBN 978-3-030-07270-4 ISBN 978-3-319-96914-5 (eBook)
https://doi.org/10.1007/978-3-319-96914-5

This Springer imprint is published by the registered company Springer Nature Switzerland AG
The registered company address is: Gewerbestrasse 11, 6330 Cham, Switzerland

Preface

Our aim in putting this book together has been to collect in one volume the range of materials systems with differing functionalities that show many of the common characteristics of geometrical frustration, where interacting degrees of freedom do not fit in a lattice or medium, and glassy behavior is accompanied by the additional presence of disorder. Ferroics include a range of materials classes with functionalities such as magnetism, polarization, orbital degrees of freedom, and strain. Frustration, due to geometrical constraints, and disorder due to chemical and/or structural inhomogeneities, can lead to glassy behavior, which has either been directly observed or inferred in a range of materials classes. These include model systems such as artificial spin ice, shape memory alloys and ferroelectrics, and electronically functional materials such as manganites. The glasses include structural glass, spin glass, relaxor ferroelectrics, and recently studied strain glass. Interesting and unusual properties are found to be associated with these glasses and they have potential for novel applications. Just as in the prototypical spin glass and structural glasses, the elements of frustration and disorder lead to non-ergodicity, history dependence, frequency-dependent relaxation behavior, and the presence of inhomogeneous nanoclusters or domains. In addition, there are new states of matter, such as spin ice, but it is still an open question whether these systems belong to the same family or universality class.

With chapters written by experts in their field and spanning experiments, theory, and simulations, the book will be of interest to a wide readership spanning areas of condensed matter physics and materials science. It is expected to be accessible to an interdisciplinary readership that includes graduate students and beginning researchers as well as experts. The chapters are intended to be partly a review with a broad perspective, partly original research, and partly delineating open issues in the field. We have organized the chapters so that we begin with a broad perspective from **David S. Sherrington** in Chap. 1, who uses the developments in spin glass theory to distill common concepts and features, observed or anticipated, among the ferroic glasses. Sherrington utilizes simple models to determine the fundamental origins within a larger picture involving spin glasses, cluster glasses and relaxors, and strain glasses. **Eric Vincent and Vincent Dupuis** in Chap. 2 bring an experimentalist's perspective to the study of spin glasses and provide an excellent introduction and review for those interested in other glasses. They consider spin glasses on laboratory time scales and discuss nonequilibrium aspects, including the aging process and its restart (known as rejuvenation), as well as the ability of a spin glass to keep memory. With its origins in frustration, the ice rule emerges as a unifying topological concept to describe a large variety of material systems. **Cristiano Nisoli** discusses in Chap. 3 how this has evolved into a powerful notion to describe diverse systems such as water ice, spin ice behavior in rare earth titanates, and systems designed to study frustration in a controlled manner, known as artificial spin ice. An example of the latter includes an array of elongated, mutually interacting, single-domain, shape-anisotropic, magnetic nano-islands arranged in different patterns. Advances in nano-fabrication, such as lithography, have allowed such systems to be readily synthesized and their properties, such as the collective magnetic and topological behavior, to be studied. As model systems with novel field-induced transitions, they provide a means to study very unusual forms of glassy behavior.

Chapter 4 by **Michael E. Manley** provides a bridge between the fundamental physics of glasses and frustration discussed in Chaps. 1–3 and the details and mechanisms at work in specific ferroic systems, such as ferroelectrics and shape memory alloys, which are discussed in the later chapters of this book. Manley discusses how the nanoregions in a ferroelectric (polar nanoregions or PNRs) can result from an interplay of nonlinearity and lattice discreteness, in modes called breathers or intrinsic localized modes. PNR can also be formed via disorder by a mechanism known as Anderson localization in which waves scattered by disorder constructively interfere in small regions and destructively interfere elsewhere. Manley describes evidence via neutron diffraction data and shows that it is more consistent with the latter mechanism. Given that in ferroic systems the competition between long-range and short-range order leads to frustration, the importance of this chapter is that it shows how the long-range order becomes localized into PNRs or why this happens specifically at the nanoscale. Broadening the theme somewhat, **Wolfgang Kleemann** in Chap. 5 considers the strain glass, cluster glass, and nanopolar glassy relaxors as based on supercritical chemical disorder, preceded by the analogous tweed-lie regions in these ferroic cases. Kleemann discusses how these glasses are the result of nontrivial interactions between point defects and the surrounding matrix, and that the mechanisms so far proposed may not be adequate to fully understand these glasses. This is in contrast to superspin glass involving an insulating matrix and interacting nonmagnetic nanoparticles. The cluster spin glass emerges from complex spin and charge frustration involving magnetic nanoparticles in dilute magnetic materials and is perhaps least understood.

The focus of Chaps. 6–10 is the study of glassy behavior in alloys, which can be magnetic. In Chap. 6, **Peter Entel et al.** review the properties of magnetic Heusler alloys, which are intermetallics that on rapid quenching acquire frozen in compositional disorder. These alloys show the magnetic shape memory effect and can be magnetocaloric with a jump-like change in magnetization as a function of decreasing temperature. They also make the point that the alloys all have the ingredients of a glass, such as intrinsic disorder and frustration as well as elastic anisotropy. The subject of Chaps. 7–10 is strain glass. **Yuanchao Ji et al.** review in Chap. 7 the origin of strain glass induced by defects, be they point defects or topological in nature, such as dislocations or precipitates. Their focus is on the results obtained from experimental measurements. In Chap. 8, **Xue and Lookman** consider phenomenological Landau models at the continuum, mesoscale level of description and show how a discrete pseudo-spin model in the sharp interface limit may be extracted. They solve the pseudo-spin model using the methods of statistical mechanics including Monte Carlo simulations. Their approach toward disorder is based on adding substitutional dopants with given chemistry and range of interaction. **Pol Lloveras et al**. and **Wang et al.** study the Landau-based models extensively in Chaps. 9 and 10, respectively. Disorder is added via the temperature-dependent quadratic term in the strain or via a stress term. The modeling studies in Chaps. 8–10 essentially calculate frequency dispersion and zero field/field cooling curves as signatures of the glassy behavior.

This book is aimed at an interdisciplinary audience and it aims to be both timely and appealing to those interested in learning about this growing field. We are grateful to all the authors for their articles as well as their support and patience during the editorial process.

Los Alamos, NM
Tsukuba, Japan

Turab Lookman
Xiaobing Ren

Contents

Contributors

M. Acet Faculty of Physics and CENIDE, University of Duisburg-Essen, Duisburg, Germany

R. Arróyave Department of Materials Science & Engineering, A&M University, College Station, TX, USA

J. M. Barandiaran BCMaterials and Department of Electricity and Electronics, University of Basque Country (UPV/EHU), Bilbao, Spain

N. M. Bruno Department of Materials Science & Engineering, A&M University, College Station, TX, USA

V. D. Buchelnikov Condensed Matter Physics Department, Chelyabinsk State University, Chelyabinsk, Russia

A. Çakır Muğla Üniversitesi, Metalurji ve Malzeme Mühendisliği Bölümü, Muğla, Turkey

T. Castán Departament de Física de la Matèria Condensada, Facultat de Física, Universitat de Barcelona, Barcelona, Catalonia, Spain

V. A. Chernenko BCMaterials and Department of Electricity and Electronics, University of Basque Country (UPV/EHU), Bilbao, Spain

Jan Dec Institute of Materials Science, University of Silesia, Katowice, Poland

T. C. Duong Department of Materials Science & Engineering, A&M University, College Station, TX, USA

Vincent Dupuis Sorbonne Universités UPMC Univ Paris 06 UMR 8234, PHENIX, Paris, France

P. Entel Faculty of Physics and CENIDE, University of Duisburg-Essen, Duisburg, Germany

S. Fähler IFW Dresden, Dresden, Germany

T. Gottschall Technical University Darmstadt, Institute of Materials Science, Darmstadt, Germany

M. E. Gruner Faculty of Physics and CENIDE, University of Duisburg-Essen, Duisburg, Germany

O. Gutfleisch Technical University Darmstadt, Institute of Materials Science, Darmstadt, Germany

A. Hucht Faculty of Physics and CENIDE, University of Duisburg-Essen, Duisburg, Germany

Yuanchao Ji Frontier Institute of Science and Technology and State Key Laboratory for Mechanical Behaviour of Materials, Xi'an Jiaotong University, Xi'an, China

I. Karaman Department of Materials Science & Engineering, A&M University, College Station, TX, USA

Wolfgang Kleemann Angewandte Physik, Universität Duisburg-Essen, Duisburg, Germany

P. Lázpita BCMaterials and Department of Electricity and Electronics, University of Basque Country (UPV/EHU), Bilbao, Spain

P. Lloveras Grup de Caracterització de Materials, Departament de Física, EEBE, Universitat Politècnica de Catalunya, Barcelona, Catalonia, Spain

Barcelona Research Center in Multiscale Science and Engineering, Barcelona, Catalonia, Spain

Turab Lookman Theoretical Division, Los Alamos National Laboratory, Los Alamos, NM, USA

S. Mankovsky Department Chemie, Ludwig-Maximilian-University Munich, Munich, Germany

M. E. Manley Material Science and Technology Division, Oak Ridge National Lab, Oak Ridge, TN, USA

Cristiano Nisoli Theoretical Division, Los Alamos National Laboratory, Los Alamos, NM, USA

A. Planes Departament de Física de la Matèria Condensada, Facultat de Física, Universitat de Barcelona, Barcelona, Catalonia, Spain

M. Porta Departament de Física Quántica i Astrofísica, Facultat de Física, Universitat de Barcelona, Barcelona, Catalonia, Spain

Shuai Ren Frontier Institute of Science and Technology and State Key Laboratory for Mechanical Behaviour of Materials, Xi'an Jiaotong University, Xi'an, China

Xiaobing Ren Frontier Institute of Science and Technology and State Key Laboratory for Mechanical Behaviour of Materials, Xi'an Jiaotong University, Xi'an, China

Center for Functional Materials, National Institute for Materials Science, Tsukuba, Ibaraki, Japan

S. Sahoo Institute of Materials Science, University of Connecticut, Storrs, CT, USA

D. Salas Department of Materials Science & Engineering, A&M University, College Station, TX, USA

L. Sandratskii Max-Planck-Institut für Mikrostrukturphysik, Halle, Germany

A. Saxena Theoretical Division, Los Alamos National Laboratory, Los Alamos, NM, USA

David Sherrington Rudolf Peierls Centre for Theoretical Physics, Clarendon Laboratory, Oxford, UK

Santa Fe Institute, Santa Fe, NM, USA

V. V. Sokolovskiy Condensed Matter Physics Department, Chelyabinsk State University, Chelyabinsk, Russia

A. Talapatra Department of Materials Science & Engineering, A&M University, College Station, TX, USA

Eric Vincent SPEC, CEA, CNRS, Université Paris-Saclay, CEA Saclay, Gif-sur-Yvette Cedex, France

Dong Wang Frontier Institute of Science and Technology and State Key Laboratory for Mechanical Behaviour of Materials, Xi'an Jiaotong University, Xi'an, China

Center of Microstructure Science, Frontier Institute of Science and Technology, State Key Laboratory for Mechanical Behavior of Materials, Xi'an Jiaotong University, Xi'an, China

Yu Wang Frontier Institute of Science and Technology and State Key Laboratory for Mechanical Behaviour of Materials, Xi'an Jiaotong University, Xi'an, China

Yunzhi Wang Center of Microstructure Science, Frontier Institute of Science and Technology, State Key Laboratory for Mechanical Behavior of Materials, Xi'an Jiaotong University, Xi'an, China

Department of Materials Science and Engineering, The Ohio State University, Columbus, OH, USA

Dezhen Xue State Key Laboratory for Mechanical Behavior of Materials, Xi'an Jiaotong University, Xi'an, China

Chapter 1
What Can Spin Glass Theory and Analogies Tell Us About Ferroic Glasses?

David Sherrington

Abstract As well as several different kinds of periodically ordered ferroic phases, there are now recognized several different examples of ferroic glassiness, although not always described as such and in material fields of study that have mostly been developed separately. In this chapter an attempt is made to indicate common conceptual origins and features, observed or anticipated. Throughout, this aim is pursued through the use of simple models, in an attempt to determine probable fundamental origins within a larger picture of greater complication, and analogies between systems in different areas, both experimental and theoretical, in the light of significant progress in spin glass understanding.

1.1 Introduction

The existence of macroscopic magnetism has been known since ancient times, with appreciation of its possible spontaneous microscopic origins coming from the mean-field theories of Weiss [1] and Stoner [2], respectively for local-moment and itinerant ferromagnets. The electrical analogue, ferroelectricity, was discovered experimentally in 1920 [3]. The subsequent recognition of antiferromagnetic and ferrimagnetic orderings is due to Néel [4]. In these conventional phases, as well as in many other subsequently discovered ferroic phases, the order is macroscopically periodic, as well as of lower symmetry than the corresponding higher temperature para-phases, which lack long range ferroic order.

The recognition of the existence of different dipolar-glassy behaviour in certain alloys, quasi-frozen locally but without periodic ferroic order, dates back some half a century in both magnetic and electrical scenarios. Initially it was thought 'just' to represent slowing down of dynamics with reduced temperature as experienced in

D. Sherrington (✉)
Rudolf Peierls Centre for Theoretical Physics, Clarendon Laboratory, Oxford, UK

Santa Fe Institute, Santa Fe, NM, USA
e-mail: David.Sherrington@physics.ox.ac.uk

T. Lookman, X. Ren (eds.), *Frustrated Materials and Ferroic Glasses*, Springer Series in Materials Science 275, https://doi.org/10.1007/978-3-319-96914-5_1

conventional glasses, but interest in the magnetic 'glasses' became more focussed with the observation, at the beginning of the 1970s, of sharp but non-divergent low-field magnetic susceptibility peaks as a function of temperature in **Au**Fe alloys [5], suggesting a conceptually new type of phase transition. In combination with evidence of local spin freezing through Mössbauer experiments and of lack of periodicity through neutron diffraction experiments, the new state became known as 'spin glass'. Attempting to understand these observations led to theoretical modelling and novel theoretical, experimental and computational methodologies [6–9] that exposed subtle new concepts and useful applications, not only in many material systems but also in many physically very different complex systems/problems, such as neural networks, hard optimization, protein-folding and also probability theory. At a model level the underlying physical origins of the behaviour are reasonably understood, although some controversies remain, and many material examples are now known; see, e.g., [10–16].

Independently, a potentially related observation was made already in the 1950s and 1960s in ferroelectric alloys [17, 18], in the form of peaks in the a.c. electrical susceptibility of the perovskite alloy $Pb(Mg_{1/3}Nb_{2/3})O_3$ (PMN), with significant frequency dependence, no ferroelectricity and no change of global symmetry, at temperatures much below those of the relatively frequency-independent ferroelectric transition in the related non-disordered compound $PbTiO_3$ (PT). This new behaviour was named 'relaxor'. The discovery of the relaxor behaviour in ferroelectric alloys[1] also sparked much interest and practical application, but its fundamental origin has remained uncertain and contested.

A third type of ferroic glass can be found in martensitic alloys, given the name 'strain glass' [19], but this was a more recent discovery, despite the fact that practical interest in martensites goes back to the nineteenth century.

In this chapter I shall try to relate these different types of ferroic glasses under a common conceptual umbrella, including both well-defined local moments and induced moments, within minimal modelling.

1.2 Experimental Indications

Before giving a theoretical discussion, it is suggestive to note some further similarities in experimental observations of different systems.

In Fig. 1.1, are shown AC susceptibilities (electrical or magnetic, as appropriate), of the original (heterovalent) relaxor PMN, the spin glass $Pt_{1-x}Mn_x$ at $x = 0.025$, and the more recently discovered homovalent relaxor $BaZr_{1-x}Ti_xO_3$ (BZT) at $x = 0.65$. They are clearly very similar, with peaks indicative of transitions or strong crossovers, with strong frequency dependence, slow to respond and glassy,

[1] We use the expression 'ferroelectric alloy' to refer to alloys which exhibit ferroelectricity (or antiferroelectricity) at appropriate concentrations and low enough temperatures.

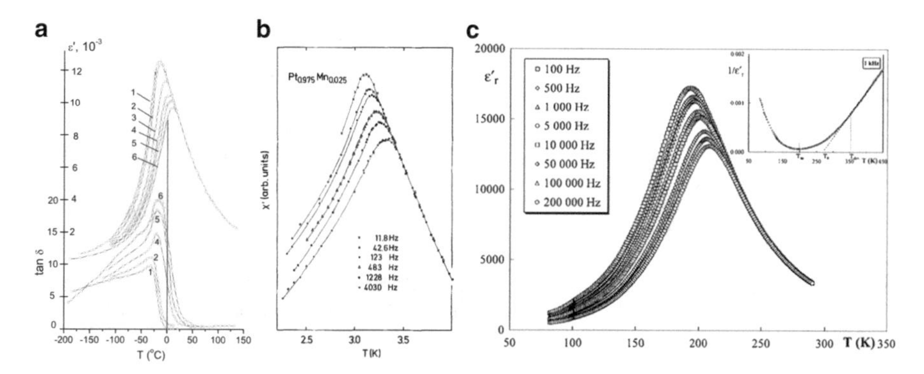

Fig. 1.1 (**a**) AC susceptibilities; heterovalent relaxor $Pb(Mg_{1/3}Nb_{2/3})O_3$ (PMN) [17], (**b**) spin glass **Pt**Mn [20] ©Springer 1983, (**c**) homovalent relaxor $BaZr_{0.35}Ti_{0.65}O_3$ (BZT) [21] ©IOPP (2004)

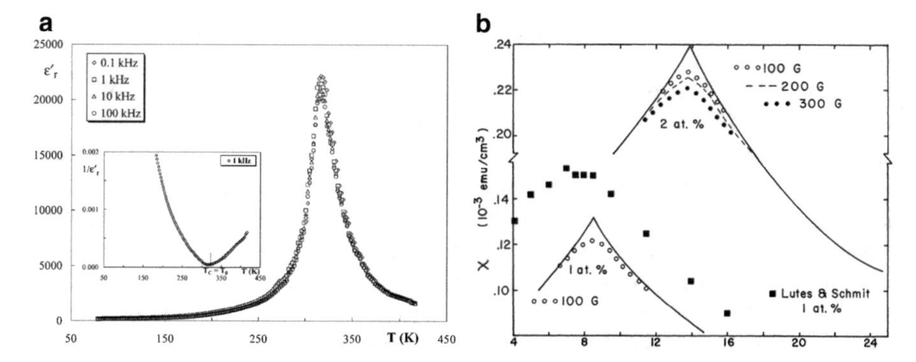

Fig. 1.2 (**a**) a.c. electrical susceptibility of ferroelectric $BaZr_{0.2}Ti_{0.8}O_3$ (BZT) at several frequencies [21] ©IOPP (2004); (**b**) low-field magnetic susceptibility of two **Au**Fe alloys under different applied fields [5] ©APS (1972)

suggesting that similar physics is at play in these experiments. Yet they are rather different in several other aspects of their physical make-ups; both PMN and BZT are ceramic (insulating) substitutional alloys with the basic average perovskite structure ABO_3, where A is an ion of charge 2+, B is an ion of charge 4+ and O has charge 2−, but with random substitution on the B sites; however, in BZT the replacement B ions also have charge 4+, hence the labelling as 'homovalent', while in PMN the replacement B ions have charges 2+ for Mg and 5+ for Nb, in ratio 1:2 to maintain the average charge, hence the description as 'heterovalent'; **Pt**Mn is a face centred cubic metallic alloy with magnetic moments only on the Mn. It is thus natural to look for conceptual common links beyond normal material appearances.

For comparison/contrast, Fig. 1.2a shows the corresponding susceptibilities of BZT at a concentration at which the alloy is ferroelectric, demonstrating no significant frequency dependence and hence no glassy slow response. Figure 1.2b shows the effects of even small applied fields in spin glass **Au**Fe, rounding the

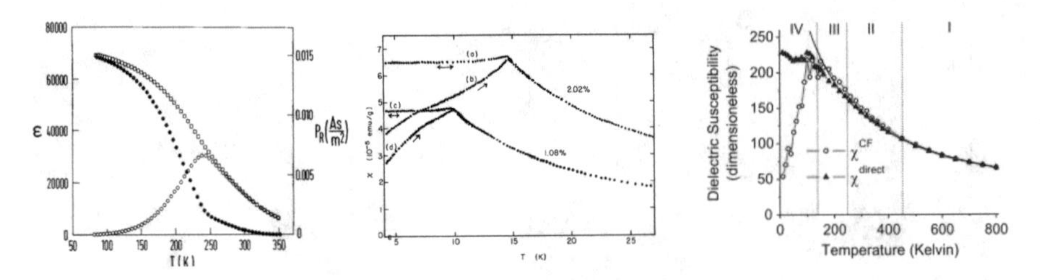

Fig. 1.3 Field-cooled (FC) and zero-field-cooled (ZFC) static susceptibility measurements; PMN [22] ©APS (1998), spin glass **Cu**Mn [23] ©APS (1979), BZT(50:50) simulation [24] ©APS (2012)

transition but also suggesting that it is sharp in the limit of zero applied field. One can also note that although the (normal) susceptibility diverges at a second-order ferromagnetic or ferroelectric transition, it does not diverge at spin glass or relaxor transitions, indicating that the global moment is not a primary order parameter for a spin glass or relaxor.

In Fig. 1.3 are shown for comparison examples of the field-cooled (FC) and zero-field-cooled (ZFC) susceptibilities for the heterovalent relaxor PMN [22] and the spin glass **Cu**Mn [23], along with results of computer simulation of analogous measures for a model of the homovalent relaxor BZT [24]. Again there are clear similarities as the temperature is reduced through that associated with the low-frequency a.c. susceptibility peak, of the continuous separation of the two kinds of susceptibility measure, cooling in the probe field (FC) and cooling without the field and then applying the field to measure (ZFC), respectively understood as probing all thermodynamic states (FC) and probing only accessible states (ZFC), the separation indicating the onset of a hierarchy of barriers.

1.3 Spin Glasses

The canonical spin glasses, such as **Au**Fe and **Cu**Mn, involve non-magnetic hosts, Au and Cu, and a finite concentration of local-moment-bearing substitutions, Fe and Mn. Paramagnetic at high temperatures, they exhibit spin glass behaviour beneath critical temperatures at lower (but finite) concentrations of magnetic ions. A similar behaviour is also found in many other systems, both metals and insulators; see, e.g., Fig. 1.4.

To model the cooperative magnetic behaviour one typically expresses the Hamiltonian as

$$H_{\mathrm{CSG}} = - \sum_{(ij)(Mag)} J(\boldsymbol{R}_{ij}) \boldsymbol{S}_i . \boldsymbol{S}_j \tag{1.1}$$

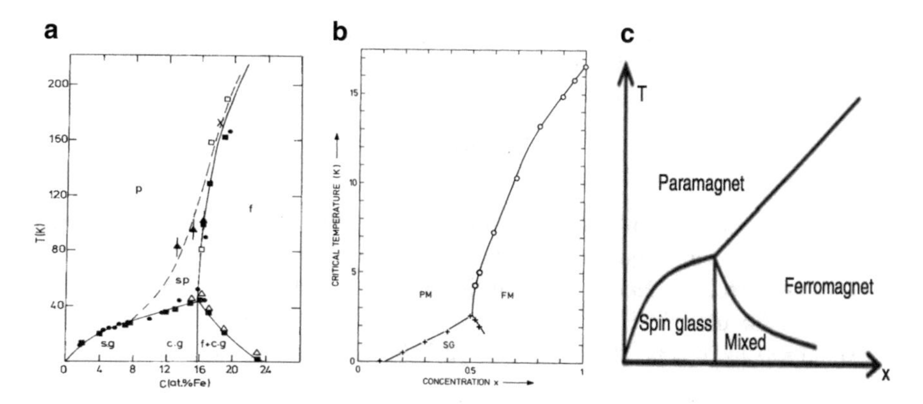

Fig. 1.4 Spin glass phase diagrams; (**a**) metal: AuFe [25] ©Taylor and Francis (1978), (**b**) semiconductor: $Eu_xSr_{1-x}S$, [26] ©APS (1979), (**c**) SK model with mean and variance of exchange distribution both scaling with concentration x. [27] ©Springer (2012)

where the $\mathbf{S}_i$ are localized spins, of fixed length but variable direction, located on the magnetic ions, $J(\boldsymbol{R})$ is a translationally invariant but spatially frustrated 'effective interaction' and the sum is over pairs of sites occupied by magnetic atoms.

For the canonical metallic systems such as **Au**Fe and **Cu**Mn, the exchange interaction between the magnetic ions is carried by the conduction electrons via the s-d coupling, resulting in the RKKY form

$$J(\boldsymbol{R}_{ij}) = \mathscr{J}^2 \chi_{ij} \tag{1.2}$$

where $\mathscr{J}$ is the coupling strength between the conduction electron spin ($\boldsymbol{s}_i$) and the local moment spin ($\boldsymbol{S}_i$) and χ_{ij} is the conduction band susceptibility between sites i and j. χ_{ij} oscillates in sign with separation R_{ij}, with wavevector $2k_F$ where k_F is the Fermi wavevector, and (in 3 dimensions) also decays in magnitude as R^{-3}. The oscillation in sign results in a competition in ordering tendencies of the spins, now known as 'frustration' [28], while the randomness of occupation of lattice sites by magnetic ions provides quenched disorder and inhomogeneity of local environments.

The RKKY interacting metal systems are, however, just one experimental example of the combination of frustration and disorder leading to spin glass behaviour. In the second example of Fig. 1.4 the material is semiconducting, the spins are on the Eu and their interaction arises from shorter-range superexchange, with frustration due to competition between nearest-neighbour and antiferromagnetic next-neighbour interactions.

It is now well-established that the combination of frustration and quenched disorder are the key ingredients for spin glass behaviour. This realization by Edwards and Anderson (EA) [6] led them to suggest in 1975 an alternative model for potentially easier but conceptually equivalent theoretical study, along with further new conceptualization and methods of analysis that ignited theoretical excitement.

In their model every site is occupied by a magnetic spin but their interactions are chosen randomly and quenched:

$$H_{\mathrm{EA}} = -\sum_{(ij)} J_{ij} \boldsymbol{S}_i . \boldsymbol{S}_j, \tag{1.3}$$

where the J_{ij} are chosen randomly from a symmetric distribution of mean zero, ensuring that no conventional periodic order is possible.

Through novel and innovative analysis EA demonstrated the existence of a new phase with random spin-freezing. They noted that a relevant order parameter to test for spin freezing, independent of overall periodic order, is

$$q_{\mathrm{EA}} = \lim_{\tau \to \infty} \overline{S_i(t) S_i(t+\tau)}, \tag{1.4}$$

where the overbar refers to an average over sites i and times t, or, equivalently

$$q_{\mathrm{EA}} = \overline{\langle S_i \rangle^2}, \tag{1.5}$$

where the $\langle . \rangle$ brackets refer to a thermodynamic average and the overbar to a site/disorder average. Thus, 'amorphous' spin freezing without ferromagnetism is signalled by non-zero q_{EA} but zero overall magnetization m, as given by

$$m = \overline{\langle S_i \rangle_i}. \tag{1.6}$$

The EA model has become an important paradigm in further theoretical study. It is normally considered as having only nearest-neighbour interactions on a simple cubic (or hypercubic) lattice. Computer simulations have demonstrated that it captures key features of real systems. An extension to allow for competition of the spin glass phase with ferromagnetism by allowing a finite mean J_0 to the interaction distribution, of standard deviation J, by Sherrington and Southern (SS) [29], showed that when J_0/J is large enough the low temperature state is a ferromagnet, while for smaller J_0/J, beneath a critical value, the low temperature state is spin glass.

The EA model with finite interaction range is not exactly soluble. However, an extension in which the distribution from which the interactions are drawn is the same for all pairs of sites, independently of their separation, the Sherrington-Kirkpatrick (SK) model [7], is soluble, although its solution is very subtle, requiring a description beyond that of a single simple order parameter [8], and has exposed several unexpected but interesting features and concepts [9]. Its solution clearly demonstrates the existence of phase transition to a glassy phase, even in an applied field, and also that its spin glass phase has a complex structure with a hierarchy of metastable states and chaotic evolution under change of global controls (such as temperature). It has stimulated much further study in many other range-free random problem scenarios. However, there remains controversy about whether all the conceptual results of the SK model studies apply to finite-ranged systems, especially those related to so-called replica-symmetry-breaking [8] and to whether a phase transition still persists in an applied field.

1.3.1 Simulations

Computer simulations of model systems have played an important role in determining whether true phase transitions exist also in systems with range-dependent interactions, using their ability to measure directly observables which are not readily accessible to conventional experimentation, such as the spin glass order parameter q_{EA} and a related spin glass susceptibility, as well as the more conventional measures such as the ferromagnetic order parameter m.

The existence of true phase transitions can be tested through sophisticated simulation studies, especially through the use of finite-size scaling and Binder plots [30]. These studies have provided clear demonstrations of spin glass phase transitions in several interesting situations, e.g. as illustrated in Fig. 1.5 for three examples; (1) spin-glass correlations in the SK model with zero mean exchange [31], (2) a nearest-neighbour Ising EA model in dimensions 3 (again with zero mean exchange) [32] and (3) a longer-range dipolar model emulating $LiHo_xY_{(1-x)}F_4$ at $x = 0.001$, a concentration at which the system is a spin glass [33].[2] Corresponding plots for ordinary magnetic correlations in these systems do not show crossings, indicating the absence of a ferromagnetic transition. The combination of these two results, crossing of the size-normalized spin glass correlation lengths together with the lack of crossing of the normal magnetic correlation lengths, lead to the deduction of a true spin glass phase transition at the crossing temperature.

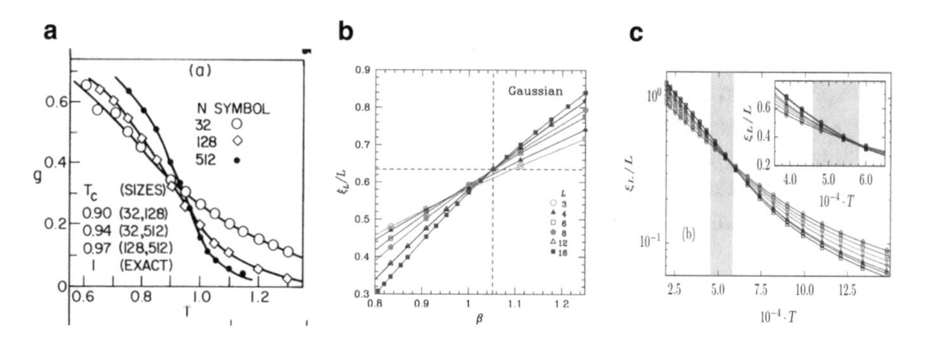

Fig. 1.5 Spin glass correlation plots for different sample sizes, with crossovers at phase transition temperatures, for (**a**) SK model [31] ©APS (1984), (**b**) three-dimensional Ising EA [32] ©APS 2006 and (**c**) $LiHo_xY_{(1-x)}F_4$;$x = 0.001$ [33]

---

[2]The phase transitions are demonstrated by the crossing of appropriate correlation measures for systems of different sizes, with scaling plots providing further confirmations and exponents.

1.3.2 Soft Spins

In analytic studies of spin glasses the Hamiltonian is often re-expressed using continuously valued 'spin fields' $\{\boldsymbol{\phi}\}$ in place of the fixed length spins $\{\boldsymbol{S}\}$. Within a simplification to one-dimensional (Ising) spins $H_{\rm EA}$ becomes

$$H_{\rm EAcont} = \sum_i (r\phi_i{}^2 + u\phi_i{}^4) - \sum_{(ij)} J_{ij}\phi_i\phi_j, \tag{1.7}$$

while the analogue for site-disorder is

$$H_{\rm CSGcont} = \sum_i (r_i\phi_i^2 + u_i\phi_i^4) - \sum_{(ij)} J(\boldsymbol{R}_{ij})\phi_i\phi_j. \tag{1.8}$$

The full hard-spin Ising case ($S = \pm 1$) results from taking the limits

$$r \to -\infty, u \to \infty, r/2u \to -1. \tag{1.9}$$

The sums are taken over only magnetic sites or, equivalently, the non-magnetic sites can be emulated by taking $r_i \to \infty$ on those sites. Note, however, that if some r_i are positive but finite then for those sites to displace there needs to be a sufficient binding energy from the interaction term to overcome the local quadratic penalty for displacements, otherwise the ground state would have $\phi = 0$. Such bootstrapping is referred to as 'induced moment'. Note that the resultant order will depend on the character of the interactions and can be either globally periodic, including ferromagnetic, or spin glass in a system with sufficient disorder and frustration.

Early experimental indications of induced moment spin glass behaviour were found in the alloys **Y**Tb and **Sc**Tb [34], in which crystal field effects lead to a singlet ground state for isolated Tb ions. For Tb concentrations less than a small but finite percentage the ground state is non-magnetic, in contrast to the corresponding alloys with non-singlet ground state Gd in place of Tb, in which the spin glass state continues to the lowest finite concentrations.

A simple extension of the EA model exhibiting induced moment spin glass behaviour was introduced in 1977 by Ghatak and Sherrington (GS) [35];

$$H_{\rm GS} = -\sum_i DS_i^2 - \sum_{(ij)} J_{ij}\boldsymbol{S}_i.\boldsymbol{S}_j \tag{1.10}$$

with the S taking values $S = 0, \pm 1$ and the $\{J\}$ again drawn randomly from a distribution of mean zero. For D less than a critical (negative) value D_c there is only a paramagnetic phase, while above there is an induced-moment spin glass phase.

1.4 Polar Glasses and Relaxors

Ferroelectric systems are often categorized as being of polar/'order-disorder' type or 'displacive' type.

In the former one envisages local electric dipolar moments well formed (but not cooperatively ordered) already in the paraelectric phase above the macroscopic ordering transition to ferroelectricity (or, if energetically preferable, to another periodic phase), in close analogy with local moment magnetism. Correspondingly, alloys with sufficient dilution of local moment units by neutral ones, together with frustrated interactions, can lead to close analogies of conventional local moment spin glasses [36, 37]. Extensions of spin glass modelling and analysis have also been developed for systems characterized by the interaction of higher-order local moments [38–40].

By contrast, in displacive ferroelectrics there are no long-lived electric moments above the transition to ferroelectricity and the charged ions fluctuate around a mean lattice structure with no overall electric moment. Rather, in such ferroelectrics, as the temperature is lowered beneath the transition temperature the time-averaged positions of charged ions displace collectively in such a manner as to yield overall ferroelectricity. The transition to ferroelectric is typically accompanied by a change in global symmetry but the ferroelectricity itself is caused by a relative distorsion of positively and negatively charged ions within the unit cells, yielding electric moments. Unless pre-empted by a first order transition, the susceptibility diverges at the transition. However, not all candidate systems with the same ionic charges and higher temperature structures do exhibit cooperative ordering; for example $BaTiO_3$ (BT) is a displacive ferroelectric while $BaZrO_3$ (BZ) is not. An energetic advantage of distorsion is needed.

Displacive ferroelectrics can be modelled by considering the displacements of the ions as variables governed by Hamiltonians including local costs, the (non-local) effects of interactions between displacements at different sites and the effects of charges on different sites, with coefficients calculable by first-principles methods, followed by computer simulations at finite temperatures.

A detailed first-principles theoretical/computational study of $BaTiO_3$ was given in [41] and demonstrated the ferroelectric transition; see also [42]. However, the conceptual principles can be seen already from a simplified model allowing only for one-dimensional displacements of the most polarizable ions:

$$H_R = \sum_i \{\kappa u_i^2 + \lambda u_i^4\} + \sum_{ij} J_{ij} u_i u_j \tag{1.11}$$

where the u_i are the displacements of the ions at sites $\{i\}$, the first (single-site) term describes the local energy costs of displacements and the last term represents the interaction energy. Clearly, this has a similar form to Eq. (1.7) and can yield an induced moment (displaced $u \neq 0$) ground state if the energy minimizing gain from the interaction term can overcome the local cost from the κ term, with a corresponding transition at a higher temperature. For κ close to zero one expects features of both displacive and order-disorder behaviour, reducing κ making it more order-disorder-like.

Here, however, the main interest is in alloys. In particular, we shall concentrate on alloys of underlying perovskite structure ABO_3 with substitutional disorder on the B sites. This disorder can be either homovalent, for which the ions on the B-sites all have the same 4+ charge as the template, or heterovalent, for which the B-ions have different charges but with the average charge of 4+.

1.4.1 Homovalent Relaxors

The homovalent alloy $Ba(Zr_{1-x}Ti_x)O_3$ (BZT) exhibits ferroelectricity at higher $x > x_{c1}$, only paraelectricity for $x < x_{c2}$, with relaxor behaviour in between [43]. The present author has argued that the relaxor state of BZT is essentially an induced moment spin glass [44]. The susceptibility measured in BZT in a relaxor region of the concentration x is shown in Fig. 1.1.

Here we shall use only a simplified model to illustrate the probable origin of the relaxor behaviour observed in BZT at intermediate concentrations [45]. We note that at para- to ferro-electric transitions of ABO_3, while the overall lattice structure stretches from cubic to tetrahedral, the B-site ions displace from the symmetric lattice positions yielding the ferroelectricity; see Fig. 1.6. Also, it is observed that in the relaxor state the overall average lattice structure remains cubic. Hence, while all the ion locations are, in principle, variable, we shall initially ignore any A and O site displacements, coupling to global strain and change in global lattice structure and concentrate on the deviations of the B-site ions from their locations on the pure perovskite ABO_3 lattice, using

$$H_R = \sum_i \{\kappa_i |\boldsymbol{u}_i|^2 + \lambda_i |\boldsymbol{u}|^4 + \gamma_i (u_{ix}^2 u_{iy}^2 + u_{iy}^2 u_{iz}^2 + u_{iz}^2 u_{ix}^2)\} + \sum_{(ij)} \sum_{\alpha\beta} J_{ij}^{\alpha\beta} u_{i\alpha} u_{j\beta} \tag{1.12}$$

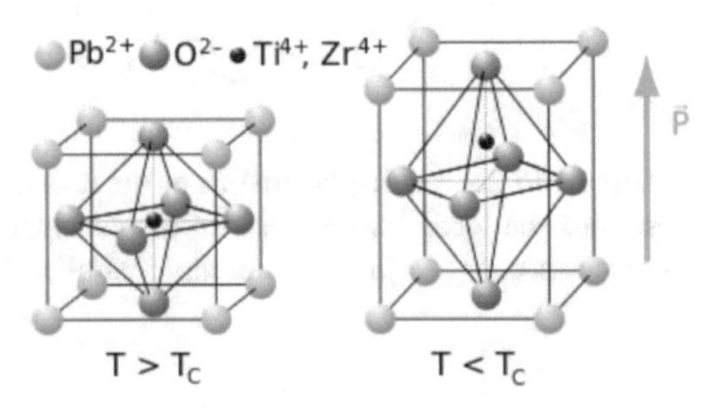

Fig. 1.6 Unit cell structure of $PbTiO_3$ above and below the ferroelectric transition temperature. $BaTiO_3$ is similar, but with smaller tetrahedral stretching

where the $\{\boldsymbol{u}_i\}$ are the displacements of the ions at B-sites $\{i\}$, the first (single-site) term describes the local energy costs of displacements, with the κ, λ and γ coefficients depending upon the types of atoms at those sites, and the last term represents the interaction energy, involving a short-range contribution due to (quantum mechanical) electronic interference of neighbouring pairs of B ions, long-range Coulomb interactions and effective interactions via the ions on A and O sites. This is immediately recognizable as a vector analogue of Eq. (1.7), also allowing for anisotropy.

For κ positive the ground state will have $u = 0$ if the interaction strength is insufficient to overcome it. This appears to be the case for Zr in $BaZrO_3$, which is everywhere paraelectric, while for Ti κ is smaller and $BaTiO_3$ is ferroelectric. Empirically, for a pure system at finite temperature one can consider Eq. (1.12) to represent instead an effective Landau free energy with temperature-dependent coefficients,

$$\kappa = \kappa_c + a(T - T_c) + \mathcal{O}(T - T_c)^2 \tag{1.13}$$

where $\kappa_c > 0$ is the critical value at which the energy cost from the local harmonic term equals the maximum energy gain from the interaction and T_c is the transition temperature.

For alloys such as BZT one can model as in Eq. (1.12) but with different κ, λ and γ on Zr and Ti sites. Given that κ^{Zr} is too great to allow order in BZ, the situation is analogous to that in Eq. (1.8), albeit without the extremes of coefficients and with r rather positive on Zr sites, reminiscent of the non-magnetic sites in conventional spin glasses but allowing for some paraelectric induction.

Dipolar interactions are frustrated, as well as long-ranged, well-known to lead to several different magnetic phases in different structures and in combination with different extra shorter-range interactions [46]; e.g. a simple cubic Ising dipolar system has an antiferromagnetic ground state, while the tetragonal $LiHoF_4$ is ferromagnetic at low temperature. It is also known from experiment and from computational studies of hard-spin dipolar models that site-dilution of dipolar sites can lead to spin glass phases in such systems; see the first two sub-figures of Fig. 1.7 [47, 48].

Hence it seems reasonable to anticipate a corresponding soft pseudo-spin glass phase in homovalently diluted frustrated ferroelectrics in appropriate parameter regions and for the observed relaxor state in BZT to be a manifestation of such a phase, the pseudo-spins being the local dipoles induced by displacement of the charged B-ions. Computer simulations of $Ba(Zr_{0.5}Ti_{0.5})O_3$ (50:50 BZT) [24] demonstrate such behaviour; see Fig. 1.3.

Note, however, that for any ordered phase the binding energy from the interaction term must be sufficient to overcome the cost of any positive κ. Hence the phase diagram for soft-spin versions of the models of [47, 48] would be expected to correspond to lowering the phase transition lines shown in the first two sub-figures

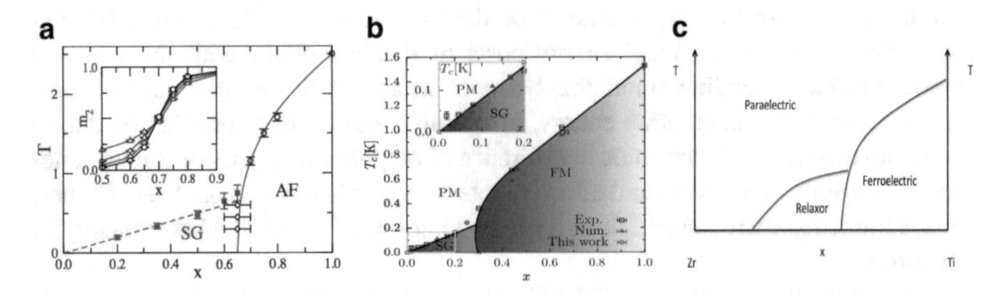

Fig. 1.7 Phase diagrams: Computer simulations of site-diluted dipolar Ising models (**a**) on a simple cubic lattice [47] ©APS (2010), (**b**) on a tetragonal lattice with also short-range antiferromagnetic interaction, with parameters based on $\mathrm{LiHo}_x\mathrm{Y}_{(1-x)}F_4$ [48], and (**c**) schematic speculation for BZT

of Fig. 1.7 by an amount of order κ^{Ti}, thereby yielding a phase diagram as indicated schematically in the third sub-figure,[3] in qualitative accord with observations [21, 43, 45].

In the model considerations above we have ignored any possible change in the basic cubic lattice structure. This is in accord with observations for relaxors. However, in para- to ferro-electric transitions there are normally observed changes in the global average lattice structure, for example in $BaTiO_3$ and $PbTiO_3$ to a tetragonal structure. This is a consequence of inclusion of global strain coupling which we have not included explicitly; see [41]. It will affect the relative energetic preferences for the ferroelectricity and relaxor and hence transition compositions at phase boundaries separating them, but the present author believes it does not affect the conceptual principles given above for the existence of pseudo spin glasses and experiment shows no change in global symmetry at the relaxor transition.[4] We have also not considered the other constituent elements explicitly, only assumed inclusion of their contributions implicitly via effective interactions between B-ions. Such extra effective interaction contributions will depend upon the ions on the A sites and is presumably at least part of the reason why $BaZr_{1-x}Ti_xO_3$ (BZT) has a relaxor phase but $PbZr_{1-x}Ti_xO_3$ (PZT) appears not to have one; while the direct B-B interactions should be similar in both alloys, the indirect interaction via the A ions will be different, with that for A=Pb more strongly ferroelectric than that for A=Ba. In fact. both experiment [49, 50] and theoretical calculations [42] show that while the Ti displacements in $BaTiO_3$ are much greater than those of the Ba ions, in $PbTiO_3$ the situation is almost inverted, the Pb displacements being greater than those of the Ti ions. Hence, in the Pb-based systems ideally one should include the

[3] Conceptually, at the simplest level, the Zr ions are analogues of the non-magnetic atoms in conventional local moment spin glasses (e.g. Cu or Au in **Cu**Mn and **Au**Fe), although in fact they should be paraelectrically displaced a small amount by the electric fields associated with the displaced Ti ions.

[4] The absence of a global strain in the relaxor state can be attributed to the lack of an overall global moment.

Pb (and O) displacements explicitly in the Hamiltonian. However, the combination of frustration and disorder should continue to allow for the possibility of a spin-glass-like relaxor state, albeit that it may not be a preferred one in PZT.

1.4.2 Heterovalent Relaxors

The original classic relaxor PMN is heterovalent, the B-site 4+ ions of the ABO_3 template being replaced by Mg 2+ ions and Nb 5+ ions in the ratio 1:2. Below we attempt to move conceptually towards a possible understanding in the light of the observations above, albeit in a discussion that is at some variance with convention.

Let us first consider in terms of the basic Hamiltonian of Eq. (1.12) but now with account needing to be taken of the fact that the B-ions are of different charges and hence that J_{ij} depends on the particular ions at i and j and not simply on their separation. Allowing also for different types of A ions we shall refer to this Hamiltonian as H^1_{AMN}. Let us also introduce a corresponding Hamiltonian $H^1_{AM^*N^*}$ for a fictitious material AM*N* in which the Mg++ and Nb+++++ of AMN are replaced by fictitious ions Mg*++++ and Nb*++++ which have the same properties as Mg++ and Nb+++++ except for their charges, which are ++++ as in the standard ABO_3 template. We next note that Mg++ has an ionic radius similar to that of Zr++++ and hence can be expected to have a similar largish κ, while Nb+++++ and Ti++++ also have similar but smaller ionic radii, suggesting similar κ and likelihood to displace. We shall assume that the B-ion replacement is random. Consequently, one might initially expect that AM*N* would have a similar phase structure to AZT at the same relative concentrations of 1:2. This would suggest that BM*N* would be a relaxor, or close to a boundary between ferroelectric and relaxor, while PM*N* would be a ferroelectric.

Hence the observation that PMN appears to show the same sort of relaxor behaviour as BZT indicates that the difference between H^1_{PMN} and $H^1_{PM^*N^*}$ is important in stabilizing the relaxor phase in PMN. This difference is given by

$$H^1_{PMN} = H^1_{PM^*N^*} + V_{Coulomb}(Z_i, Z_j, R_{ij}) - V_{Coulomb}(Z^0_i, Z^0_j, R_{ij}) \quad (1.14)$$

where $V_{Coulomb}(\tilde{Z}_i, \tilde{Z}_j, \tilde{R}_{ij})$ is the Coulomb energy associated with charges $\tilde{Z}_i$ and $\tilde{Z}_j$ separated by a distance $\tilde{R}_{ij}$, the $\{Z_i\}$ are the actual charges at sites $\{i\}$ while Z^0_i is the charge at site i accounted for in PM*N* (i.e. for B-sites, $Z^0 = 4+$, for A-sites $Z^0 = 2+$ and for O-sites $Z^0 = 2-$), and

$$R_{ij} = |\mathbf{R}^0_i + \mathbf{u}_i - \mathbf{R}^0_j - \mathbf{u}_j|. \quad (1.15)$$

Expanding, the perturbation component compared with PM*N* includes terms both linear and bilinear in the displacements [51]. The coefficients of the linear terms can be viewed as effective fields and the bilinear terms as effective extra 'exchange' interactions. The effective fields at any site i depend upon the types of ions on all sites $j \neq i$. Given that the B-site interactions are (quasi-)random, so are the effective fields.

Let us concentrate now on the possible effects of including the random fields, which have been considered as driving forces for relaxor behaviour in PMN, particularly since the work of [52]; for more recent discussion, see [53, 54].

Microscopically random magnetic fields are difficult to produce so there is little experiment to compare directly with in magnetic systems; rather, diluted antiferromagnets have been studied in uniform fields, emulating ferromagnets in random uniaxial $\pm h$ fields; in the context of relaxor analogies, see [55].

The problem of the statistical physics of a system controlled by the Hamiltonian of Eq. (1.14) is not soluble exactly and raises many questions. One relates to whether a system with a spin-glass transition in the absence of applied fields should continue to exhibit a sharp transition in the presence of such field(s). It is accepted that the range-free SK model (with spins of any dimension) has an ergodic-non-ergodic spin glass transition even in the presence of uniform or randomly chosen local fields [56–59]. On the other hand, there remains controversy about the effects of fields in short-range spin glasses, with many authors arguing that they destroy sharp spin glass transitions, on the basis of both theoretical arguments and computer simulations, but still without a clear accepted answer [60–63]. Most computer simulations have been performed on Ising EA-like model systems with interactions drawn randomly from symmetric (zero-mean) distributions, whereas in the relaxor systems there are biases in the overall effective interactions, as demonstrated by the existence of ferroelectric phases in appropriate concentration regimes. Most of the simulated models have also had short-range nearest-neighbour interactions or are on one-dimensional structures employed to emulate short-range systems in different dimensions.

It is generally accepted that random fields have a detrimental effect on tendencies for ferromagnetism and that for sufficient strength they suppress ferromagnetism. Thus, the effective random fields in PMN can be expected to act to reduce the ferroelectric tendency anticipated above in PM*N*. An approximate Ising analogue of interactions in PMN has, in fact, been studied in computer simulations of Ising spins based on magnetic (Ho) sites of the diluted alloy $\mathrm{LiHo}_x\mathrm{Y}_{(1-x)}F_4$ [64] and of an EA model with non-zero mean exchange [65], each in the presence of random fields; see Fig. 1.8. These simulations also indicate what their authors call a ‘quasi-spin-glass’ in not-too-large random fields, including the existence of parameter regions where the quasi-spin-glass is preferred to the ferromagnet in sufficient finite random fields, even though at lower random fields the opposite is the case and for higher fields the system is paramagnetic. It is tempting to wonder whether PMN might lie in such a region, hence relaxor. However, more study is needed, particularly of the transition/crossover from paramagnet to (quasi-)spin glass; currently, there is no computational study indicating a sharp transition from paraelectric to relaxor, as suggested by extrapolation to zero frequency of the a.c. susceptibility observed experimentally in PMN.

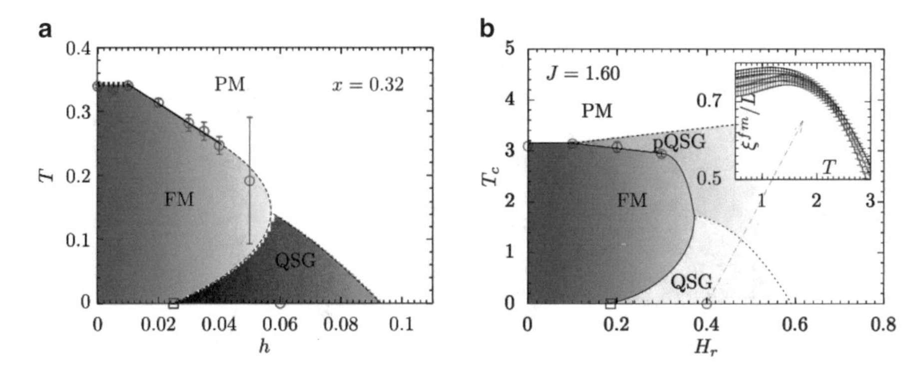

Fig. 1.8 Phase diagrams: Computer simulations of (**a**) diluted Ising model based on $LiHo_xY_{(1-x)}F_4$ plus Gaussian distributed quenched random fields of standard deviation h [64] ©APS (2013). (**b**) n.n. EA/SS Ising model with finite mean $J_0 = 1$ and variance J with Gaussian-distributed random fields of standard deviation H_r [65]

We also note that computer simulational study of an Ising dipolar system on a simple cubic lattice has indicated[5] that the sharp spin glass transition seen in zero applied field is removed in a uniform field [66]. Note, however, that random $\pm h$ fields cannot be simply gauged away into a uniform field h in systems with non-zero mean exchange, as they can in the usually studied models with zero mean. Furthermore, the magnitudes of effective fields in PMN are also randomly multivalued.

It should also be recalled that the displacements in real relaxor systems are not one-dimensional, but are 3-vector. It was realized long ago that vector spin versions of the SK model in a uniform field would exhibit a spin glass transition in a transverse direction as the temperature is lowered [58], but with only weak non-ergodicity in the longitudinal direction until a lower crossover temperature [59]. It seems probable that the first of these transition temperatures will persist even for short-range interactions. It has been observed experimentally [67]. It is also of probable relevance that the effective dipolar interaction in displacive systems is not anisotropic as in the Ising cases of Refs. [64] and [65], in which the dipoles are restricted to lie in the z-direction. Rather it has the more general isotropic form [41]; $[\mathbf{u}_i \cdot \mathbf{u}_j - 3(\hat{\mathbf{R}}_{ij} \cdot \mathbf{u}_i)(\hat{\mathbf{R}}_{ij} \cdot \mathbf{u}_j)]/|\mathbf{R}_{ij}|^3$.

As already noted, others have claimed that the relaxor peak observed in PMN is driven dominantly by the random fields [52, 53].[6] A recent simulational study inspired by PMN has also indicated in favour of this [68], using a model similar to Eq. (1.14) with the only disorder attributed to the random field terms; *i.e.* with H^1_{PM*N*} calculated with parameters averaged over the M and N and ignoring

[5]Via a study of the size dependence of the spin glass correlation length, showing no Binder crossover.

[6]See also [54].

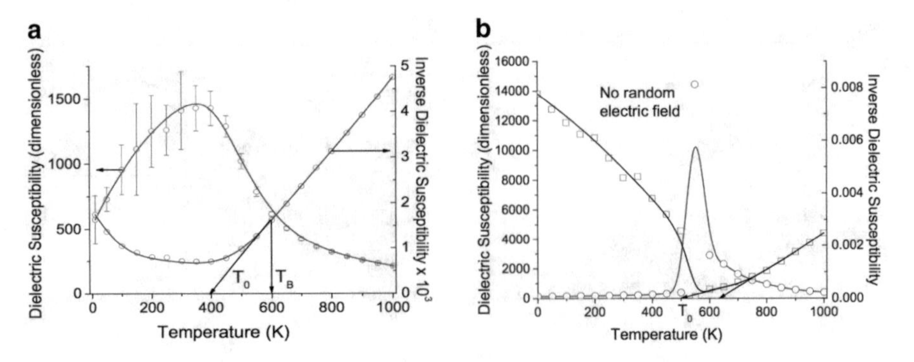

Fig. 1.9 Susceptibilities of model of PMN based on averaged interactions, (**a**) with random fields, (**b**) without random fields [68] ©APS (2015)

the extra site-interaction terms. Figure 1.9 shows the results obtained for the susceptibility both with and without inclusion of the random fields. Taken in combination these results are suggestive that relaxor/glass-like behaviour might be possible in the combination of frustrated interactions and disorder of either dilution or random fields. However, as yet, there is no convincing finite-size scaling demonstration of a true thermodynamic transition in the case of purely random field disorder, even when the non-disordered system has a ferro transition. There remains also uncertainty in the statistical mechanics community as to whether there can be a frozen spin glass phase driven purely by random fields without exchange frustration, although it has been proven not to be a thermodynamically stable state in a system of one-dimensional spins with only ferromagnetic (or zero) exchange interactions [69].

1.4.3 Polar Nanoregions

Another observed feature of displacive relaxors is that of the appearance of polar nanoregions (PNRs)[70, 71] already at temperatures higher than those of the susceptibility peaks, beneath a higher so-called 'Burns temperature' characterized by the onset of deviations from Curie behaviour [72]. A commonly expressed conceptualization is that relaxor behaviour is a consequence of interaction of such PNRs, but specific details are not clarified. Here we indicate how such PNRs and both ferroelectric and relaxor phase transitions can be expected as a consequence of a simple extension of the modelling above. The initial discussion will be restricted to a simple mean field consideration and, for simplicity, within a picture allowing only for one-dimensional deviations, but allowing for spatial inhomogeneity.

We start with the homovalent case. Thus we consider minimization of a Landau-type free energy

$$F_R = \sum_i \{\tilde{\kappa}_i(T)u_i^2 + \tilde{\lambda}_i(T)u_i^4\} - \sum_{(ij)} \tilde{J}(R_{ij}, T)u_i u_j. \tag{1.16}$$

where the coefficients are now temperature-dependent, the $\{u\}$ at minimum are now the mean-field values and allowance is made for different u_i at different sites i.

Minimizing with respect to the $\{u_i\}$ yields the self-consistency relation

$$\tilde{\kappa}_i(T)u_i - \sum_j \tilde{J}(R_{ij})u_j = -2\tilde{\lambda}_i u_i^3. \tag{1.17}$$

Of particular interest are non-zero solutions and phase transitions as a consequence of reducing the $\tilde{\kappa}(T)$ with reducing T. This Eq. (1.17) always allows solutions $\{u = 0\}$, corresponding to undisplaced paraelectricity, but interest is in possible solutions $\{u \neq 0\}$. These only occur for small enough $\tilde{\kappa}$.

For a pure ferroelectric all the u_i have the same value, given by

$$u = \left\{ \left[\sum_j \tilde{J}(R_{ij}, T) - \tilde{\kappa}(T) \right] / \tilde{\lambda}(T) \right\}^{1/2}, \tag{1.18}$$

from which we see that there is a critical temperature T_c given by

$$\tilde{\kappa}(T_c) = \sum_j \tilde{J}(R_{ij}, T_c). \tag{1.19}$$

For $T < T_c$ the system is ferroelectric whereas for $T > T_c$ it is paraelectric.

In a general alloy, however, the solutions u_i for different sites i will vary. Equation (1.17) must have a (real) solution at each site i and, in principle, can be either localized or extended/percolating. Localized solutions would represent internally ordered nanoregions, while the onset of extended solutions would signify a phase transition. A suggestive conceptual guide to the character of such solutions can be visualized by comparing with the (linear) Anderson localization equation [73]

$$\epsilon_i \psi_i + \sum_j t_{ij} \psi_j = E \psi_i. \tag{1.20}$$

with the identifications

$$\{\epsilon_i\} = \{\tilde{\kappa}_i\} \quad ; \quad \{t_{ij}\} = -\{\tilde{J}_{ij}\}. \tag{1.21}$$

Figure 1.10 shows a schematic density of states $\rho(E)$ for the Anderson model in a situation where the lower band edge is positive.. Correspondingly the only solution to Eq. (1.17) is $u = 0$. However, if the temperature is reduced so the mean ϵ is decreased sufficiently for the lower band edge to reduce below zero, then solutions of Eq. (1.17) with $u \neq 0$ exist. For a pure system with no ϵ-disorder, all the states of Eq. (1.20) are extended, with the lower band edge state having the highest symmetry, resulting in a phase transition to a state of similar symmetry for Eq. (1.19). This

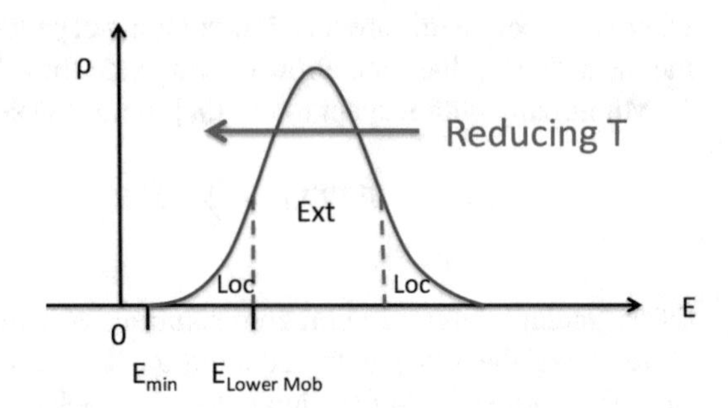

Fig. 1.10 Schematic density of states of an Anderson model with local energy disorder, showing localized and extended regions. The arrow indicates movement of the whole figure relative to its vertical axis on decreasing the mean local energy ϵ while maintaining its relative distribution

corresponds to the ferroelectric transition as found in BT, whereas in BZ $\tilde{\kappa}^{Zr}$ is never small enough for $\rho(E)$ to reach $E = 0$.

However, if the Anderson model coefficients, such as the ϵ, are disordered then states at the outer regions of $\rho(E)$ are localized, with the mapping leading to internally correlated but not cooperatively frozen clusters, identifiable as the observed PNRs, while for a true thermodynamic phase transition an extended state solution is required. Hence it is (crudely) suggestive that the temperature must be lowered further until the lower mobility edge, separating localized and extended states, crosses $E = 0$. This consideration suggests that lowering T in the model system of Eq. (1.12) will lead first to finite internally ordered nanoregions, growing in number and size as T is lowered, followed by a true thermodynamic transition at a lower temperature. The onset of PNRs is expected at a temperature region close to the phase transition of the pure ferroelectric host. The cooperatively ordering phase transition is expected to be to ferroelectric at higher concentrations of ferroelectric B-ions, passing to relaxor/pseudo-spin-glass at intermediate concentrations, and failing to reach cooperative order at too low concentrations. This is illustrated schematically in Fig. 1.11, where the solid lines indicate phase transitions but the dotted and dashed lines are heuristic indications of onset and visibility of PNRs.[7] This prediction including PNRs is in qualitative accord with experimental observations [43].

[7]Conceptually one can view the situation in a substitutional alloy as follows: (1) quenched statistical fluctuations in the locations of the ions on the underlying lattice will lead to a range of clusterings of the more potentially displaceable ions (Ti in BZT), with regions both denser and less dense than the average concentration; (2) For clusters to displace-order internally the energy lowering gained through interaction must overcome the local free energy penalties; (3) such internal correlation will first occur on clusters that are close in structure to the pure ferroelectric one (BT for BZT); (4) this can always occur in principle at a temperature close to that of the pure ferroelectric, but will become rarer as the concentration of potentially ferroelectric ions reduces; (5) as the temperature is lowered the decrease in the effective $\kappa(T)$ will lead to the internal mean-field stabilization of larger clusters, until eventually there will be clusters that percolate throughout the whole system; (6) the character of the final low temperature macroscopically cooperative state will be determined by minimizing the free energy, which in a disordered and frustrated system can be either globally periodic or spin glass-like.

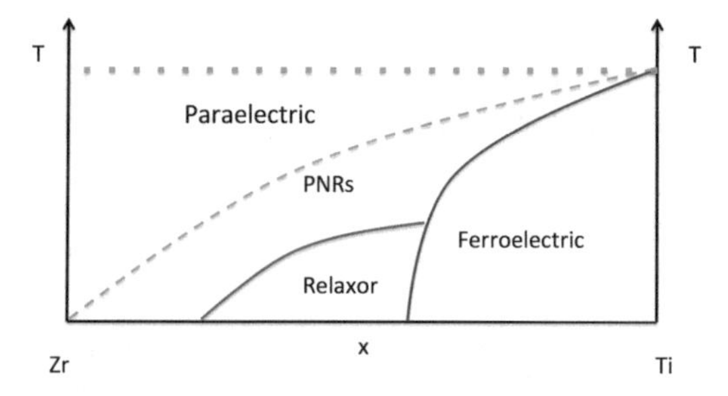

Fig. 1.11 Schematic 'phase diagram' for BZT expected in the light of heuristic considerations. Solid lines denote true phase transitions. The dotted line indicates the onset of PNRs in the picture discussed. The dashed line is a speculative illustration of crossover for the onset of significant visibility of PRNs

For heterovalent alloys it is necessary also to take account of the effective random field terms in the Hamiltonian, which yield corresponding linear contributions to the mean-field free energy of Eq. (1.16) and consequently terms of zero-order in the $\{u\}$ in Eq. (1.17) and hence induced displacements even at higher temperatures; e.g. even without any interaction terms the presence of the extra charges on Mg++ and Nb+++++ ions randomly distributed on B-sites of PMN would lead to a corresponding quasi-spherical distribution of Pb deviations from their mean-lattice positions, as discussed in [74], in qualitative accord with observations [75]. Statistical clusterings of effective fields of similar orientations can be expected to lead to nucleation of polar nanoregions, even without frustration in the site-to-site interactions. Although the suggestive quasi-mapping to the Anderson equation suggested above will no longer be applicable, the concept of relating the 'transition' to an 'edge' separating localized and extended solutions should remain qualitatively valid.

In principle one could also change description further by considering the PNR as 'superspins' with effective interactions between them, with eventually a percolating coherence between them marking the relaxor transition. This conceptualization was used in the context of itinerant spin glasses in the early 1970s [76] and has become popular in considerations of relaxors; for a recent discussions, see e.g. [53, 54, 77, 78].

Of course, to be fully representative of even the soft-Ising-like model of Eq. (1.12) for a homovalent alloy, one needs to go beyond the simple mean-field form used above.

1.5 Itinerant Spin Glasses

In fact, some of the suggestions above in Sect. 1.4 were conceptually pre-empted many decades ago by theoretical considerations of itinerant spin glasses, in which the magnetism resides with conduction electrons [76, 79, 80]. These studies were not pursued but are probably worthy of resurrection here in the light of [44]. Again, we follow the philosophy of using a simple model and approximations for illustration.

A simple Hubbard model for a transition metal alloy is given by the Hamiltonian

$$H_{HA} = \sum_{ij;s=\uparrow,\downarrow} t_{ij} a_{is}^{\dagger} a_{js} + \sum_{i;s=\uparrow,\downarrow} V_i a_{is}^{\dagger} a_{is} + \sum_i U_i \hat{n}_{i\uparrow} \hat{n}_{i\downarrow} \tag{1.22}$$

where the $a, a^{\dagger}$ are site-labelled d-electron annihilation and creation operators, $\hat{n}_{is} = a_{is}^{\dagger} a_{is}$ and, in general, the t_{ij}, V_i and U_i depend upon the type of atoms at sites *i,j*. We are concerned with cases in which the electron density is such that the conduction band is only partially filled and the alloys are metallic.

This can be transformed into a form analogous to that of Eq. (1.12) with the variables local magnetization and charge fluctuations.

We first re-write $\hat{n}_{i\uparrow}\hat{n}_{i\downarrow}$ in terms of complete squares using the identity

$$\hat{n}_{i\uparrow}\hat{n}_{i\downarrow} = \frac{1}{4}\{\hat{n}_i^2 - \hat{m}_i^2\} \tag{1.23}$$

where

$$\hat{n}_i = \hat{n}_{i\uparrow} + \hat{n}_{i\downarrow}; \qquad \hat{m}_i = \hat{n}_{i\uparrow} - \hat{n}_{i\downarrow}. \tag{1.24}$$

For easier conceptualization of the possible magnetic consequences, with minimal more peripheral distractions, we further simplify by assuming that the charge fluctuations are of lesser importance and take their contribution to be absorbed into the V_i and furthermore set these all V_i equal and hence ignorable. Further re-writing in a symmetric notation, we are left with

$$H = \sum_{ij,\sigma} t_{ij} a_{i\sigma}^{\dagger} a_{j\sigma} - \frac{1}{4} \sum_i U_i \hat{\mathbf{S}}_i . \hat{\mathbf{S}}_i \tag{1.25}$$

where

$$\hat{\mathbf{S}}_i = a_{is}^{\dagger} \boldsymbol{\sigma}_{s,s'} a_{is'}. \tag{1.26}$$

The quadratic form of the $\hat{\mathbf{S}}$-term in Eq. (1.25) enables the use of an 'inverse completion of a square' procedure [81–86] to effectively 'linearize' the Hamiltonian in $a_{is}^{\dagger} a_{js'}$ through the introduction of an auxiliary magnetization field variable $\hat{m}$, conjugate to $\hat{\mathbf{S}}$.

One can then further 'integrate out' the original electron operators in favour a description in terms purely of magnetization variables [86]. Further taking the static approximation yields an effective Hamiltonian in local magnetisation variables[8]; to fourth order,

$$H_m = \sum_i (1 - U_i \chi_{ii})|\hat{\boldsymbol{m}}_i|^2 - \sum_{ij;i\neq j} U_i^{1/2} U_j^{1/2} \chi_{ij} \hat{\boldsymbol{m}}_i.\hat{\boldsymbol{m}}_j$$
$$- \sum_{ijkl;\alpha\beta\gamma\delta} (U_i U_j U_k U_l)^{1/2} \Pi_{ijkl}^{\alpha\beta\gamma\delta} \hat{m}_i^\alpha \hat{m}_j^\beta \hat{m}_k^\gamma \hat{m}_l^\delta, \tag{1.27}$$

where χ is the static band susceptibility function of the bare system (with only the t term), Π is a corresponding bare 4-point function and we have dropped the higher order contributions.

A further change of variables

$$\hat{\boldsymbol{M}}_i = U_i \hat{\boldsymbol{m}}_i \tag{1.28}$$

immediately brings this to a form reminiscent of Eq. (1.12):

$$H_M = \sum_i (U_i^{-1} - \chi_{ii})|\hat{\boldsymbol{M}}_i|^2 - \sum_{ij;i\neq j} \chi_{ij} \hat{\boldsymbol{M}}_i.\hat{\boldsymbol{M}}_j$$
$$- \sum_{ijkl;\alpha\beta\gamma\delta} (\Pi_{ijkl}^{\alpha\beta\gamma\delta} \hat{M}_i^\alpha \hat{M}_j^\beta \hat{M}_k^\gamma \hat{M}_l^\delta, \tag{1.29}$$

with local self-energy weight $(U_i^{-1} - \chi_{ii})$ the analogue of κ in Eq. (1.12). Minimization of H_m or H_M gives the equation for the magnetizations $\{\boldsymbol{m}\}$ in mean field approximation, the itinerant magnetic analogue of the relaxor Eq. (1.17).

A simple consideration of a system with two components A and B with $U_A = 0$ but $U_B > 0$ immediately demonstrates the following well-known mean field results: (1) pure A is only paramagnetic; (2) pure B is ferromagnetic only if $(1 - U_B \sum_j \chi_{ij}) \equiv (1 - U_B \chi(q = 0)) < 0$, the Stoner criterion [2], otherwise paramagnetic, (3) a single B substituted in an A-host will only carry a mean-field moment if $(1 - U_B \chi_{ii}) \equiv (1 - U_B \int_q \chi(q)) < 0$, the Anderson condition [87].

For metallic systems χ_{ij} oscillates in sign as a function of separation, so there is frustration in the effective interactions of Eq. (1.29). Hence, a more concentrated alloy with a sufficient finite non-zero density of B atoms can, in principle, exhibit either ferromagnetism or another periodic order, while beneath a critical concentration x_c and with sufficient frustration, can exhibit spin-glass order.

If the Anderson local moment criterion is satisfied at B-sites, then the situation is essentially the same as in the conventional hard spin case discussed in Sect. 1.3.

However, if the Anderson criterion is not satisfied (equivalent to $\kappa > 0$), then a sufficiently strong potential energy lowering due to coherently-acting mutual

[8]Note that here the $\hat{m}$ are auxiliary field variables, not the actual equilibrium magnetizations.

magnetization fluctuation freezing at different sites is needed to bootstrap a macroscopic magnetically ordered phase, overcoming the $(1 - U^B \chi_{ii})|\mathbf{m}_i|^2$ local fluctuation penalties on a percolating network, otherwise the system would be paramagnetic. For the case of a high concentration of B this phase is still essentially Stoner's itinerant ferromagnetism. But for an intermediate concentration of B the spontaneous cooperative phase can be a spin glass, bounded by a lower critical concentration separating it from the (Pauli-type) paramagnet and an upper critical concentration separating it from the ferromagnet. Hertz [80] provided the first theory[9] and introduced the term 'Stoner glass' to refer to the itinerant spin glass.

Furthermore, the same considerations concerning the formation of PNRs as discussed for displacive ferroelectric alloys should apply to the formation of bootstrapped super-spin nano-clusters due to quenched statistical fluctuations in the locations of the B atoms, in such itinerant magnetic alloys, even above a transition temperature for spin glass or ferromagnet [76, 79].[10] Their 'visibility' would however depend on their effective dynamical lifetimes, not discussed here but surely much shorter than for ferroelectric PNRs.

The sequence paramagnet/spin glass/ferromagnet was already observed in the early days of experimental spin glass physics; e.g. in **Rh**Co alloys [88]. It seems highly probable that a similar phenomenological explanation should apply in other metallic spin glass alloys, especially those labelled as 'cluster glasses', and it could prove interesting to review them in this light. A full theoretical treatment would require going beyond the simple mean-field theory presented above, as well as beyond the other simplifying assumptions employed above, but hopefully it could already provide a useful starting perspective complementary to those currently employed.

1.6 Strain Glass

In this section we consider another analogue of spin glasses, of glassy strain distortions in martensitic alloys [19, 27, 89, 101] and given the name 'strain glass'.

Martensitic materials, see, e.g., [90, 91], exhibit first-order structural transitions from higher to lower symmetry phases as temperature is lowered. An example is from high temperature cubic austenite to a lower temperature phase of alternating twin planes of complementary tetragonal character, epitomized by TiNi which in its pure higher-temperature state is an ordered compound of rocksalt structure. Our interest here will be particularly in when this compound is atomically disordered, for example by randomly altering the balance of Ti and Ni or by replacing some of these atoms by another element (e.g. Fe).

---

[9] Using a different formulation than presented here.

[10] These early papers discussed the formation of clusters and anticipated that their interactions would then yield the spin glass state, much as in later considerations of relaxors.

Although, in principle, it could be modelled microscopically in terms of atomic displacements, as was done for relaxor problem of Sect. 1.4, or phenomenologically at a Landau-Ginsburg level in terms of continuous-valued deviatoric strains, here we shall simply employ a crude discrete pseudo-spin mean-field modelling for illustration. Furthermore, again for simplicity, we shall consider a two-dimensional version in which the local cell structure is describable in terms of a variable S_i which takes the values 0, +/−1 corresponding, respectively, to a square and two orthogonal rectangular structures. The local part of an effective free energy is then given by

$$F_L = \sum_i D_i S_i^2 \tag{1.30}$$

where the $\{i\}$ label cells, so that if $D_i > 0$ then minimizing F_L yields $S_i = 0$, while if $D_i < 0$ it yields degeneracy $S_i = \pm 1$. There are also intercell interaction terms

$$F_{int} = -\sum_{ij} S_i V_{ij} S_j \tag{1.31}$$

arising from both short-range neighbouring similarity effects and from longer range St. Venant's compatibility constraints that give (in 2 dimensions) [92]

$$V_{ij}^{StV} \propto -\cos(4\theta(\mathbf{R}_{ij}))/|\boldsymbol{R}_{ij}|^2. \tag{1.32}$$

where $\theta(\mathbf{R})$ is the angle subtended by $\mathbf{R}$ at a Cartesian axis of the cubic lattice.

Temperature is emulated by taking the D to be temperature dependent, reducing with reducing temperature. Without the interaction term the S-values at the minimum of the free energy will change from $S = 0$ to $S = \pm 1$ when their corresponding D change from positive to negative. When the interaction is included cooperative bootstrapping will result in $S = \pm 1$ at sites where the resultant free energy minimization can overcome the local D, with the favoured state given by the relative signs of the $\{S_i\}$ that yield the lowest free energy. For a pure system one gets the martensite phase of diagonally alternating stripes of $S = +1$ and $S = -1$. For a system with quenched randomness of site occupation one can expect a corresponding quenched randomness of the $\{D_i\}$, along with quenched effective random fields. Clearly this represents a scenario similar to that discussed above for relaxors and spin glasses, namely that for sufficient quenched disorder the frustration, arising from the antiferroelastic interactions at values of θ where the cosine is negative, leads to an expectation of strain glass, as has been observed [93, 94]; see Fig. 1.12.[11] Both the cases of quenched D [89] and of the effective random fields [95], arising from local substitutions, have been proposed separately as the disorders responsible strain glass behaviour. The reality probably includes both.

[11] In this case the analogue of the ferromagnet is the martensitic stripe phase.

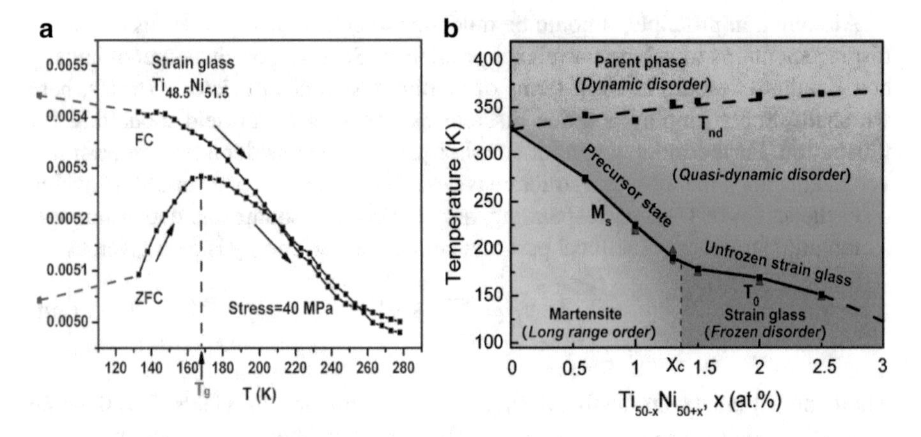

Fig. 1.12 $Ni_{50-x}Ti_{50+x}$: (**a**) FC/ZFC evidence for strain glass [93] ©APS (2007). (**b**) Phase diagram [94] ©APS (2010)

1.7 Conclusion

In this chapter I have tried to demonstrate similarities in the potential for ferroic glass behaviour in several different types of system, magnetic, ferroelectric and martensitic/ferrolastic alloys, both metallic and insulating, using a combination of simple modelling and analogies, experimental, theoretical/mathematical, computational and conceptual, and with a particular consideration of 'induced moment' and continuously displaceable systems.

The key ingredients to permit such glassy behaviour appear to be frustrated interactions and quenched disorder, as has often been expressed before. Probing these relationships has provided possible explanations for phenomena such as the onset of non-ergodicity and slow dynamics. Cluster effects such as those known as polar nano-regions (PNR) in displacive relaxors are considered in analogy with localization phenomena.

The simple analogies considered here suggest further conceptual transfers between different ferroic materials and further experimental investigations; for further discussion concerning characteristic experimental aspects of spin glasses, see [96]; for a complementary recent discussion of relaxors, see [54].

These comparisons have also highlighted some remaining questions, particularly concerning the issue of the role of quenched random fields. BZT has no quenched random fields but PMN has significant such fields, yet the susceptibility measurements look very similar to one another.

As noted earlier, there is controversy in the spin glass community as to whether a true spin glass phase transition can continue to exist in the presence of an applied field. Even without a non-analyticity it would not necessarily mean that the peak in the pure zero-field susceptibility cannot continue in a finite-field, in a more rounded form, as indeed was clear already in the early important small-field experiments of

Cannella and Mydosh [5]; see Fig. 1.2b. There has been much interest in the random field Ising model (RFIM), normally with short-range ferromagnetic interactions, without demonstration of a spin glass, and indeed it has been proven not to be thermodynamically stable for purely ferromagnetic or zero interactions [69]. There have been theoretical suggestions that in a system with higher spin dimension there could be a 'spin glass' state driven by the random fields [97] but there has been no observed evidence of a sharp transition to such a phase in a magnetic system.

Although there have been many experimental demonstrations of spin glass behaviour in frustrated and quench-disordered three-dimensional systems of three-dimensional (Heisenberg) spins, there is still some debate about theory [15]. There are no simple experimental methods to apply three-dimensional random magnetic fields. On the other hand, the relaxor systems discussed above have displacement variables able to orient in the full three-dimensional space and in heterovalent relaxor alloys, such as PMN, the effective random fields are also spread throughout the three-dimensional orientation space and are of significant strength, yet the peaks in the susceptibility are quite sharp. Both the classic spin glass and the relaxor examples have long range interaction frustration.

It is thus tempting to wonder whether the criterion of frustrated interaction and quenched disorder as the key ingredients for spin glass/relaxor/strain glass behaviour might apply independently whether the disorder arises from site-disorder, bond-disorder or random fields, or a combination, preventing simple homogenous and smoothly varying optimal compromises, and also whether one needs to go beyond one-dimensionality of the 'pseudospins', but more work is required to help decide.

Finally, let me note that my aim has not been to describe quantitatively or completely the systems that I have discussed, but rather through simple extraction and comparisons, to try to draw links and to expose contrasts and remaining puzzles and uncertainties, in the hope that they might stimulate work that might not have been obvious within the confines of just sub-classes of systems. I should also point out that I am not the first to propose that either relaxors or martensitic alloys might be considered as pseudo-spin glasses (see, e.g., [98–100]), but I hope my small contribution can be stimulating in moving towards a greater understanding.

References

1. P. Weiss, L'hypothèse du champ moléculaire et la propriété ferromagnétique. J. Phys. Theor. Appl. **6**, 661–690 (1907)
2. E.C. Stoner, Collective electron ferromagnetism. Proc R. Soc. A **165**, 372 (1938)
3. J. Valasek, Piezoelectric and allied phenomena in Rochelle salt. Phys. Rev. **15**, 537 (1920)
4. L. Néel, Propriétées magnétiques des ferrites; Ferrimagnétisme et antiferromagnétisme. Ann. Phys. Paris **3**, 137 (1948)
5. V. Cannella, J.A. Mydosh, Magnetic ordering in gold-iron alloys. Phys. Rev. B **6**, 4220 (1972)
6. S.F. Edwards, P.W. Anderson, Theory of spin glasses. J. Phys. F. **5**, 965 (1975)

7. D. Sherrington, S. Kirkpatrick, Solvable model of a spin-glass. Phys. Rev. Lett. **35**, 1792 (1975)
8. G. Parisi, Infinite number of order parameters for spin-glasses. Phys. Rev. Lett. **43**, 1754 (1979)
9. M. Mézard, G. Parisi, M.-A. Virasoro, *Spin Glass Theory and Beyond* (Word Scientific, Singapore, 1987)
10. K. Binder, A.P. Young, Spin glasses: experimental facts, theoretical concepts and open questions. Rev. Mod. Phys. **58**, 801 (1986)
11. K.H. Fischer, Hertz J.A., *Spin Glasses* (Cambridge University Press, Cambridge, 1991)
12. J.A. Mydosh, *Spin Glasses: An Experimental Introduction* (Taylor and Francis, Philadephia, 1993)
13. H. Nishimori, *Statistical Physics of Spin Glasses and Information Processing* (Oxford University Press, New York, 2001)
14. J.A. Mydosh, Spin glasses: redux: an updated experimental/materials survey. Rep. Prog. Phys. **78**, 052501 (2015)
15. H. Kawamura, T. Taniguchi, Spin glasses, in *Handbook of Magnetic Materials*, ed. by K.H.J. Buschov (Elsevier BV, Amsterdam, 2015)
16. D. Panchenko, *The Sherrington-Kirkpatrick Model* (Springer, New York, 2013)
17. G.A. Smolenskii, V.A. Isupov, Dokl. Acad. Nauk SSSR **97**, 653 (1954)
18. G.A. Smolenskii, V.A. Isupov, A.I. Agranovskaya, S.N. Popov, Ferroelectrics with diffuse phase transitions. Fiz.Tverd Tela **2**, 2906 (1960); [Sov. Phys. Solid State **2**, 2584 (1961)]
19. S. Sarkar, X. Ren, K. Otsuka, Evidence for strain glass in the ferroelastic-martensitic system $Ti_{50-x}NI_{50+x}$. Phys. Rev. Lett. **95**, 205702 (2005)
20. G.V. Lecomte, H. von Löhneysen, E.F. Wassermann, Frequency dependent magnetic susceptibility and spin glass freezing in PtMn Alloys. Z. Phys. B **50**, 239 (1983)
21. A. Simon, J. Ravez, M. Maglione, The crossover from a ferroelectric to a relaxor state in lead-free solid solutions. J. Phys. Condens. Matter **16**, 963 (2004)
22. A. Levstik, Z. Kutnjak, C. Filipič, R. Pirc, Glassy freezing in relaxor ferroelectric lead magnesium niobate. Phys. Rev. B **57**, 11204 (1998)
23. S. Nagata, P.H. Keesom, H.R. Harrison, Low-dc-field susceptibility of *Cu*Mn Spin glass. Phys. Rev. B **19**, 1633 (1979)
24. A.R. Akbarzadeh, S. Prosandeev, E.J. Walter, A. Al-Barakaty, L. Bellaiche, Finite-temperature properties of Ba(Zr, Ti)O_3 relaxors from first principles. Phys. Rev. Lett. **108**, 257601 (2012)
25. B.R. Coles, B. Sarkissian, R.H. Taylor, The role of finite magnetic clusters in Au-Fe alloys near the percolation concentration. Phil. Mag. B **37**, 489 (1978)
26. H. Maletta, P. Convert, Onset of ferromagnetism in $Eu_xSr_{1-x}S$ near x=0.5. Phys. Rev. Lett. **42**, 108 (1979)
27. D. Sherrington, Understanding glassy phenomena in materials, in *Disorder and Strain Induced Complexity in Functional Materials*, vol. 177, ed. by T. Kakeshita, T. Fukuda, A. Saxena, A. Planes (Springer, Berlin, 2012)
28. G. Toulouse, Theory of frustration effect in spin-glasses. Commun. Phys. **2**, 115 (1977)
29. D. Sherrington, B.W. Southern, Spin glass versus ferromagnet. J. Phys. F **5**, L49 (1975)
30. K. Binder, Finite size scaling analysis of Ising model block distribution functions. Z. Phys. B **43**, 119 (1981)
31. R.N. Bhatt, A.P. Young, Search for a transition in the 3-dimensional ±1 spin-glass. Phys. Rev. Lett. **54**, 924 (1985)
32. H.G. Katzgraber, M. Koerner, A.P. Young, Universality in three-dimensional Ising Spin glasses: a Monte Carlo study. Phys. Rev. B **73**, 224432 (2006)
33. J.C. Andresen, H.G. Katzgraber, V. Organesyan, M. Schechter, Existence of a thermodynamic spin-glass phase in the zero-concentration limit of anisotropic dipolar systems. Phys. Rev. X **4**, 041016 (2014)
34. B.V.B. Sarkissian, B.R. Coles, Spin-glass to Overhauser-alloy transitions in Y-rare-earth and Sc-rare-earth alloys. Commun. Phys. **1**, 17 (1976)

35. S.K. Ghatak, D. Sherrington, Crystal field effects in a general S Ising spin glass. J. Phys. C **10**, 3149 (1977)
36. U.T. Hochli, K. Knorr, A. Loidl, Orientational glasses. Adv. Phys. **1990**, 405 (1990)
37. B.E. Vugmeister, M.D. Glinchuk, Dipole glass and ferroelectricity in random-site electric dipole systems. Revs. Mod. Phys. **62**, 993 (1990)
38. P.M. Goldbart, D. Sherrington, Replica theory of the uniaxial quadrupolar glass. J. Phys. C **18**, 1923 (1985)
39. D. Sherrington, Potts and related glasses. Prog. Theor. Phys. Japan Supp. **87**, 180 (1986)
40. K. Binder, J.D. Reger, Theory of orientational glasses models, concepts, simulations. Adv. Phys. **41**, 547 (1992)
41. W. Zhong, D. Vanderbilt, K.M. Rabe, Phase transitions in $BaTiO_3$ from first principles. Phys. Rev. Lett. **73**, 1861 (1994); First-principles theory of ferroelectric phase transitions for perovskites: the case of $BaTiO_3$. Phys. Rev. B **52**, 6301 (1995)
42. R.D. King-Smith, D. Vanderbilt, First-principles investigation of ferroelectricity in perovskite compounds. Phys. Rev. B **49**, 5828 (1994)
43. T. Maiti, R. Guo, A.S. Bhalla, Structure-property phase diagram of $BaZr_xTi_{1-x}O_3$ system. J. Am. Ceram. Soc. **91**, 1769 (2008)
44. D. Sherrington, BZT: a soft pseudospin glass. Phys. Rev. Lett. **111**, 227601 (2013)
45. W. Kleemann, S. Miga, J. Dec, J. Zai, Crossover from ferroelectric to relaxor and cluster glass in $BaTi_{(1-x)}Zr_xO_3 (x = 0.25 - 0.35)$ studied by non-linear permittivity. Appl. Phys. Lett. **102**, 232907 (2013)
46. J.M. Luttinger, L. Tisza, Theory of dipole interaction in crystals. Phys. Rev. **70**, 954 (1946)
47. J.J. Alonso, J.F. Fernandez, Monte Carlo study of the spin-glass phase of the site-diluted dipolar Ising model. Phys. Rev. B **81**, 064408 (2010)
48. J.C. Andresen, H.G. Katzgraber, V. Organesyan, M. Schechter, Existence of a thermodynamic spin-glass phase in the zero-concentration limit of anisotropic dipolar systems. Phys. Rev. X **4**, 041016 (2014)
49. A.W. Hewat, Structure of rhombohedral ferroelectric barium titanate. Ferroelectrics **6**, 215 (1974)
50. G. Shirane, R. Pepinsky, B.C. Frazer, X-ray and neutron diffraction study of ferroelectric $PbTiO_3$. Acta Crystallogr. **9**, 131 (1955)
51. D. Sherrington, Relaxors, Spin, Stoner and cluster glasses. Phase Trans. **88**, 202 (2015)
52. V. Westphal, W. Kleemann, M.D. Glinchuk, Diffuse phase transitions and random-field-induced domain states of the "Relaxor" ferroelectric $PbMg_{1/3}Nb_{2/3}0_3$. Phys. Rev. Lett. **68**, 847 (1992)
53. W. Kleemann, Relaxor ferroelectrics: cluster glass ground state via random fields and random bonds. Phys. Stat. Sol. B **251**, 1993 (2014)
54. W. Kleeman, Relaxor ferroelectrics and related cluster glasses, in *Frustrated Materials and Ferroic Glasses*, ed. by T. Lookman, X. Ren. Springer Series in Materials Science, vol. 275 (Springer, Cham, 2018). https://doi.org/10.1007/978-3-319-96914-5_5
55. R.A. Cowley, S.N. Gvasaliya, S.G. Lushnikov, B. Roessli, G.M. Rotaru, Relaxing with relaxors. Adv. Phys. **60**, 229 (2011)
56. J.R.L. de Almeida, D.J. Thouless, Stability of the Sherrington-Kirkpatrick solution of a spin glass model. J. Phys. A **11**, 983 (1978)
57. A. Sharma, A.P. Young, de Almeida-Thouless line in vector spin glasses. Phys. Rev. E **81**, 061115 (2010)
58. M. Gabay, G. Toulouse, Coexistence of spin-glass and ferromagnetic orderings. Phys. Rev. Lett. **47**, 201 (1981)
59. D.M. Cragg, D. Sherrington, M. Gabay, Instabilities of an m-vector spin glass in a field. Phys. Rev. Lett. **49**, 158 (1982)
60. H.G. Katzgraber, A.P. Young, Probing the Almeida-Thouless line away from the mean-field model. Phys. Rev. B **72**, 184416 (2005)
61. D. Larsen , H.G. Katzgraber , M.A. Moore, A.P. Young, Spin glasses in a field: three and four dimensions as seen from one space dimension. Phys. Rev. B **87**, 024414 (2013)

62. M. Baity-Jesi et al. (JANUS), The three-dimensional Ising spin glass in an external magnetic field: the role of the silent majority. J. Stat. Mech. **2014**, P05014 (2014)
63. R.R.P. Singh, A.P. Young, de Almeida-Thouless instability in short-range Ising spin glasses. Phys. Rev. E **96**, 012127 (2017)
64. J.C. Andresen, C.K. Thomas, H.G. Katzgraber, M. Schechter, Novel disordering mechanism in ferromagnetic systems with competing interactions. Phys. Rev. Lett. **111**, 177202 (2013)
65. J.C. Andresen , H.G. Katzgraber, M. Schechter, Random-field-induced disordering mechanism in a disordered ferromagnet: between the Imry-Ma and the standard disordering mechanism (2017). ArXiv 1706.07904
66. J.F. Fernandez, Evidence against an Almeida-Thouless line in disordered systems of Ising dipoles. Phys. Rev. B **82**, 144436 (2010)
67. D. Petit, L. Fruchter, I.A. Campbell, Ordering in Heisenberg Spin glasses. Phys. Rev. Lett. **88**, 207206 (2002)
68. A. Al-Barakaty, S. Prosandeev, D. Wang, B. Dkhil, L. Bellaiche, Finite-temperature properties of the relaxor $PbMg_{1/3}Nb_{2/3}O_3$ from atomistic simulations. Phys. Rev. B **91**, 214117 (2015)
69. F. Krzakala, F. Ricci-Tersenghi, L. Zdeborovà, Elusive Spin-glass phase in the random field Ising model. Phys. Rev. Lett. **104**, 207208 (2010)
70. L.E. Cross, Relaxor ferroelectrics. Ferroelectrics **76**, 241 (1987)
71. I.K. Jeong, T.W. Darling, J.K. Lee, T. Proffen, R.H. Heffner, J.S. Park, K.S. Hong, W. Dmowski, T. Egami, Local lattice dynamics and the origin of the relaxor ferroelectric behavior. Phys. Rev. Lett. **94**, 147602 (2005)
72. G. Burns, F. Dacol, Crystalline ferroelectrics with glassy polarization behavior. Phys. Rev. B **28**, 2527 (1983)
73. P.W. Anderson, Absence of diffusion in certain random lattices. Phys. Rev. **109**, 1492 (1958)
74. D. Sherrington, $Pb(Mg_{1/3}Nb_{2/3})O_3$: a minimal induced-moment soft pseudo-spin glass perspective. Phys. Rev. B **89**, 064105 (2014)
75. S.B. Vakhrushev, N.M. Okuneva, Evolution of structure of $PbMg1/3Nb_{2/3}O_3$ in the vicinity of the burns temperature. AIP Conf. Proc. **626**, 117 (2002)
76. D. Sherrington, K. Mihill, Effects of clustering on the magnetic properties of transition metal alloys. J. Phys. Colloq. **35**, C4–199 (1974)
77. W. Kleemann, J. Dec, S. Miga, The cluster glass route of relaxor ferroelectrics. Phase Trans. **88**, 234 (2015)
78. W. Kleemann, J. Dec, Ferroic superglasses: polar nanoregions in relaxor ferroelectric PMN versus CoFe superspins in a discontinuous multilayer. Phys. Rev. B **94**, 174203 (2016)
79. D. Sherrington, K. Mihill, Magnetic ordering in transition metal alloys. Proc. Int. Conf. Mag. (Moscow 1973) **1**, 283 (1974)
80. J.A. Hertz, The Stoner glass. Phys. Rev. B **19**, 4796 (1979)
81. S.F. Edwards, The nucleon Green function in pseudoscalar meson theory. I. Proc. R. Soc. A **232**, 371 (1955)
82. I.M. Gel'fand, A.M. Yaglom, Integation in functional spaces and its applications in quantum physics. Uspekhi Mat. Nauk, **11**, 77 (1956); J. Math. Phys. **1**, 48 (1960)
83. R.L. Stratonovich, A method for the computation of quantum distribution functions. Doklady Acad. Nauk SSSR **115**, 1097 (1957); Sov. Phys. Doklady **2**, 416 (1958)
84. J. Hubbard, Calculation of partition functions. Phys. Rev. Lett. **3**, 77 (1959)
85. D. Sherrington, A new method of expansion in the quantum many-body problem III. The density field. Proc. Phys. Soc. **91**, 285 (1967)
86. D. Sherrington, Auxiliary fields and linear response in Lagrangian many body theory. J. Phys. C **4**, 401 (1971)
87. P.W. Anderson, Localized magnetic states in metals. Phys. Rev. **124**, 41 (1961)
88. B.R. Coles, A. Tari, H.A. Jamieson, Onset of ferromagnetism in alloys at low temperatures, in *Low-Temperature Physics-LT13*, vol. 2, ed. by K.D. Timmerhaus, W.J. O' Sullivan, E.F. Hammel (Plenum, New York, 1974), p. 414
89. D. Sherrington, A simple spin glass perspective on martensitic shape-memory alloys. J. Phys. Condens. Matter **20**, 304213 (2008)

90. K. Bhattacharya, *Microstructure of Martensite* (Oxford University Press, Oxford, 2003)
91. K. Otsuka, C.M. Wayman (eds.), *Shape Memory Materials* (Cambridge University Press, Cambridge, 1998)
92. T. Lookman, S.R. Shenoy, K.O. Rasmussen, A. Saxena, A.R. Bishop, Ferroelastic dynamics and strain compatibility. Phys. Rev. B **67**, 0241114 (2003)
93. Y. Wang, X. Ren, K. Otsuka, A. Saxena, Evidence for broken ergodicity in strain glass. Phys. Rev. B **76**, 132201 (2007)
94. Z. Zhang, Y. Wang, D. Wang, Y. Zhou, K. Otsuka, X. Ren, Phase diagram of $\mathrm{Ti}_{50-x}Ni_{50+x}$: crossover from martensite to strain glass. Phys. Rev. B **81**, 224102 (2010)
95. D. Wang, Y. Wang, Z. Zhang, X. Ren, Modeling abnormal strain states in ferroelastic systems: the role of point defects. Phys. Rev. Lett. **105**, 205702 (2010)
96. E. Vincent, V. Dupuis, Spin glasses: experimental signatures and salient outcomes, in *Frustrated Materials and Ferroic Glasses*, ed. by T. Lookman, X. Ren. Springer Series in Materials Science, vol. 275 (Springer, Cham, 2018). https://doi.org/10.1007/978-3-319-96914-5_2
97. M. Mézard, R. Monasson, Glassy transition in the three-dimensional random-field Ising model. Phys. Rev. B **50**, 7199 (1994)
98. D. Viehland, J.F. Li, S.J. Jang, L.E. Cross, M. Wuttig, Glassy polarization behavior of relaxor ferroelectrics. Phys. Rev. B **46**, 8013 (1992)
99. R. Pirc, R. Blinc, Spherical random-bond-random-field model of relaxor ferroelectrics. Phys. Rev. B **60**, 13470 (1999)
100. S. Kartha, T. Castan, J.A. Krumhansl, J.P. Sethna, Spin-glass nature of tweed precursors in martensitic transformations. Phys. Rev. Lett. **67**, 3630 (1991)
101. Y. Ji, S. Ren, D. Wang, Y. Wang, X. Ren, Strain glasses, in *Frustrated Materials and Ferroic Glasses*, ed. by T. Lookman, X. Ren. Springer Series in Materials Science, vol. 275 (Springer, Cham, 2018). https://doi.org/10.1007/978-3-319-96914-5_7

Chapter 2
Spin Glasses: Experimental Signatures and Salient Outcomes

Eric Vincent and Vincent Dupuis

Abstract Within the wide class of disordered materials, spin glasses occupy a special place because of their conceptually simple definition of randomly interacting spins. Their modelling has triggered spectacular developments of out-of-equilibrium statistical physics, as well analytically as numerically, opening the way to a new vision of glasses in general. "Real" spin glasses are disordered magnetic materials which can be very diverse from the chemist's point of view, but all share a number of common properties, laying down the definition of generic spin glass behavior. This paper aims at giving to nonspecialist readers an idea of what spin glasses are from an experimentalist's point of view, describing as simply as possible their main features as they can be observed in the laboratory, referring to numerous detailed publications for more substantial discussions and for all theoretical developments, which are hardly sketched here. We strived to provide the readers who are interested in other glassy materials with some clues about the potential of spin glasses for improving their understanding of disordered matter. At least, arousing their curiosity for this fascinating subject will be considered a success.

2.1 Introduction

We are surrounded by disordered materials, in which the atoms or molecules are disposed at random. This is the case of window glass, but also of plastics, polymers, foams, gels, granular media, etc. Although being random when seen microscopically, they have controllable and reproducible properties at the macroscopic scale. Their modelling is a challenge for the material scientist as well as for the physicist.

E. Vincent (✉)
SPEC, CEA, CNRS, Université Paris-Saclay, CEA Saclay, 91191 Gif-sur-Yvette Cedex, France
e-mail: eric.vincent@cea.fr

V. Dupuis
Sorbonne Universités UPMC Univ Paris 06 UMR 8234, PHENIX, 75005 Paris, France

T. Lookman, X. Ren (eds.), *Frustrated Materials and Ferroic Glasses*, Springer Series in Materials Science 275, https://doi.org/10.1007/978-3-319-96914-5_2

Within the wide class of disordered materials, spin glasses appear as a remarkably simple archetype, because they can be defined in very simple terms. A spin glass is a set of interacting magnetic moments (originating from spins), in which the interactions are *randomly distributed in sign* (and possibly in magnitude). We easily represent ourselves ferromagnets (forming our permanent magnets), which are constituted of *positively* interacting moments, tending to all align in the same direction and produce a macroscopic magnetization. We also know antiferromagnets, in which the moments are in a *negative sign* interaction that drives them to anti-alignment, establishing a set of two intricated ferromagnetic sublattices oriented in opposite directions.

The case of spin glasses can be simply described as a mixture of both ferro- and antiferromagnetic situations. The magnetic moments (or spins) are in *random sign* interactions, that is, each moment experiences contradicting constraints from its neighbors, which are either ferromagnetically or antiferromagnetically interacting with it. This situation of contradicting influences has been termed *frustration*. No simple symmetric configuration of the set of spins corresponds to an equilibrium state with a clear minimum of energy. On the opposite, the numerous possible spin arrangements with comparable energy yield a huge number of metastable states. Finding the absolute minimum is thus extremely difficult and, from a practical point of view, a spin glass is virtually always out of equilibrium.

In a spin glass, the disorder is contained in the set of the magnetic interactions, which is fixed. Contrary to this situation of *frozen* disorder, in usual glasses the molecules are located at random positions that are evolving with time. The spin glass problem is conceptually simpler, it has allowed rich, far-reaching theoretical developments[1] [1] and numerical investigations (see for instance the recent work [2] of the Janus collaboration, and references therein). Yet, both classes of systems share a lot of similitudes, and the spin glass has been progressively identified as a powerful model for the description and understanding of various glassy systems.

Disordered systems in which a cooperative behavior is developing below a characteristic temperature are sometimes called "ferroic materials," a wide class of materials that constitute the subject of the book in which the present paper on spin glasses is a chapter. Interesting examples are martensitic alloys with shape memory effects [3–5], and relaxor ferroelectrics, on which some light can now be shed thanks to the analogy with certain spin glass models [6–8].

2.2 What Is a Spin Glass Made of?

The first spin glass materials identified were nonmagnetic metals (Au, Ag, Pt...) in which a few percents of magnetic atoms (Fe, Mn...) were dispersed at random

[1]See numerous references in "Spin glasses and random fields", A.P. Young Editor, Series on Directions in Condensed Matter Physics Vol. 12, World Scientific (1998).

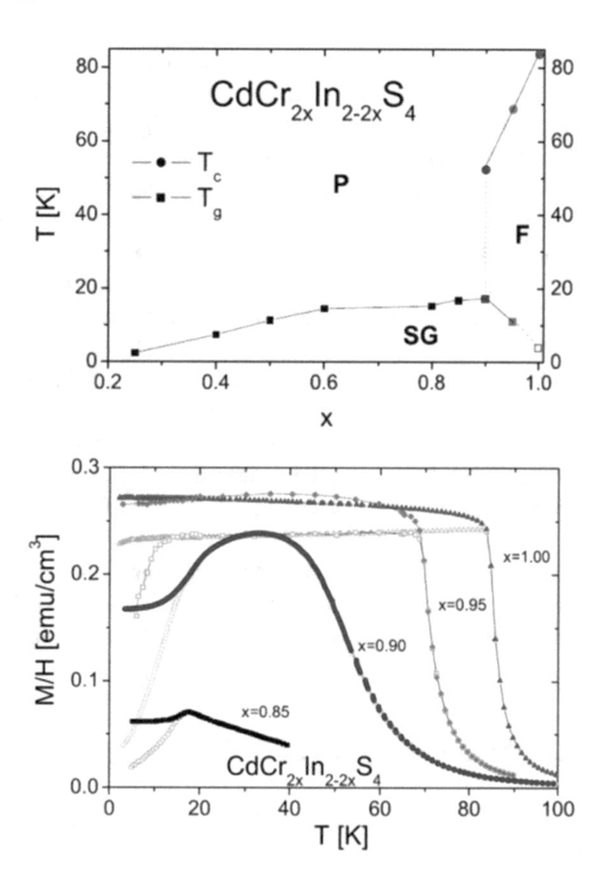

Fig. 2.1 *Top*: Phase diagram of the $CdCr_{2x}In_{2(1-x)}S_4$ thiospinel compound, as a function of the dilution parameter x, showing the measured transition points between paramagnetic (P), ferromagnetic (F), and spin glass (SG) phases (lines are guides for the eye) [14–16]. *Bottom*: Magnetization (normalized to the field) as a function of temperature for five samples of the compound with various dilutions (the colors of the curves refer to the colors of the points in the phase diagram) [15]. The measurement follows the usual **ZFC** and **FC** procedures: for **ZFC**, cooling in zero field, applying the field at low temperature, then measuring upon increasing slowly the temperature, for **FC**, measuring upon slowly cooling in the field (the same curve is obtained upon reheating)

[9]. In Cu:Mn3% for example, the Mn magnetic atoms are separated by random distances, and the oscillating character of their RKKY interaction with respect to distance makes their coupling constants take a random sign [10, 11]. Examples of spin glasses have also been found within insulating compounds [12]. Interestingly, although chemically very different, these various compounds have been found to show a common general behavior that is now understood as generic for spin glasses [13].

An example in which a number of spin glass properties have been observed in detail is the Indium diluted Chromium thiospinel $CdCr_2xIn_{2(1-x)}S_4$, with superexchange magnetic interactions between the (magnetic) Cr^{3+} ions [14, 16]. The phase diagram is shown in Fig. 2.1 (top) [15].

Let us first examine the $x = 1$ compound, which is a ferromagnet with $T_c = 85$ K. The nearest-neighbor interactions are ferromagnetic and dominant for $x = 1$, but the next-nearest ones are antiferromagnetic. Hence, there is some frustration even in the pure Cr compound. Characteristic variations of magnetization as a function of temperature are shown in Fig. 2.1 (bottom), they are measured along the usual ZFC and FC procedures (see caption of Fig. 2.1 for explanation). Starting from high temperatures, a rise-up of magnetization from the paramagnetic phase to the ferromagnetic plateau is clearly observed when approaching $T_c = 85$ K. Below T_c, an irreversible behavior is found, signed up by a separation of the FC and ZFC curves. The irreversibility, in which different geometrical arrangements of the ferromagnetic domains and walls are realized according to the temperature/field procedure, is probably due to some defects in the sample.

In $CdCr_{2x}In_{2(1-x)}S_4$, when a fraction $(1-x)$ of the (magnetic) Cr ions is substituted by (nonmagnetic) In ions, some nearest-neighbor *ferromagnetic* links are suppressed, and the effect of next-nearest *antiferromagnetic* interactions is enhanced [14, 16]. The balance that globally favors ferromagnetism in the absence of In-dilution is disturbed. This is illustrated in Fig. 2.1 (bottom) in the case of the $x = 0.95$ and $x = 0.90$ samples, for which the ferromagnetic plateau becomes rounded. Meanwhile, the onset of irreversibility is shifted towards lower temperatures, indicating the appearance of a different, re-entrant spin glass phase at low temperatures (see phase diagram in top of Fig. 2.1) [15].

For increasing dilution, below $x \leq 0.85$, the ferromagnetic phase disappears. The transition occurs directly from the high-temperature paramagnetic phase to a disordered low-temperature phase, which presents all characteristic features of a spin glass, as will be explained below with various spin glass examples.

2.3 What Happens at T_g?

2.3.1 Dynamical Aspects of the Transition

When a structural glass is cooled down from its liquid phase, it fails to crystallize and becomes a supercooled liquid. The increase of relaxation times when cooling to the glass temperature is so abrupt that the supercooled liquid rapidly starts to behave as a good solid at all accessible experimental time scales [17–19]. For the so-called *fragile* glasses, which are the most common case, the viscosity (proportional to a typical relaxation time τ) of the supercooled liquid increases faster than the Arrhenius law corresponding to simple thermal slowing down over a barrier E:

$$\tau = \tau_0 \exp\left(E/k_B T\right),$$

τ_0 being a microscopic time. This is pictured in the left part of Fig. 2.2 [20], where the viscosity data from various glasses is presented. In this plot of the viscosity versus inverse temperature, the Arrhenius behavior corresponds to a straight line, and most glasses show an upward curvature.

In a spin glass, there is also an abrupt increase of the relaxation times when cooling to the glass temperature. We know it quantitatively from the precise study of the magnetic ac susceptibility. At a fixed frequency $\omega/2\pi$, the ac susceptibility of a spin glass as a function of temperature presents a maximum at $T_f(\omega)$ that can even be very sharp [9]. $T_f(\omega)$ can be understood as the temperature at which the

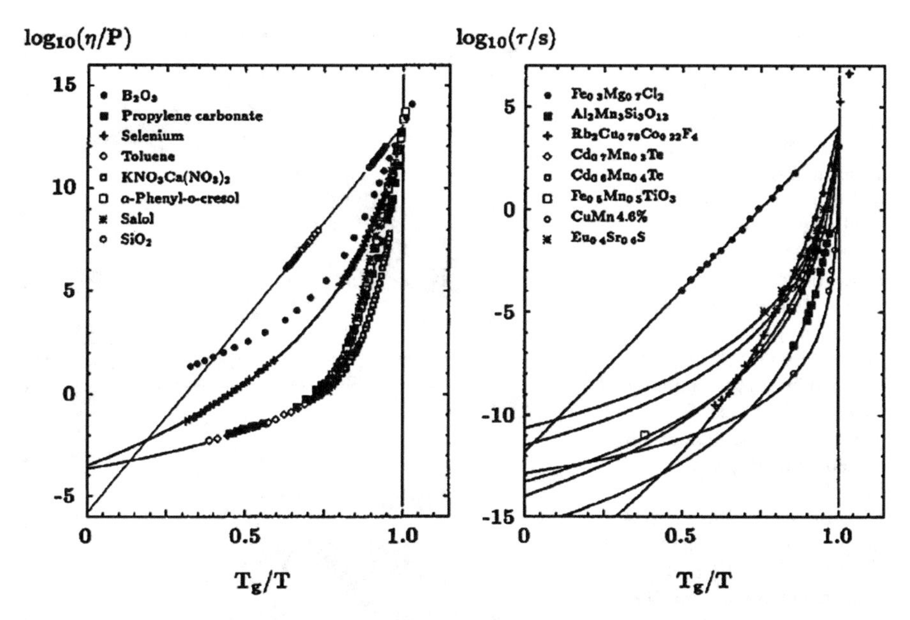

Fig. 2.2 (from [20]) *Left*: Plot of the logarithm of the viscosity η of different glass-forming systems vs. T_g/T, the glass temperature T_g being defined such that $\eta = 10^{13}$ Poise at T_g. Only SiO_2 follows an Arrhenius behavior (straight line in this plot, "strong" glass). The more common "fragile" glasses show a viscosity increase towards low temperatures that is faster than Arrhenius. *Right*: Plot of the logarithm of a typical relaxation time τ of different spin glasses vs. T_g/T, the glass temperature T_g being defined such that $\tau = 10^4$ s at T_g. Most standard spin glasses appear as "fragile," according to the classification of glasses. In more details, spin glasses usually obey critical dynamic scaling (see text), and for structural glasses the question of criticality implies further investigations [21–23]

spin glass becomes frozen *at the experimental probing time scale* τ *equal to the inverse of the frequency,* $\tau = 2\pi/\omega$. This peak temperature is frequency dependent, and it systematically shifts to lower temperatures for decreasing frequencies, as can be seen in Fig. 2.3 [24]. The important point is to quantitatively examine the scale of this frequency dependence, and to determine whether the peak temperature tends to a finite value in the limit of vanishing frequencies [25].

The shift of T_f (ω) with ω in spin glasses can be regarded as an increase of the value of a typical relaxation time $2\pi/\omega$ for decreasing temperature T_f (ω). It can be presented in the same kind of Arrhenius plot of time versus inverse temperature as is used for glasses. This is shown in the right part of Fig. 2.2. Both Arrhenius plots for glasses and spin glasses look very similar. In both cases the increase of the relaxation times with decreasing temperature is faster than an Arrhenius law (upward curvature) [20].

Numerically, the curves in Fig. 2.2 can be well fitted to the Vogel-Fulcher law [26].

$$\tau = \tau_0 \exp\left(E/k_B(T - T_0)\right),$$

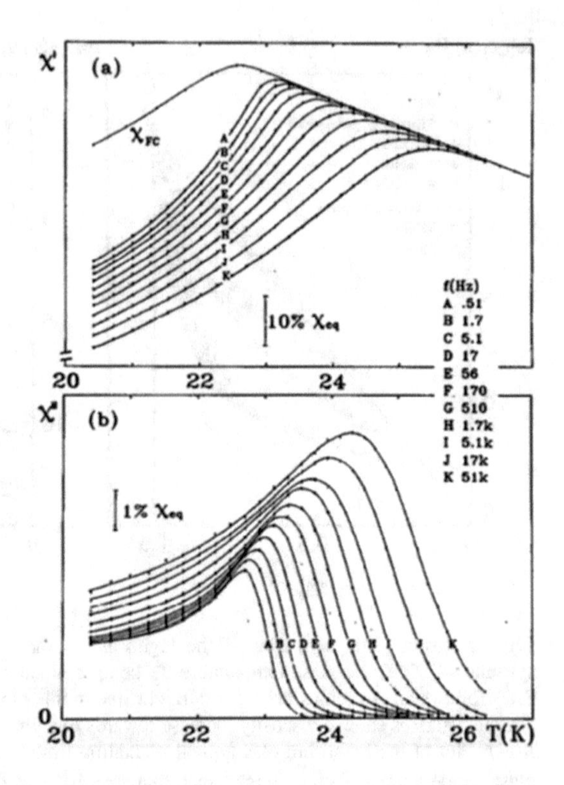

Fig. 2.3 (from [24]) Complex *ac* susceptibility $\chi(\omega) = \chi'(\omega) + \chi''(\omega)$ of the amorphous metallic spin glass $(Fe_xNi_{1-x})B_{16}P_6Al_3$, versus temperature, at different frequencies (0.51 Hz to 51 kHz) of the applied oscillating field (amplitude 50 mOe). (**a**) $\chi'(\omega)$. The field-cooled susceptibility χ_{FC} at an applied field of 50 mOe is also plotted. 10% of the equilibrium susceptibility ($\approx\chi_{FC}$) is indicated. (**b**) $\chi''(\omega)$. 1% of the equilibrium susceptibility ($\approx\chi_{FC}$) is indicated. The frequency-dependent freezing temperature T_f (ω) can be defined equivalently as the temperature of the $\chi'(\omega)$ peak or as that of the $\chi''(\omega)$ inflection point

where T_0 is an adjustable parameter. But, for spin glasses, nothing particular is happening at T_0, to which no physical interpretation is usually given (the situation is different for structural glasses[2]). It is instructive to simply consider that the departure of the data from the Arrhenius law corresponds to a modification of the Arrhenius law by the introduction of a *temperature-dependent* effective energy barrier $E_{\text{eff}}(T)$:

$$\tau = \tau_0 \exp\left(E_{\text{eff}}(T)/k_BT\right).$$

In these terms, the upward curvature of the data in Fig. 2.2 means an increase of $E_{\text{eff}}(T)$ for decreasing temperatures, which can be considered as a signature of the development of correlations when approaching T_g from above, both in glasses and in spin glasses.

In glass-forming liquids such as glycerol, an increase of the number of correlated molecules has now been identified at the approach of T_g [21–23]. This increase remains limited to a relatively modest extent of a few tens of molecules before dynamical arrest, but it paves the way to a new understanding of the glass transition

[2]In structural glasses, it has been found that T_0 from the Vogel-Fulcher law is close to the Kauzman temperature T_K, at which the measured configurational entropy extrapolates to zero. And, in the "Random First Order Transition" scenario, T_K is the critical point at which the size of the glassy ordered domains diverges [18].

in terms of an increase, even though limited before freezing, of a correlation length at the approach of the glass transition [27].

In spin glasses, the increase of a relaxation time $\tau = 2\pi/\omega$ for decreasing temperatures (data in Fig. 2.2 right, obtained from measurements like those presented in Fig. 2.3) can be well fitted considering a divergence of a correlation length ξ at the approach of a transition at T_g,

$$\xi = \xi_0[(T_f(\omega) - T_g)/T_g]^{-\nu}$$

(ν being the usual exponent for the correlation length in a phase transition), and using the dynamic scaling hypothesis.

$$\tau \propto \xi^z$$

(z is thus defined as a dynamical exponent), which yields the critical dynamics scaling law:

$$\tau = \tau_0[(T_f(\omega) - T_g)/T_g]^{-z\nu} \text{ [28].}$$

The exponent $z\nu$ is found to have a rather high value, ranging from 5 to 11 in the various samples (see, for example, [24, 29, 30]).

2.3.2 A Thermodynamic Phase Transition

There are other classes of experiments in spin glasses which support the idea of a thermodynamic phase transition at the zero-frequency limit T_g of the freezing temperature T_f (ω). In a ferromagnet, the order parameter is the spontaneous magnetization, and the order parameter susceptibility is the usual magnetic susceptibility. In a spin glass, some "glassy order" takes place, yielding to an apparently random arrangement of the spins with no visible macroscopic symmetry. The low-temperature phase can be characterized by the Edwards-Anderson order parameter [31], which corresponds to an average of the squared moduli of the spins, and the order parameter susceptibility is the nonlinear magnetic susceptibility [32–34].

The magnetic susceptibility χ can be expanded in even powers of the magnetic field H:

$$\chi = \chi_0 - a_3H^2 + a_5H^4 \ldots,$$

χ_0 being the linear susceptibility. The coefficients of the nonlinear terms are all diverging at T_g, with critical exponents corresponding to the specific spin glass order parameter [32–34]:

$$a_3 \propto (T - T_g)^{-\gamma}, a_5 \propto (T - T_g)^{-(\beta + \gamma)}, \text{ etc.}$$

Their determination implies rather extensive measurements of the magnetic susceptibility as a function of the field, at various temperatures close to T_g, and careful extrapolation at zero field. Figure 2.4 shows the example of Ag:$Mn_{0.5\%}$ [35]. The nonlinear part of the susceptibility is plotted as a function of the field. In this figure, a significant increase of the slope of the curves at the origin for decreasing values of the temperature T towards T_g is very clearly visible. Such

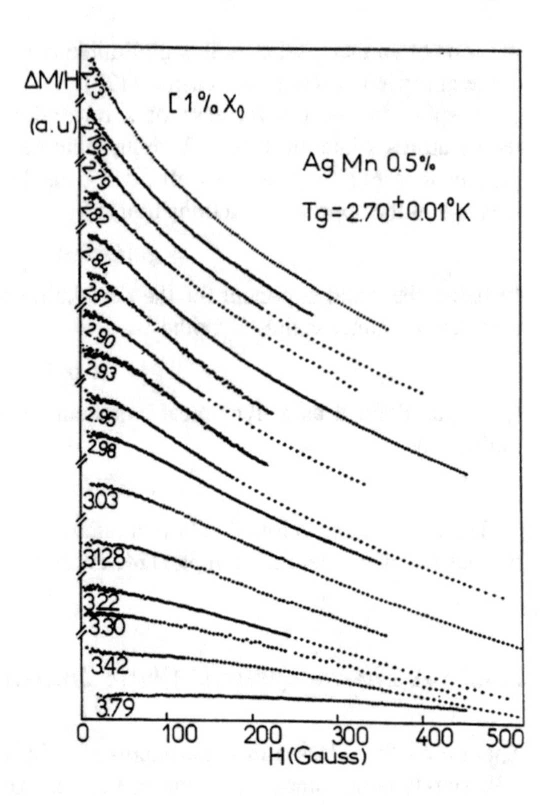

Fig. 2.4 (from [35]) Nonlinear susceptibility (obtained as the difference ΔM between the total magnetization and its field-linear part, divided by the applied magnetic field H), as a function of field H, at different temperatures approaching the critical temperature $T_g = 2.70$ K. The relative origins on the Y-axis are arbitrary. 1% of the linear susceptibility χ_0 is indicated. As $T \to T_g$, a sharp increase of the slope at the origin is clearly visible. Below $1.1T_g = 2.97$ K, the low field behavior of the nonlinear susceptibility is seen to become singular instead of being quadratic

evidences of a static critical behavior from the nonlinear susceptibility, in addition to dynamic critical behavior determined from the *ac* susceptibility, have been obtained in numerous different spin glass samples (see, for example, [35–38], and numerous references in [13]).

2.3.3 Spin Glass Transition: Open Questions

The spin glass transition in real samples is now widely considered as a thermodynamic phase transition, in agreement with mean-field spin glass models [1, 39]. Still, a few interesting questions are worth being mentioned on the subject. In the mean-field theory [1], which is equivalent to infinite dimension d, a true phase transition is indeed obtained, which persists in the presence of a magnetic field, as well for scalar Ising [40] as for vector Heisenberg [41] spins. In $d = 3$, a phase transition is expected for Ising spins, but not for Heisenberg spins. However, many evidences of a phase transition are found in real $d = 3$ Heisenberg-like samples (e.g., [36], see references in [13]). A very plausible explanation can be found in the scenario proposed by Kawamura of chirality driven phase transition of spins, a mechanism that has been detailed and argued both numerically and analytically

[42–44]. The agreement between the values of the critical exponents found in this scenario and those observed in the experiments is remarkable [13]. Experimentally, direct access to the observation of chirality freezing is difficult, but some pioneering measurements of the anomalous Hall effect in spin glasses have already given very interesting results [45–47].

At the spin glass transition, critical exponents which vary from Ising to Heisenberg situations have been measured in a wide series of $d = 3$ samples with variable spin anisotropy [13, 38]. Experiments on spin glass *thin films* have allowed to study the crossover from $d = 3$ to $d = 2$, a situation in which the transition is expected to take place at $T = 0$, and which allows interesting studies of the growth of the spin glass correlation length under constrained conditions (see recent results in [48, 49], and older references therein).

In the mean-field theory of spin glasses [1], an *infinite number of different pure states* is obtained, yielding a very interesting and complex picture of the spin glass phase. This would imply that in a real sample, after cooling from the paramagnetic phase, many domains with different types of spin glass order should coexist and compete. A phase diagram with a transition line is obtained as a function of the magnetic field [40, 41]. On the other hand, in very different approaches, scaling theories of the spin glass behavior have been developed for Ising spins on the basis of phenomenological arguments [50–52]. In the droplet model [50, 52], there are simply two (spin reversal symmetric) pure states, and *the phase transition is expected to be destroyed by any magnetic field*. Let us comment briefly on these important questions.

1. The question of a multiple or single nature of the ground state in the spin glass is very difficult to address experimentally. Some (indirect) arguments in favor of multiple pure states are discussed below (Sect. 2.5.3). Also, following a theoretical suggestion stating that the correlation of the conductance fluctuations in two spin states should be a direct function of their overlap [53], a new experimental approach using transport measurements on mesoscopic samples has been developed. Magneto-resistance traces are observed, which are reproducible until the sample is heated well over T_g. They are likely to be correlated to frozen spin configurations which strikingly persist under high field cycling [54]. Measurements of the universal conductance fluctuations in mesoscopic spin glasses are a real challenge, and have not yet given full results, but in principle they are a promising way to obtain information on the nature of the pure states.
2. The vanishing of the phase transition in the presence of a magnetic field has been reported in a study of the dynamic scaling properties of a $Fe_{0.5}Mn_{0.5}TiO_3$ single crystal [55], which is a good example of a short-range Ising spin glass [56]. Interestingly, very recent experiments on the $Dy_xY_{1-x}Ru_2Si_2$ show that the phase transition persists in a field in this Ising *but long-range* (RKKY) system [57]. For Heisenberg-like spin glasses, data from torque measurements bring robust evidence for a true spin glass ordered state which survives under high applied magnetic fields [38].

Thereby, important questions concerning the nature of the spin glass transition are still open. They are also a hot topic for structural glasses, in which the nonlinear susceptibility is now understood as playing a similar role as in spin glasses [27].

An important experimental program has now allowed investigating, by the means of nonlinear dielectric susceptibility measurements, the growth of correlations at the approach of the transition and in the glassy phase during aging [21–23].

In experiments on spin glasses, no true thermodynamic equilibrium state can be reached at laboratory time scales. What we see in experiments probing the spin glass state is essentially out-of-equilibrium properties [58]. A wide panel of rich results have been obtained, of which we highlight in Sects. 2.4 and 2.5 some of the most prominent features.

2.4 Slow Dynamics and Aging

2.4.1 DC Experimental Procedures

We present in Fig. 2.5 a typical measurement of the magnetization as a function of temperature in a spin glass [59], performed along the usual ZFC-FC procedures (see caption of Fig. 2.1). The sample is here a $Fe_{0.5}Mn_{0.5}TiO_3$ single crystal, which is a good example of a spin glass of Ising type, due to strong easy axis anisotropy [56].

The curves are very similar to those presented in Fig. 2.1 for the $x = 0.85$ diluted thiospinel sample, they illustrate the general features of simple *dc* magnetization measurements on spin glasses. Above T_g, the magnetization M follows a characteristic Curie-Weiss law.

$$M \propto C/(T - \theta)$$

which is characteristic of a paramagnetic phase (C is a constant proportional to the square of the individual magnetic moments, and θ is a temperature proportional to the energy of the interactions). Below T_g, the temperature behavior becomes

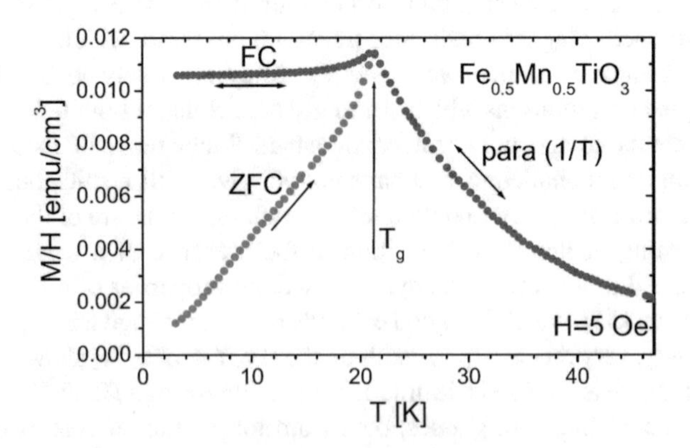

Fig. 2.5 (from [59]) Magnetization divided by the applied field, as a function of temperature, measured along the usual ZFC-FC procedures (see caption of Fig. 2.1) on a $Fe_{0.5}Mn_{0.5}TiO_3$ single crystal [56] along the *c*-axis direction. The onset of irreversibility is seen at T_g in the separation of the ZFC-FC curves, which above T_g are identical with a $1/T$-like paramagnetic behavior

irreversible. The magnetization is history-dependent, and a splitting of the ZFC and FC curves is observed. Such a splitting can be found in various magnetic systems in which some freezing occurs for any reason (e.g. non-interacting magnetic nanoparticles, see a review in [60]). What is characteristic here of the collective behavior related to the spin glass transition is the (approximate) flatness of the FC curve observed here below T_g (also obtained as a characteristic feature in mean-field models [1], as emphasized in [61]). When going from the paramagnetic region to low temperatures, the magnetization increase suddenly stops.

In the FC state, the magnetization value can be considered to a first approximation to be at equilibrium (this is usually true within 1%), and the FC curve can be measured upon cooling or as well heating in the presence of the field. On the contrary, the ZFC curve is out of equilibrium, because the application of the field has been made at low temperature, in the frozen phase. The value of the ZFC magnetization depends on the time spent at each temperature. After cooling in zero field and applying the field, the ZFC(t) magnetization slowly increases as a function of time, most probably towards the FC value (that is, however, never attained in experimental time scales).

An example is shown in Fig. 2.6 [24], in which we see that the relaxation curves are influenced by another time parameter: the waiting time. In the procedure used to measure these relaxations, the sample is rapidly cooled in zero field from above T_g to $T < T_g$ (quench), and it is kept at temperature T during a given *waiting time* t_w,

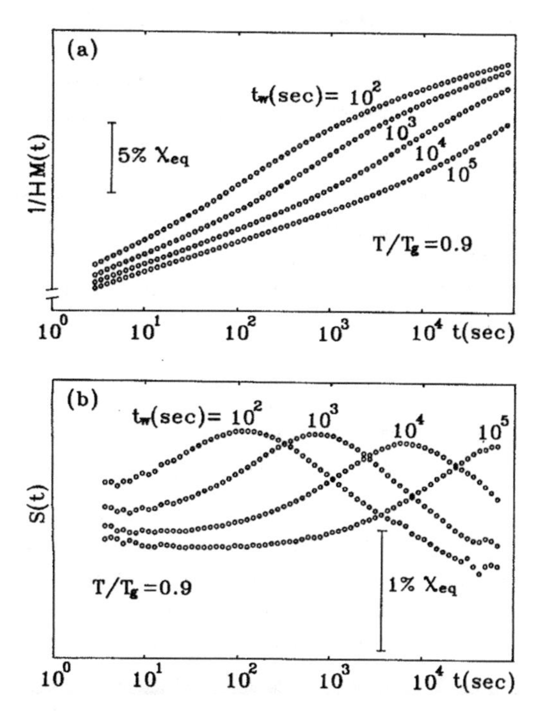

Fig. 2.6 (from [24]) Zero field cooled susceptibility [$(I/H)M(t)$] and corresponding relaxation rate [$S(t) = (I/H)dM/d\ \ln t$] at different waiting times ($t_w = 10^2$, 10^3, 10^4, and 10^5 s) plotted versus time t (log scale), for the amorphous metallic spin glass $(Fe_xNi_{1-x})B_{16}P_6Al_3$ at $T = 20.3$ K ($T/T_g = 0.9$), $H = 0.1$ Oe. *Top*: [$(I/H)M(t)$], and *Bottom*: $S(t)$. 1% of the equilibrium susceptibility ($\approx \chi_{FC}$, see Fig. 2.3 on the same sample) is indicated

after which a small field is applied (defining $t = 0$). Then the relaxation is measured as a function of the observation time t (elapsed since the field change).

The relaxation curves in Fig. 2.6 (top) display two essential features of spin glass dynamics:

1. The magnetization relaxation following a field change is slow, roughly logarithmic in time (glassy state).
2. It strongly depends on the waiting time: the longer t_w, the slower the relaxation. This waiting time dependence is called *aging*.

Hence, time translation invariance is lost in the slow dynamics of the spin glass: the relaxation dynamics depend on *both* t_w *and* t, not only on t, the dynamics is nonstationary [62].

The logarithmic derivative $dM/d \log t$ of these magnetization curves can be given an interesting physical interpretation, as proposed by L. Lundgren at the beginning of the 80's [63]. The derivatives of the curves of the top part of Fig. 2.6 are shown in the bottom part. They are bell shaped, and, remarkably, their broad maximum (inflection point of the magnetization curves) occurs after a time t of the order of t_w itself.

The physical interpretation is the following. As can be seen on the $\log t$ scale, the relaxations are slower than exponential. They can be modelled as a sum of exponential decays $\exp(-t/\tau)$, the decay times τ being distributed according to a distribution function $g(\tau)$ which is defined in this way, and represents the effective density of relaxation times. Taking the derivative $dM/d \log t$ introduces a term t/τ $\exp(-t/\tau)$ in the integrand of the sum over the τ distribution. This term is sharply peaked around $t = \tau$. Approximating this peaked function by a δ-function, we estimate the integral by taking the value of the integrand for $\tau = t$, and obtain [63].

$$dM_{tw}/d \log t \propto g_{tw}(\tau = t).$$

Here M_{tw} and g_{tw} are labelled by t_w to emphasize that each relaxation curve, taken for a given t_w, gives access (through its logarithmic time derivative) to the density of relaxation times that represents the dynamics of the spin glass at a time of the order of t_w after the quench. Thus, each derivative $dM_{tw}/d \log t$ gives an estimate of the density $g_{tw}(\tau = t)$, and as t_w increases $g_{tw}(\tau)$ shifts towards longer times. This gives a physical picture of the two important features listed above:

1. The effective relaxation times are widely distributed (glassy state).
2. This distribution function, peaking around $\tau = t_w$, shifts towards longer times with increasing t_w: this is the phenomenon of aging.

The mirror experiment of the above ZFC relaxations can be performed as well, it gives access to the same information, provided that the amplitude of the field change remains small (far below the limit of linearity, typically 1–10 Oe). In this mirror procedure, one starts from a FC state at temperature $T > T_g$. After cooling the spin glass from the paramagnetic phase to a temperature $T < T_g$, and waiting a time t_w at T, the field is turned to zero. Then the remnant magnetization (called "thermo-remnant" magnetization, TRM) slowly decays [64].

Both relaxations observed by ZFC and TRM mirror procedures are symmetric: ZFC(t) + TRM(t) = FC (this relation holds even if a slight relaxation of the FC magnetization occurs, FC ≡ FC(t) [65]). All these curves present an inflection point at $t \sim t_w$. When plotted as a function of t/t_w, the curves can be almost superimposed. In a first approximation, we can thus consider that the relaxation curves obey a t/t_w scaling. When examined in more details, however, some systematic departures from t/t_w scaling are observed, and can be taken into account very precisely by more refined procedures ("subaging," [64, 66, 67]).

The same phenomenon of aging has been known for a long time in the mechanical properties of a wide class of materials called "glassy polymers" [68, 69]. When a piece of, e.g., PVC is submitted to a mechanical stress, its response (elongation, torsion ...) is logarithmically slow. And the response depends on the time elapsed since the polymer has been quenched below its freezing temperature. Like in spin glasses, for increasing aging time the response becomes slower (called "physical aging", as opposed to "chemical aging"). The t_w-dependence of the dynamics of glassy polymers has been expressed as a scaling law [68] that could be applied to the case of spin glasses ([64], see also [70]). Numerous other glassy materials show similar aging phenomena, although not necessarily obeying precisely a t/t_w scaling (see, for example, [68, 69, 71, 72]). Numerical simulations of packed hard spheres provide us with very powerful toy models of simple glasses [73].

2.4.2 AC Experimental Procedures

Slow dynamics and aging in the spin glass phase can also be observed by *ac* susceptibility measurements, in which a small *ac* field (~1 Oe) is applied all along the measurement [58, 64]. Again, the starting point consists in cooling the spin glass from above T_g, down to a given $T < T_g$ at which the *ac* response is measured as a function of the time elapsing, which is the "age" of the system (equivalent to $t_w + t$ in the *dc* procedures).

Figure 2.7 [59] shows the time evolution of both components χ' and χ'' of the ac susceptibility. We find here the same features as observed in *DC* experiments:

1. The *ac* response is delayed, as seen from the existence of an out-of-phase susceptibility χ''. χ'' is zero above T_g in the paramagnetic phase, and rises up as the sample is cooled into the spin glass phase (as already visible in Fig. 2.3).
2. The susceptibility relaxes down, signing up the occurrence of aging. This relaxation is visible on both χ' and χ'', but is more important in relative value $\Delta\chi/\chi$ in the out-of-phase component χ''.

In Fig. 2.7, the relaxations of χ' *and* χ'' are shown for different frequencies ($\omega/2\pi$), and plotted as a function of $(\omega/2\pi).t$ for reasons that will soon become clear. Their asymptotic (stationary) values $\chi'_{\text{eq}}(\omega)$ and $\chi''_{\text{eq}}(\omega)$ in the infinite time limit can be determined by a fit of the decaying part to $(\omega.t)^{-b}$ (the exponent b is found in the range 0.15–0.20 in various samples [59]).

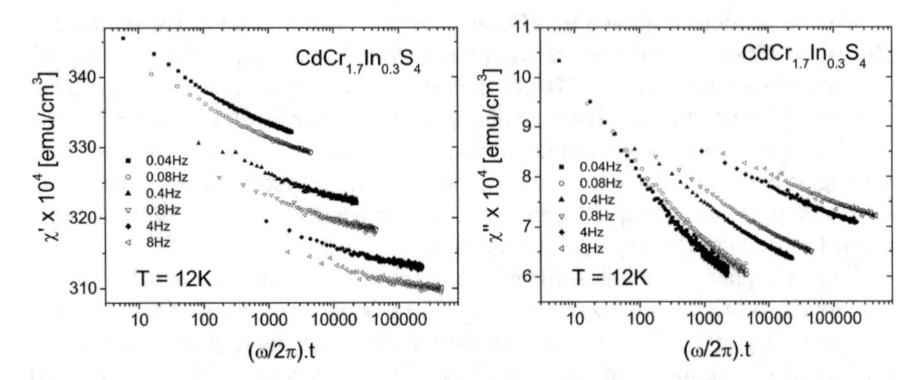

Fig. 2.7 (from [59]) Relaxation of both components of the *ac* susceptibility during the time t following the quench from above T_g down to $T = 12$ K $= 0.7T_g$, as a function of the product of the frequency $\omega/2\pi$ times t. *Left*: in-phase component χ'. *Right*: out-of-phase component χ''

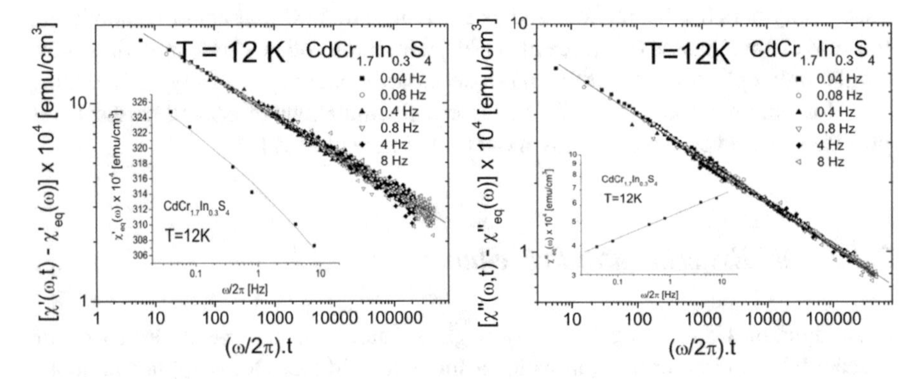

Fig. 2.8 (from [59]) Relaxation of both components of the *ac* susceptibility, same data as in Fig. 2.7, but after subtraction of the equilibrium part, log-log scale. The inserts show the fitted equilibrium values. The relaxations at different frequencies $\omega/2\pi$ merge onto a unique curve (power law) as a function of the reduced variable $(\omega/2\pi).t$

In Fig. 2.8 [59], the asymptotic values have been subtracted, and the remaining decaying part is represented on a log scale, which emphasizes the power law behavior (straight lines in this log-log plot). Remarkably, the decay part of the curves measured at different frequencies are all superimposed when plotted as a function of $(\omega/2\pi).t$, as well for χ' as for χ''.

This $\omega.t$ scaling can be related to the t/t_w scaling of the *dc* data in the following way. In an *ac* measurement, the time $2\pi/\omega$ can be considered as a typical observation time, which plays the same role as t in the *DC* relaxation procedures. On the other hand, the total age of the system is here the time t along which the *ac* relaxation is measured after cooling, that is equivalent to $t_w + t$ in the *DC* experiment. Hence,

$$\omega.\, t \approx (1/t)(t + t_w) = 1 + t_w/t,$$

the present $\omega.t$ scaling is equivalent to the t/t_w scaling of the *dc* experiments [58, 63, 64, 66].

In structural and polymer glasses, similar *ac* procedures have also been used for the study of aging effects. For instance, in [74], the dielectric constant ε of glycerol is found to show a strong relaxation, with a frequency dependence which has qualitatively the same trend as $\omega.t$ scaling. And, more recently, detailed measurements of the third harmonics dielectric susceptibility in glycerol [22] have revealed an increase of the size of the glassy domains with the aging time.

Beyond the study of the response of the spin glass to small *dc* and *ac* excitations, the dynamics can also be explored by measuring the spontaneous magnetic fluctuations (magnetic noise) in the absence of any excitation. This is a difficult experiment, because contrary to the case of the ferromagnet the amplitude of the spontaneous noise is very small in spin glasses. Nevertheless, such measurements could be performed [62, 64, 75, 76]. They brought very interesting information on the violation of the fluctuation-dissipation relations in the aging regime, which allow comparisons with some important features of spin glass theories [77, 78] (see also Figs. 5–7 in [61]).

2.5 Aging, Rejuvenation, and Memory Effects

It is well known that the state of a structural glass is very much dependent on the speed at which it has been cooled, a slower cooling bringing smaller values of the enthalpy and specific volume that are closer to equilibrium values [17, 68, 79]. This view of glasses was the starting point of a new class of experiments in spin glasses. We explored how the aging behavior could be influenced by the temperature history, having in mind that well-suited cooling procedures might perhaps bring the spin glass into a strongly aged state, which otherwise would require astronomical waiting times to be established [80, 81]. These experiments have brought important surprises [58].

2.5.1 Temperature Step Experiments

Figure 2.9 presents the result of an experiment in which a small negative temperature cycle is performed during the relaxation of the *ac* susceptibility [81]. After a normal cooling ($\sim$100 s from $1.3\,T_g$ to $0.7\,T_g$), the spin glass is kept at constant temperature $T = 12\ K = 0.7\ T_g$ for $t_1 = 300$ min, during which aging is visible in the strong relaxation of χ''. Then, the temperature is lowered one step further from $T = 12\ K$ to $T - \Delta T = 10\ K$. What is then observed is not a slowing down of the relaxation, but on the contrary a jump of χ'' and a restart. Such a restart upon further cooling was termed *rejuvenation*, because the relaxation of χ'' behaves as if aging was starting anew at $T - \Delta T$. Apparently, there is no influence of former aging at T.

The question one may naturally ask is whether this renewed relaxation corresponds to a *full rejuvenation* of the sample. The answer is no. Let us first point out

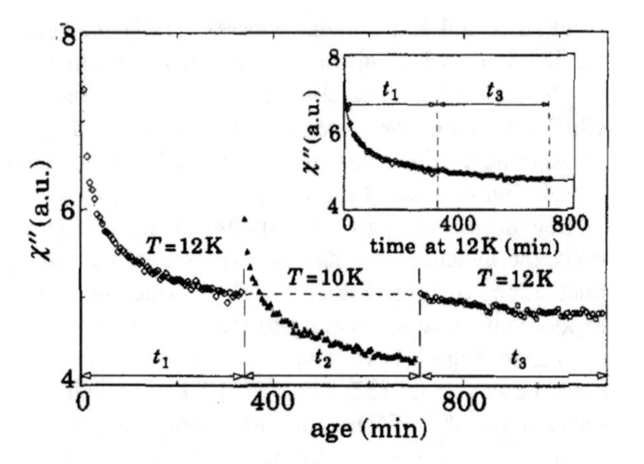

Fig. 2.9 (from [81]) Relaxation of the out-of-phase susceptibility χ'' during a negative temperature cycle of amplitude $\Delta T = 2$ K (frequency 0.01 Hz), showing aging at 12 K, rejuvenation at 10 K, and memory at 12 K. The inset shows that, despite the *rejuvenation* at 10 K, both parts at 12 K are in continuation of each other (*memory*). The sample is the $CdCr_{2x}In_{2(1-x)}S_4$ thiospinel spin glass ($T_g = 16.7$ K)

that, for observing such a restart, the temperature interval ΔT must obviously be sufficiently large, here $\Delta T \geq 2$–3 *K*. And still, the time window explored in this experiment is limited, therefore we do not know very much about the *overall* state of the spin glass, which involves relaxation processes on a very wide time scale.

The final part of the experiment brings the answer to the question. After aging during $t_2 = 300$ min at $T - \Delta T = 10$ *K*, when the temperature is turned back to $T = 12$ *K*, the χ'' relaxation restarts exactly from the point that was attained at the end of the stay at the original temperature T. It goes in precise continuity of the former one, as if nothing of relevance at T had happened at T-ΔT. As shown in the inset of Fig. 2.9, this can be checked by shifting the third relaxation to the end of the first one: they are in continuity, and can be superposed on the reference curve which is obtained in a simple aging at T. Hence, during aging at $T - \Delta T$ and despite the strong associated χ''-relaxation, the spin glass has kept a *memory* of previous aging at T. This memory is retrieved when heating back to T.

This negative temperature cycle experiment pictures in a spectacular manner the phenomenon of rejuvenation and memory in a spin glass. However, examination of the situation in more details shows that it should not be considered too simply. We see in Fig. 2.10 the results of negative temperature cycle experiments performed with various values of ΔT ([59], but see also [82]).

For $\Delta T = 1$ *K*, the beginning of the third part relaxation shows a transient spike, which lasts for ~5000 s before the curve merges with those, obtained for higher ΔT's, that are in continuity with the relaxation at T. Thus, for a smaller ΔT than that corresponding to a "full" memory effect, there is indeed some contribution at T from aging at $T - \Delta T$ that may be divided in two parts:

1. An *incoherent* contribution (spike), extending over rather long but finite times (3–5000 s). For smaller ΔT, the observed "transient spike" decreases, and finally vanishes.

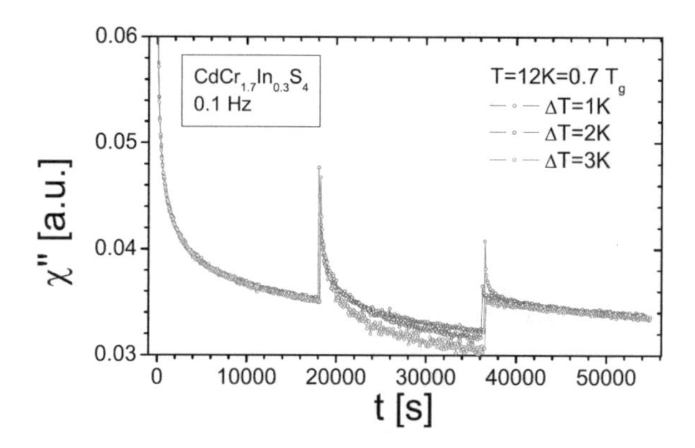

Fig. 2.10 (data from [59]) Relaxation of the out-of-phase susceptibility χ'' during negative temperature cycles of different amplitudes, ranging from $\Delta T = 1$ K (upper curve, with the prominent spike) to $\Delta T = 3$ K (lower curve, no spike and full memory). Same sample as in Fig. 2.9, but the frequency is here 0.1 Hz, instead of 0.01 Hz (Here, the points can be taken more rapidly, and a small upturn is visible for $T = 2$ K: full memory is only obtained for $T = 3$ K)

2. A coherent contribution to aging at T as an additional aging time t_{eff} (called "cumulative process" below), in such a way that the third relaxation must be shifted by $(t_2 - t_{\text{eff}})$ to be in continuity with the first part.

Details on the results in this regime of intermediate ΔT values, together with their discussion in terms of a Random Energy Model, can be found in [82] (see also [83]). Many sets of temperature step experiments of this kind have been performed, by the Saclay group (see references in [58, 59]) and also by the Uppsala group (see, for example, [84]), with similar results, even though sometimes discussed in slightly different terms.

2.5.2 *Memory Dip Experiments*

The ability of the spin glass to keep a *memory* despite *rejuvenation* has been further explored in experiments with multiple temperature steps. The first "memory dip" experiments, suggested by P. Nordblad, were developed in collaboration between the Uppsala and Saclay groups [85, 86]. An example of a "multiple dip experiment" is shown in Fig. 2.11 [59, 87, 88].

This is an *ac* experiment in which the sample is cooled by 2 K steps of duration half an hour down to 4 K, and then reheated continuously (sketch in the inset of Fig. 2.11). Figure 2.11 shows χ'' as a function of temperature during this procedure, starting from $T > T_g$ where $\chi'' = 0$ (paramagnetic phase). χ'' rises up when crossing $T_g = 16.7$ *K*, and when the cooling is stopped at 14 K, the relaxation of χ'' due to aging is recorded during ½ h (successive points at the same temperature in the figure). Upon further cooling by another 2 K step, a χ'' jump of rejuvenation

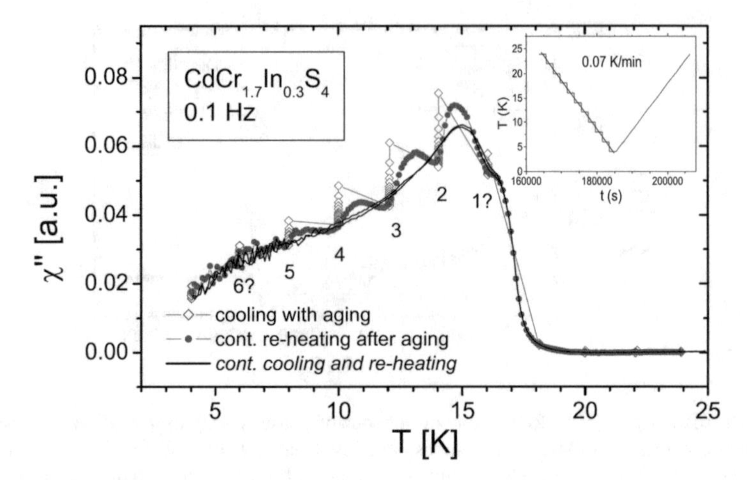

Fig. 2.11 (from [59, 87]) An example of multiple rejuvenation and memory steps. The sample was cooled by 2 K steps, with an aging of time of 2000 s at each step (open blue diamonds). Continuous reheating at 0.001 K/s (full red circles) shows memory dips at each temperature of aging. The solid black line shows a reference measurement with continuous cooling/heating and no steps. The inset is a sketch of the procedures

is found, and the relaxation due to aging takes place. At each new cooling step, rejuvenation and aging are seen, and this happens ~6 times in the experiment of Fig. 2.11.

In the second part of the experiment, the sample is reheated continuously at a slow rate (0.001 *K/s*, equal to the average cooling rate). Amazingly, apart from the rather noisy low-*T* region, the memory of each of the aging stages performed during cooling is revealed in shape of "memory dips" in χ'' (*T*), tracing back the lower value of χ'' which was attained at each of the aging temperatures. Thus, the spin glass is able to keep the simultaneous memory of several (up to 5–6!) successive aging periods performed at lower and lower temperatures. Increasing the temperature afterwards reveals the memories (and meanwhile erases them).

One can think of a certain type of aging in terms of domain growth dynamics, of the type occurring in a ferromagnet. Aging by domain growth is a naturally "cumulative" process, in the sense that aging continues additively during the various parts of the experimental procedure, from one temperature to the other, as long as $T < T_g$. This cumulative process corresponds to the coherent contribution to aging observed in small temperature step experiments. But it is difficult to imagine how rejuvenation and memory effects may arise in this scheme. In the droplet model [50, 52], they are related to "temperature chaos" effects [89]. Discussions on the possible relevance of this scenario to experiments can be found in [87, 90, 91], and also [48, 49].

In some spin glass experiments like the one that we present now in Fig. 2.12 [59], the dual aspect of aging dynamics in terms of coherent (cumulative) and

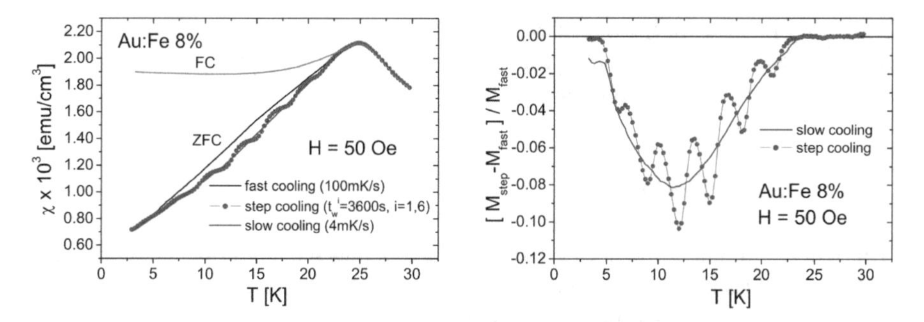

Fig. 2.12 (from [59]) *Left*: Effect of various cooling procedures on the ZFC magnetization of the Au:Fe8% spin glass. Comparison of fast and slow cooling, with and without stops. *Right*: difference with the magnetization obtained after fast cooling

incoherent (rejuvenation and memory) contributions appears very clearly. In this *dc* type procedure, proposed by the Uppsala group [92], the sample is cooled in zero field with various thermal histories, and after applying the field at low temperature the magnetization is measured while increasing the temperature continuously at fixed speed (small steps of 0.1 K/min).

On one hand, we can observe the effect of a slow cooling in comparison with that of a fast cooling: the slow-cooled curve lies below the fast one in the whole temperature range. There is indeed a cooling rate effect in spin glasses, provided that one chooses an appropriate procedure to bring it to evidence. On the other hand, memory effects can be demonstrated by stopping the cooling at several distinct temperatures and waiting. The magnetization measured during reheating after this step-cooling procedure shows clear dips at all temperatures at which the sample has been aging. These effects are emphasized in the right part of Fig. 2.12 where we have plotted the difference between the curves obtained after a specific cooling history and the reference one obtained after a fast cooling. Sharp oscillations (memory dips) show up on top of a wide bump (cumulative aging).

2.5.3 Discussion

Thus, aging effects in spin glasses can be described as a combination of rejuvenation and memory effects, which are strongly temperature specific, with some more classical cooling rate effects [70]. Structural glasses are usually considered to be dominated by cooling rate effects [17, 18]. However, experiments in various glassy systems have been designed these last years to search for possible rejuvenation and memory effects. Interesting examples of such phenomena can be found for instance in [93–95] with PMMA and in [96, 97] with gelatine. New experimental ways have been developed more recently for the investigation of aging effects in colloids and soft matter, like microrheology techniques using optical traps [98, 99]. There is now a growing interest for the out-of-equilibrium properties of *active* colloidal systems (chemically powered colloids [100], biomolecular motors [101], Janus particles with asymmetric surface coating [102], etc.)

2.5.3.1 Hierarchical Picture

The rejuvenation and memory effects reveal a strong sensitivity of the aging state of the spin glass to temperature changes, which have very different effects when the temperature is decreased or increased. The Saclay group proposed to account for these phenomena in terms of a hierarchical organization of the metastable states as a function of temperature, as pictured in Fig. 2.13, which we now explain ([80, 103], see details in [58, 66]).

In this scheme, the effect of temperature variations is represented by a modification of the free-energy landscape of the metastable states (not only a change in the transition rates between them). At fixed temperature T, aging corresponds to the slow exploration of the numerous metastable states (at level T in Fig. 2.13). When going from T to $T-\Delta T$, the free-energy valleys subdivide into smaller ones, separated by new barriers (level $T-\Delta T$ in Fig. 2.13). *Rejuvenation* arises from the transitions that are now needed to equilibrate the population rates of the new sub-valleys: this is a new aging stage. For large enough ΔT (and in the limited experimental time window), the transitions can only take place between the sub-valleys inside the main valleys, in such a way that the population rates of the main valleys are untouched, keeping the *memory* of previous aging at T. Hence the memory can be retrieved when reheating and going back to the T-landscape.

This tree picture may seem somewhat naïve when described in these qualitative terms. It is, however, able to reproduce many features of the experiments when discussed in more details (see discussions of experimental results in [58, 66]).

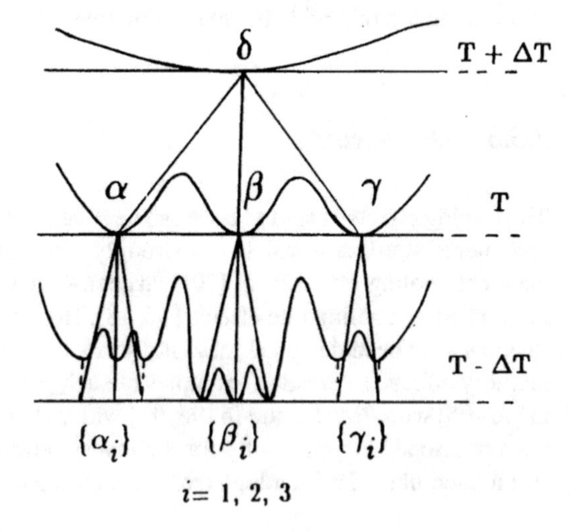

Fig. 2.13 Schematic picture of the hierarchical structure of the metastable states as a function of temperature ([80, 103], see details in [58, 66])

Indeed, quantitative theoretical models have been derived along such a hierarchical scheme, in terms of developments of Bouchaud's Trap Model [104, 105] and also Derrida's Random Energy model [82, 106, 107].

The hierarchical organization of the *metastable* states as a function of *temperature* is, of course, reminiscent of the hierarchical organization of the *pure* states (as a function of their *overlap*) that is obtained in the mean-field theory of the spin glass with full replica symmetry breaking [1]. It has been shown that rejuvenation and memory effects can indeed be expected in the dynamics of this model [108]. Detailed analysis of the temperature growth of the free-energy barriers involved in temperature variation experiments has suggested that the hierarchical organization of the *metastable* states as a function of temperature can indeed be extrapolated to a hierarchical organization of the *pure* states [109] (see however the discussion of a different barrier analysis in [87]).

Also, from another point of view [88], rejuvenation effects can be expected from the fact that in a frustrated system effective interactions can be defined, which are found to be temperature dependent in many cases [110]. Still, memory effects require considering other processes in addition [111].

2.5.3.2 A Correlation Length for Spin Glass Order

Aging can also be considered as the slow establishment of a "spin glass order" [50–52]. Starting from a random configuration obtained when cooling the spin glass from the paramagnetic state (like the structurally liquid state of a glass after quench), the spins will locally optimize their respective orientations over longer and longer length scales, which define a *time growing correlation length*. This correlation length could be determined, although a bit indirectly, in various sets of experiments [112–114].

The rejuvenation and memory effects have important implications in terms of these length scales. Let us consider that during aging at T the correlation length of the spin glass order grows up to a certain L_T. When going to $T-\Delta T$, rejuvenation implies that new reorganization processes take place. But, in order to keep the memory of what happened at T, these new processes at $T-\Delta T$ should occur on smaller length scales $L_{T-\Delta T} < L_T$. In practice, the independence of aging at length scales $L_{T-\Delta T}$ and L_T can be realized by a strong separation of the corresponding *time* scales τ:

$$\tau(L, T - \Delta T) >> \tau(L, T).$$

This necessary separation of the aging length scales with temperature has been coined "temperature-microscope effect" by Bouchaud [87]. In experiments like those from Figs. 2.11 and 2.12, at each temperature stop aging should take place at well-separated length scales

$$L_n < \cdots < L_2 < L_1,$$

as if the magnification of the microscope was varied by orders of magnitude at each temperature step. This hierarchy of embedded length scales as a function of temperature is a *real space* equivalent of the hierarchy of metastable states in the *configuration space* (Fig. 2.13) [87].

Spin glass numerical simulations have allowed the exploration of the microscopic organization of the spins, and investigated the properties of the correlation length of the spin glass state (see, for example [115], and references therein). But, due to frustration, the evolutions towards equilibrium are very slow, which implies time-consuming computations. The experiments on real spin glasses are typically exploring the 10^0–10^5 s time range, which in units of the paramagnetic spin flip time $\tau_0 \sim 10^{-12}$ s corresponds to 10^{12}–10^{17} τ_0. Taking $\tau_0 = 1$ MC step for comparison, the first numerical simulations were exploring up to $\sim 10^7$ Monte-Carlo (MC) steps, a rather short-time regime compared with the experiments. In the Janus and Janus II projects [2], dedicated supercomputers have been designed, which allow computation up to $\sim 10^{11}$ MC steps. A wide set of numerical results is now available in a time range that is close to that of spin glass experiments, yielding considerable progresses in their interpretation [2].

Intermediate time range of this dynamics could be explored in experiments on the glassy state formed by interacting magnetic nanoparticles (see [116], and references therein). The microscopic flip time of the "super spins" born by the nanoparticles (10^4–10^6 ferromagnetically coupled spins in each nanoparticle) is much longer than 10^{-12} s (depending on the temperature and on the size of the nanoparticles, τ_0 can range from 10^{-10} s up to milliseconds), and in τ_0 units the explored experimental time window can be close to that of simulations. The results concerning the time growth of the correlation length ξ of the glassy order during aging in *experiments* (spin and super spin glass) and *simulations* are in overall agreement [2, 112–116]. The general trend of this growth is a slow power law

$$\xi \propto (t/\tau_0)^{aT/Tg},$$

with $a \sim 0.15$, going up to a few tens of atomic distances for spin glasses in the laboratory time window.

2.6 Conclusions

In this paper, we have tried to emphasize some general experimental features of the disordered magnetic systems that are known as spin glasses. The spin glass state develops at and below a well-defined temperature T_g, above which the spins form a paramagnetic phase. At T_g, the nonlinear susceptibility diverges, and the transition to the spin glass state presents most characteristics of a thermodynamic transition. However, since relaxation times are diverging at T_g, in an experiment the equilibrium phase cannot be established. Starting from a frozen random configuration inherited from the paramagnetic phase, a "spin glass order" is progressively established at longer and longer range, and this slow evolution is accompanied by aging phenomena that show up in various dynamical properties, like a waiting time dependence of the *ac* susceptibility or of the magnetization relaxation in *dc* field variation procedures.

Lowering the temperature during aging causes a restart of aging processes (*rejuvenation*), while the *memory* of previous aging at higher temperatures can be kept, and retrieved when reheating. The effect of temperature changes can be seen as a combination of these rejuvenation and memory effects with more common *cumulative* effects. In structural and polymer glasses, and many other glassy systems, the dominant scenario is the continuation of aging from one temperature to another in terms of *cumulative* processes, but rejuvenation and memory processes can often be found, even though with weaker importance. Hence, spin glasses appear as glassy systems in which rejuvenation and memory effects are more pronounced than in others, but they can yet be used as powerful models for glasses in general, because of their rather simple theoretical formulation in terms of a system of randomly interacting spins. Certain classes of spin glass theoretical models (with p-spin interactions) have even been found to reproduce rather precisely [117, 118] the properties of structural glasses as modelled by mode coupling theory [119].

Since the early pioneering studies, spin glasses have continuously benefitted from active exchanges between experiments and theory [9, 31, 39]. They have been the opportunity of important conceptual breakthroughs in statistical physics [1] and these developments have shed new light on the problem of glasses in general [18, 27]. New results allowing a better microscopic understanding of the transition in structural glasses are now being obtained [21–23]. The out-of-equilibrium properties of spin glasses have inspired a lot of original experimental investigations of glassy systems in general. After several decades, many crucial questions on the nature of the glassy state are still open, and the important experimental, analytical and numerical efforts that are presently deployed offer very promising perspectives of new progress on these fascinating questions.

Acknowledgements The authors are grateful to J.-P. Bouchaud, H. Bouchiat, H. Kawamura, F. Ladieu, S. Miyashita, S. Nakamae, P. Nordblad, Y. Tabata, T. Taniguchi, and H. Yoshino for recent discussions and helpful suggestions concerning this manuscript. EV is thankful to the University of Tokyo for kind hospitality during the most part of the writing.

References

1. M. Mezard, G. Parisi, M.A. Virasoro, *Spin Glass Theory and Beyond* (World Scientific, Singapore, 1987)
2. M. Baity-Jesi, E. Calore, A. Cruz, L.A. Fernandez, J.M. Gil-Narvion, A. Gordillo-Guerrero, D. Iñiguez, A. Maiorano, E. Marinari, V. Martin-Mayor, J. Monforte-Garcia, A. Muñoz-Sudupe, D. Navarro, G. Parisi, S. Perez-Gaviro, F. Ricci-Tersenghi, J.J. Ruiz-Lorenzo, S.F. Schifano, B. Seoane, A. Tarancon, R. Tripiccione, D. Yllanes, Phys. Rev. Lett. **118**, 157202 (2017)
3. X. Ren, in *Disorder and Strain-Induced Complexity in Functional Materials*, Springer Series in Materials Science, ed. by T. Kakeshita et al.. Strain glass and strain glass transition, vol 148 (Springer, New York, 2012)

4. D. Wang, Y. Wang, Z. Zhang, X. Ren, Phys. Rev. Lett. **105**, 205702 (2010)
5. Y. Zhou, D. Xue, T. Ya, X. Ding, S. Guo, K. Otsuka, J. Sun, X. Ren, Phys. Rev. Lett. **112**, 025701 (2014)
6. D. Sherrington, in *Disorder and Strain-Induced Complexity in Functional Materials*, Springer Series in Materials Science, ed. by T. Kakeshita et al.. Understanding glassy phenomena in materials, vol 148 (Springer, New York, 2012)
7. D. Sherrington, Phys. Rev. Lett. **111**, 227601 (2013)
8. D. Sherrington, Phys. Status Solidi B **1**, 1–15 (2014)
9. V. Cannella, J.A. Mydosh, Phys. Rev. B **6**, 4220 (1972)
10. J.A. Mydosh, *Spin Glasses, an Experimental Introduction* (Taylor & Francis, London, 1993)
11. J.A. Mydosh, Rep. Prog. Phys. **78**, 052501 (2015)
12. H. Maletta, W. Felsch, Phys. Rev. B **20**, 1245 (1979)
13. H. Kawamura, T. Taniguchi, in *Handb. Magn. Mater*, ed. by K. H. J. Bischow. Spin glasses, vol 24 (Elsevier, Amsterdam, 2015)
14. M. Alba, J. Hammann, M. Noguès, J. Phys. C **15**, 5441 (1982)
15. V. Dupuis, E. Vincent, M. Alba, J. Hammann, Eur. Phys. J. B. **29**, 19 (2002)
16. J.L. Dormann, A. Saifi, V. Cagan, M. Noguès, Phys. Status Solidi (b) **131**, 573 (1985)
17. M.D. Ediger, P. Harrowell, J. Chem. Phys. **137**, 080901 (2012)
18. A. Cavagna, Phys. Rep. **476**, 51 (2009)
19. L. Berthier, G. Biroli, Rev. Mod. Phys. **83**, 587 (2011)
20. J. Souletie, D. Bertrand, J. Phys. I (France) **1**, 1627 (1991)
21. C. Crauste-Thibierge, C. Brun, F. Ladieu, D. L'Hôte, G. Biroli, J.-P. Bouchaud, Phys. Rev. Lett. **165703**, 104 (2010)
22. C. Brun, F. Ladieu, D. L'Hôte, G. Biroli, J.-P. Bouchaud, Phys. Rev. Lett. **109**, 175702 (2012)
23. S. Albert, T. Bauer, M. Michl, G. Biroli, J.-P. Bouchaud, A. Loidl, P. Lunkenheimer, R. Tourbot, C. Wiertel-Gasquet, F. Ladieu, Science **352**, 1308 (2016)
24. P. Svedlindh, P. Granberg, P. Nordblad, L. Lundgren, H.S. Chen, Phys. Rev. B **35**, 268 (1987)
25. J.L. Tholence, Physica **108B**, 1287 (1981)
26. J. Souletie, J.L. Tholence, Phys. Rev. B **32**, 516 (1985)
27. J.-P. Bouchaud, G. Biroli, Phys. Rev. B **72**, 064204 (2005)
28. K. Binder, A.P. Young, Phys. Rev. B **29**, 2864 (1984)
29. N. Bontemps, J. Rajchenbach, R.V. Chamberlin, R. Orbach, Phys. Rev. B **30**, 6514 (1984)
30. E. Vincent, J. Hammann, Solid State Comm. **58**, 57–62 (1986)
31. S.F. Edwards, P.W. Anderson, J. Phys. F **5**, 965 (1975)
32. M. Suzuki, Prog. Theor. Phys. **58**, 1151 (1977)
33. J. Chalupa, Solid State Comm. **22**, 315 (1977)
34. M. Suzuki, S. Miyashita, Physica **106A**, 344 (1981)
35. H. Bouchiat, J. Phys. (France) **47**, 71 (1986)
36. E. Vincent, J. Hammann, J. Phys. C **20**, 2759 (1987)
37. L.P. Lévy, Phys. Rev. B **38**, 4963 (1988)
38. D. Petit, L. Fruchter, I.A. Campbell, Phys. Rev. Lett. **88**, 207206 (2002)
39. D. Sherrington And S. Kirkpatrick, Phys. Rev. Lett. 35, 1972 (1975)
40. J.R.L. de Almeida, J.R.L and D.J. Thouless. J. Phys. A **11**, 983 (1978)
41. M. Gabay, G. Toulouse, Phys. Rev. Lett. **47**, 201 (1981)
42. H. Kawamura, Phys. Rev. Lett. **68**, 3785 (1992)
43. D.X. Viet, H. Kawamura, Phys. Rev. Lett. **102**, 027202 (2009)
44. H. Kawamura, J. Phys. Soc. Jpn. **79**, 011007 (2010)
45. T. Taniguchi, K. Yamanaka, H. Sumioka, T. Yamazaki, Y. Tabata, S. Kawarazaki, Phys. Rev. Lett. **93**, 246605 (2004)
46. P. Pureur, F. Wolff Fabris, J. Schaf, I.A. Campbell, Europhys. Lett. **67**, 123 (2004)
47. T. Taniguchi, J. Phys. Condens. Matter **19**, 145213 (2007)
48. S. Guchhait, R.L. Orbach, Phys. Rev. B **92**, 214418 (2015)
49. Q. Zhai, D.C. Harrison, D. Tennant, E. Dan Dalhberg, G.G. Kenning, R.L. Orbach, Phys. Rev. B **95**, 054304 (2017)

50. D.S. Fisher, D.A. Huse, Phys. Rev. B **38**, 373 (1988)
51. G.J.M. Koper, H.J. Hilhorst, J. Phys. (Paris) **49**, 429 (1988)
52. D.S. Fisher, D.A. Huse, Phys. Rev. B **38**, 386 (1988)
53. D. Carpentier, E. Orignac, Phys. Rev. Lett. **100**, 057207 (2008)
54. T. Capron, G. Forestier, A. Perrat-Mabilon, C. Peaucelle, T. Meunier, C. Baüerle, L.P. Lévy, D. Carpentier, L. Saminadayar, Phys. Rev. Lett. **111**, 187203 (2013)
55. J. Mattsson, T. Jonsson, P. Nordblad, H. Aruga Katori, A. Ito, Phys. Rev. Lett. **74**, 4305 (1995)
56. H. Aruga Katori, A. Ito, J. Phys. Soc. Jpn. **63**, 3122 (1994)
57. Y. Tabata, T. Waki, H. Nakamura. Phys. Rev. B **96**, 184406 (2017)
58. E. Vincent, Lect. Notes Phys. **716**, 7 (2007)
59. V. Dupuis, PhD Thesis, Orsay University (France) (2002), https://tel.archives-ouvertes.fr/tel-00002623/document
60. J.L. Dormann, D. Fiorani, E. Tronc, in *Adv. Chem. Phys.*, ed. by I. Prigogine, A. Rice, vol XCVIII (J. Wiley & Sons, Inc, Hoboken, 1997)
61. G. Parisi, PNAS **103**, 7948 (2006)
62. P. Refregier, M. Ocio, J. Hammann, E. Vincent, J. Appl. Phys. **63**, 4343 (1988)
63. L. Lundgren, P. Svedlindh, O. Beckman, Phys. Rev. B **26**, 3990 (1982)
64. M. Alba, J. Hammann, M. Ocio, P. Refregier, H. Bouchiat, J. Appl. Phys. **61**, 3683 (1987)
65. P. Nordblad, L. Lundgren, L. Sandlund, J. Magn. Magn. Mater. **92**, 228 (1990)
66. E. Vincent, J. Hammann, M. Ocio, J.-P. Bouchaud, L.F. Cugliandolo, in *Complex Behavior of Glassy Systems*, Lecture Notes in Physics, ed. by M. Rubi, vol 492 (Springer, Berlin, 1997), p. 184 (preprint available as arXiv:cond-mat/9607224)
67. D. Parker, F. Ladieu, J. Hammann, E. Vincent, Phys. Rev. B **74**, 184432 (2006)
68. L.C.E. Struik, *Physical Aging in Amorphous Polymers and Other Materials* (Elsevier, Houston, 1978)
69. W.K. Waldron Jr., G.B. McKenna, M.M. Santore, J. Rheol. **39**, 471 (1995)
70. J. Hammann, E. Vincent, V. Dupuis, M. Alba, M. Ocio, J.-P. Bouchaud, J. Phys. Soc. Jpn. **69**(Suppl.A), 206 (2000)
71. M. Cloitre, R. Borrega, L. Leibler, Phys. Rev. Lett. **85**, 4819 (2000)
72. H. Tanaka, S. Jabbari-Farouji, J. Meunier, D. Bonn, Phys. Rev. E **71**, 021402 (2005)
73. Y. Jin, H. Yoshino, Nat. Commun. **8**, 14935 (2016)
74. R.L. Leheny, S.R. Nagel, Phys. Rev. B **57**, 5154 (1998)
75. D. Hérisson, M. Ocio, Phys. Rev. Lett. **88**, 257202 (2002)
76. D. Hérisson, M. Ocio, Eur. Phys. J. B. **40**, 283 (2004)
77. L.F. Cugliandolo, J. Kurchan, J. Physics A **27**, 5749 (1994)
78. L.F. Cugliandolo, J. Kurchan, L. Peliti, Phys. Rev. E **55**, 3898 (1997)
79. S.L. Simon, G.B. McKenna, J. Chem. Phys. **107**, 8678 (1997)
80. P. Refregier, E. Vincent, J. Hammann, M. Ocio, J. Phys. (France) **48**, 1533 (1987)
81. F. Lefloch, J. Hammann, M. Ocio, E. Vincent, Europhys. Lett. **18**, 647 (1992)
82. M. Sasaki, V. Dupuis, J.-P. Bouchaud, E. Vincent, Eur. Phys. J. B. **29**, 469 (2002)
83. M. Sasaki, K. Nemoto, J. Phys. Soc. Jpn. **70**, 1099 (2001)
84. P.E. Jönsson, H. Yoshino, P. Nordblad, Phys. Rev. Lett. **89**, 097201 (2002)
85. K. Jonason, E. Vincent, J.P. Bouchaud, P. Nordblad, Phys. Rev. Lett. **81**, 3243 (1998)
86. K. Jonason, P. Nordblad, E. Vincent, J. Hammann, J.-P. Bouchaud, Eur. Phys. J. B. **13**, 99 (2000)
87. J.-P. Bouchaud, V. Dupuis, J. Hammann, E. Vincent, Phys. Rev. B **65**, 024439 (2001)
88. S. Miyashita, E. Vincent, Eur. Phys. J. B. **22**, 203 (2001)
89. A.J. Bray, M.A. Moore, Phys. Rev. Lett. **58**, 57 (1987)
90. H. Yoshino, A. Lemaître, J.-P. Bouchaud, Eur. Phys. J. B. **20**, 367–395 (2001)
91. P.E. Jönsson, R. Mathieu, P. Nordblad, H. Yoshino, H. Aruga Katori, A. Ito, Phys. Rev. B **70**, 174402 (2004)
92. R. Mathieu, P. Jönsson, D.N.H. Nam, P. Nordblad, Phys. Rev. B **63**, 092401 (2001)
93. L. Bellon, S. Ciliberto, C. Laroche, Europhys. Lett.**51**, 551 (2000)
94. K. Fukao, A. Sakamoto, Phys. Rev. E **71**, 041803 (2005)

95. L. Bellon, S. Ciliberto, C. Laroche, Eur. Phys. J. B **25**, 223 (2002)
96. A. Parker, V. Normand, Soft Matter **6**, 4916 (2010)
97. O. Ronsin, C. Caroli, T. Baumberger, Phys. Rev. Lett. **103**, 138302 (2009)
98. P. Jop, J. Ruben Gomez-Solano, A. Petrosyan, S. Ciliberto, J. Stat. Mech., **2009**, P04012 (2009)
99. S. Jabbari-Farouji, D. Mizuno, M. Atakhorrami, F.C. MacKintosh, C.F. Schmidt, E. Eiser, G.H. Wegdam, D. Bonn, Phys. Rev. Lett. **98**, 108302 (2007)
100. J. Palacci, C. Cottin-Bizonne, C. Ybert, L. Bocquet, Phys. Rev. Lett. **105**, 088304 (2010)
101. Y. Sumino, K.H. Nagai, Y. Shitaka, D. Tanaka, K. Yoshikawa, H. Chaté, K. Oiwa, Nature **483**, 448 (2012)
102. T. Mano, J.-B. Delfau, J. Iwasawa, M. Sano, PNAS **114**, E2580 (2017)
103. V.S. Dotsenko, M.V. Feigel'man, L.B. Ioffe, Sov. Sci. Rev. A. Phys. **15**, 1–250 (1990)
104. J.P. Bouchaud, J. Phys. I (France) **2**, 1705 (1992)
105. J.P. Bouchaud, D.S. Dean, J. Phys. I (France) **5**, 265 (1995)
106. B. Derrida, Phys. Rev. B **24**, 2613 (1981)
107. B. Derrida, E. Gardner, J. Phys. C **19**, 2253 (1986)
108. L.F. Cugliandolo, J. Kurchan, Phys. Rev. B **60**, 922 (1999)
109. J. Hammann, M. Lederman, M. Ocio, R. Orbach, E. Vincent, Physica A **185**, 278 (1992)
110. S. Miyashita, Progr. Theor. Phys. **69**, 714 (1983)
111. S. Tanaka, S. Miyashita, Progr. Theor. Phys. Suppl. **157**, 34 (2005)
112. Y.G. Joh, R. Orbach, J.J. Wood, J. Hammann, E. Vincent, Phys. Rev. Lett. **82**, 438 (1999)
113. F. Bert, V. Dupuis, E. Vincent, J. Hammann, J.-P. Bouchaud, Phys. Rev. Lett. **92**, 167203 (2004)
114. Y.G. Joh, R. Orbach, G.G. Wood, J. Hammann, E. Vincent, J. Phys. Soc. Jpn. **69**(Suppl.A), 215 (2000)
115. L. Berthier, A.P. Young, Phys. Rev. B **71**, 314429 (2005)
116. S. Nakamae, C. Crauste-Thibierge, D. L'Hôte, E. Vincent, E. Dubois, V. Dupuis, R. Perzynski, Appl. Phys. Lett. **101**, 242409 (2012)
117. J.-P. Bouchaud, L. Cugliandolo, J. Kurchan, M. Mezard, Physica A **226**, 243 (1996)
118. T. Castellani, A. Cavagna, J. Stat. Mech., **2005**, P05012 (2005)
119. W. Gotze, L. Sjogren, Rep. Prog. Phys. **55**, 241 (1992)

Chapter 3
Frustration(s) and the Ice Rule: From Natural Materials to the Deliberate Design of Exotic Behaviors

Cristiano Nisoli

Abstract The ice rule has a long, interesting history, one that proved most influential to thermodynamics, physical chemistry, statistical mechanics, magnetism, material science, and soft matter. First introduced to solve the mystery of the residual entropy in water ice, it has motivated an entire set of exactly solvable problems in statistical mechanics and applied mathematics. It was then recognized in exotic magnets at low temperature, and designed in new artificial frustrated systems, both magnetic and colloidal. As new classes of artificial ice rule materials are being presented, dedicated geometries are proposed to generate new, exotic collective behaviors, often not found in natural systems. There, a deeper understanding of the origin of the ice rule from different forms of frustration can be exploited for design of unusual phases.

3.1 Introduction

The story of the ice rule begins with the brilliant intuition of Linus Pauling within the limited context of water ice. It has then evolved to a powerful topological notion to conceptualize the effects of frustration in a variety of interesting systems, ranging from chemistry to magnetism to soft matter to superconductors. It often leads to a so-called "ice manifold" a quasi-disordered set of low-energy configurations wherein new interesting phenomena can take place, and sometimes even be designed in artificial realizations. These systems reveal common properties in their diversity, as well as different behaviors in their commonality. Their properties are often emergent from the collective dynamics of discrete underlying degrees of freedom. We shall see how such exotic states and behaviors can now be designed in artificial, collectively interacting materials at the nano or micro-scale.

C. Nisoli (✉)
Theoretical Division, Los Alamos National Laboratory, Los Alamos, NM, USA
e-mail: cristiano@lanl.gov

T. Lookman, X. Ren (eds.), *Frustrated Materials and Ferroic Glasses*, Springer Series in Materials Science 275, https://doi.org/10.1007/978-3-319-96914-5_3

The scope of this chapter is not to provide an exhaustive, diachronic review of these systems, but rather to compare different materials described by the ice rule, and conceptualize their similarities and differences. It should not be considered as a reference for the systems introduced, but rather as an attempt to provide perspective over disparate materials, and as such it should be read in its entirety. In our choice of the material we do not deny our intellectual bias—indeed we declare it. As reviews of frustration and of the ice rule in such systems are plentiful, we will devote quite some space on the most recent developments of *artificial* ice-rule materials, whose phenomenology is not just richer than their "natural" counterparts, but can in fact be designed at will, characterized at the constituent level, and accessed at desirable ranges of temperature and field.

In particular, we will begin in Sect. 3.2 by introducing fundamental concepts in many different yet well-known and relatively simpler systems, both natural and artificial. Then in Sect. 3.3 we will try to conceptualize some theoretical themes that are common to such materials, to provide some understanding on the diverse origin on the ice rule from different conceivable forms of frustration. Finally, in Sect. 3.4 we show how this deeper insight can lead to more complexity in behavior in deliberately manufactured, artificial ice rule systems.

3.2 Common Systems

3.2.1 Water Ice

Water is a most common and vital substance, fundamental for Life on the planet, and of course for our everyday life. It is therefore rather remarkable how mysterious water still is to science, and in how many ways. The study of its chameleonic behavior has generated famous controversies and debates as many of its physical properties often do not lead themselves to simple modeling (see the very exhaustive website of M. Chaplin at http://www1.lsbu.ac.uk/water/ and the references therein).

One of the early mysteries of water was the apparent residual entropy of water ice. In the 1930s Giaque and Ashley [1, 2] performed a series of carefully conducted calorimetric experiments, deduced the entropy of ice at very low temperature, and found that it was not zero despite the ordered, crystalline structure. An explanation was provided by Linus Pauling a few years later [3], and it goes as follows.

Ice comes in about eighteen crystalline forms [4], depending on metastability, but all of them involve oxygen atoms residing at the center of tetrahedra, sharing four hydrogen atoms with four nearest neighbor oxygen atoms (Fig. 3.1). As in all cases the water molecules are hydrogen-bonded to each other, two of such hydrogens are covalently bonded to the oxygen of their molecule, and two will realize hydrogen bonds with oxygens of different molecules. One way to say that is that two are "in," two are "out" of the tetrahedron whose center is occupied by the oxygen atom, and this is the so-called ice-rule introduced by Bernal and Fowler [6]. Each tetrahedron

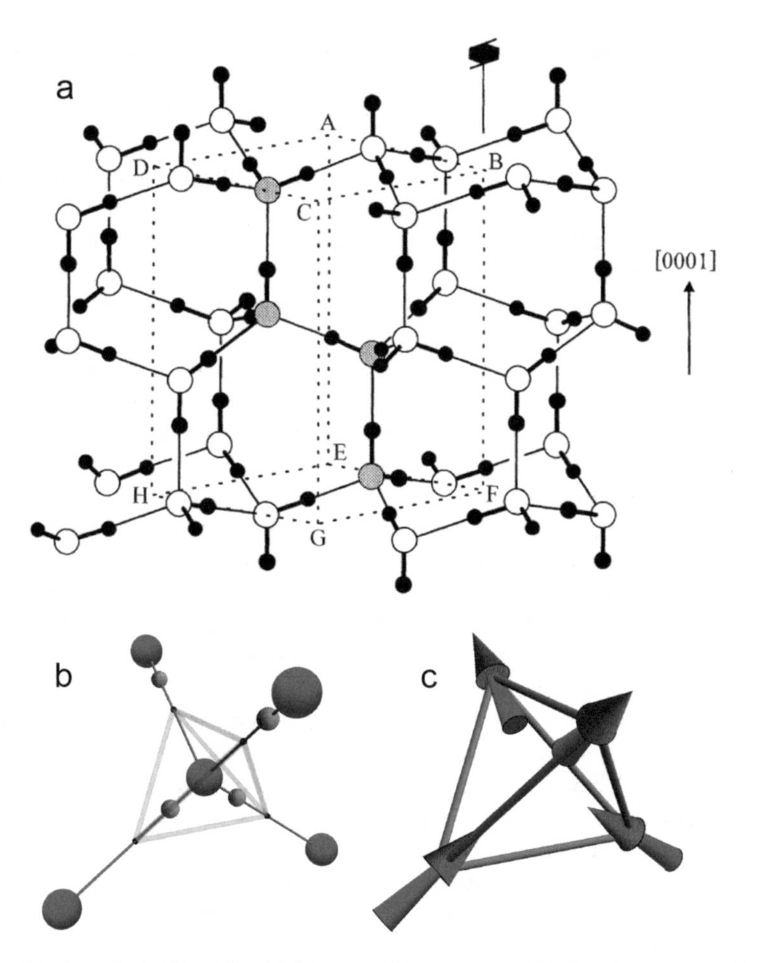

Fig. 3.1 The ice rule in Water Ice. (**a**) The crystalline structure of Ih ice shows proton disorder in the allocation of hydrogen atoms among oxygen centered tetrahedra (image from [4]). (**b**) In water ice oxygen atoms sit at the center of tetrahedra, connected to each other by a hydrogen atom. Two of such protons are close (covalently bonded) to the oxygen at the center, two are further away, close to two of the four neighboring oxygens. (**c**) One might replace this picture with spins pointing in or out depending on whether the proton is in or out of the tetrahedron. Then two spins point in, two point out. This corresponds to the disposition of magnetic moments on pyrochlore spin ices, rare earth titanates whose magnetic ensemble does not order at low temperature because of frustration and, much like water ice, has non-zero low temperature entropy density (figures **b**, **c** from [5])

has therefore 6 admissible configurations out of the $2^4 = 16$ ideally possible, and the collective degeneracy of the ice grows exponentially in the number of tetrahedra N as W^N. This leads to a non-zero entropy per tetrahedron $s = k_B \ln W$.

In one of the most precise and felicitous back-of-the envelope estimates in the history of statistical mechanics, Pauling counted such degeneracy starting from the allowed configurations of each tetrahedron, obtaining $W = 3/2$. This value was later found to be remarkably close to both the experimental value and to the numerically evaluated value ($W = 1.50685 \pm 0.00015$ [7]). We can anticipate here that the same kind of heuristic estimate can be successfully applied to other ice-rule models, including two-dimensional ones, such as the six-vertex model [8–12], or the hexagonal ice (see later) and the results are similarly accurate. And yet, as we shall see, the Pauling approach would fail if applied to new, vertex-frustrated ice-rule materials described later in this chapter [13–17], where vertices are instead ordered, and degeneracy comes from the collective inability to place all of them in an ordered state.

The works of Jaques, Pauling, and others pointed to the reality of exotic states of cosntrained disorder in the most common and vital substance on earth. The idea is however more general, and of broader application. As Fig. 3.1 shows, one can associate to the ice rule a spin model: classical binary spins are assigned to the bonds between molecules, and they point toward the proton. Then the ice rule dictates that two spins point in, two point out, as two protons are close and two are away. One immediately suspects some underlying level of mathematical abstraction and generality beyond the chemistry of water, which indeed was understood and explored decades later in statistical mechanics. The ice rule immediately resembles a precept for "charge" cancellation—for a proper definition of a topological charge (see Sect. 3.3)—within each tetrahedron, which in turn can be generalized as a "vertex" in about any geometry, not necessarily regular, as we will explain in Sect. 3.3.1. Violations of the ice rule can also be formalized and generalized in similar fashion. They are already present in water ice where they are known as Bjerrum defects [18], but in magnetic materials they can lead to magnetic monopoles as fractionalized excitations, as we shall see later [5, 19].

Before the seventies these approaches had motivated the introduction by Lieb, Wu, Rys, and others of simplified models of mathematical physics, known as vertex models. These are two-dimensional spin models where different energies are ascribed to different vertex configurations, and which could in many cases be solved exactly, typically via transfer matrix methods [8–12]. The six-vertex model in particular [8] only admits ice ruling obeying vertices on a square lattice, and was meant to represent a solvable, two-dimensional equivalent of water ice. Beside helping clarifying the physics of ice, these models provided new phenomenology (for instance, the Rys F-model, a six-vertex model with lifted degeneracy, contains an infinitely continuous transition [9]), but also were often shown to be equivalent to other, outstanding problems of statistical mechanics, and thus initiated an independent, theoretical line of research in mathematical physics. Moreover, the same set of ideas proved later of application behind the chemistry of water, to magnetic materials.

3.2.2 *Spin Ice: Rare Earth Titanates*

Beside water ice, ice-like systems have received renewed interest in the 1990s, when unusual behaviors were discovered in the low temperature regime of rare earth titanates such as $Ho_2Ti_2O_7$ and $Dy_2Ti_2O_7$. In these substances, the magnetic cations Ho^{3+} and Dy^{3+} carry a very large magnetic moment, $\mu \sim 10\mu_B$. When temperature lowered below 200 K for $Ho_2Ti_2O_7$ and 300 K for $Dy_2Ti_2O_7$, these moments can be modeled as binary, classical Ising spins constrained to point along the directions of the lattice bonds which form a pyrochlore lattice (Fig. 3.2), mutually interacting as magnetic dipoles. Thus, these materials exhibit a net ferromagnetic interaction between nearest neighbor spins, yet when temperature is lowered, no ordering transition is present. It was suggested that strong frustration impeded ordering.

It was then noted by Harris et al. [23] and confirmed experimentally by Ramirez et al. [20] that, similar to protons in water ice, the magnetic moments of these spin ice materials reside on a lattice of corner-sharing tetrahedra (Fig. 3.2c), and they are constrained to point either directly toward or away from the center of a tetrahedron

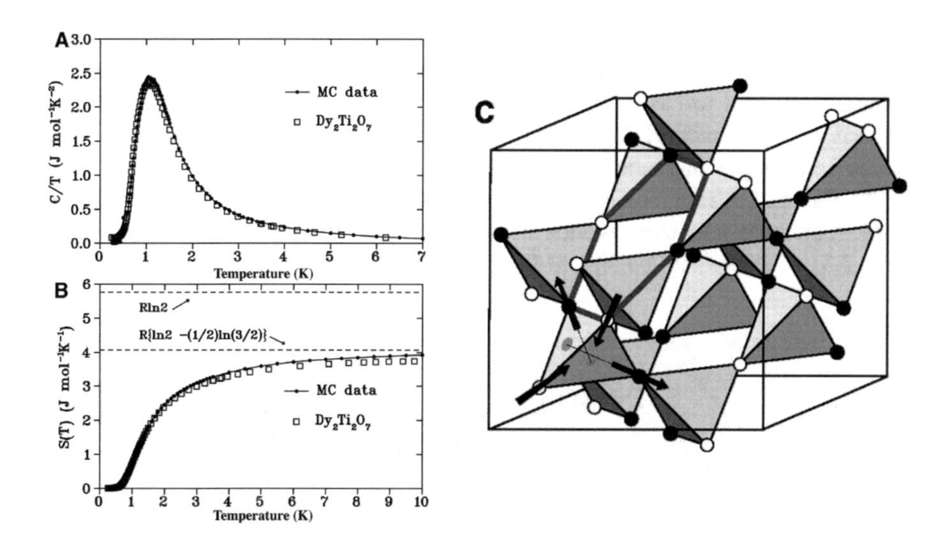

Fig. 3.2 The ice rule in Spin Ice. (**a**) Specific-heat and (**b**) entropy data for $Dy_2Ti_2O_7$ [20] obtained by integrating the specific heat, compared with Monte Carlo simulation results for the dipolar spin ice model [21]. These curves demonstrate the residual entropy of spin ice materials. If we ascribe zero entropy to zero temperature, the entropy per spin converges at large temperature to a value that is inferior to the expected $R \ln 2$, the entropy per spin of uncorrelated moments. The difference is very close to $(R/2)\ln(3/2)$ the Pauling estimate for water-ice residual entropy. The broad bump in the specific heat corresponds to the crossover into the ice state. (**c**) Spin ice materials $Ho_2Ti_2O_7$ and $Dy_2Ti_2O_7$ arrange in a pyrochlore lattice of corner-sharing tetrahedra, occupied by the magnetic rare-earth ions. The magnetic Ising moments of cations occupy the corners of the tetrahedra and are the equivalents of the proton displacement vectors in Fig. 3.1b. Figures from [22]

(Fig. 3.1c). The resulting ferromagnetic interaction favors the ice-rule described in the previous section.

These materials provided important model systems of constrained disorder and its exotic state, inclusive of novel field-induced phase transitions and unusual forms of glassiness. Many of these effects, as we shall see later on, are mostly related to an early and practical example of a classical topologically ordered state (see Sect. 3.3) as spin ice harbors a new fractionalization phenomenon in its low-energy dynamics: emergent magnetic monopoles [5, 19].

3.2.3 Artificial Spin Ice

Between the late 1990s and the early 2000s, the exploration of the magnetic state of nanodots, nanoislands, and in general various geometries at the nanoscale was already a mature effort [24]. Such studies concentrated on how to obtain interesting magnetic textures—and transitions among them—in different, properly shaped, lithographically fabricated nanostructures. However, a different direction emerged around 2006 [25–27]. It considered employing *relatively simple* nanostructures, which could lead to easily modeled binary degrees of freedom, and concentrate instead on their *mutual interaction* to generate possibly interesting collective states from simple magnetic building blocks.

To this end, arrays of elongated, mutually interacting, single-domain, shape-anisotropic, magnetic nano-islands arranged along a variety of different geometries (Fig. 3.3) were chosen: the size of the (typically, NiFe alloys, of size 200 × 80 ×

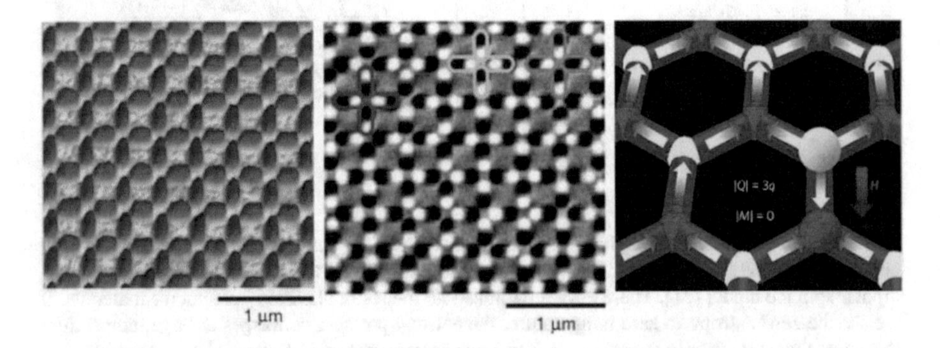

Fig. 3.3 Artificial spin ice in its most common geometries. Left: Atomic force microscopy image of square ice showing its structure (figures from [25]). Center: a magnetic force microscopy image of square ice, showing the orientation of the islands' magnetic moments (north poles in black, south poles in white); Type-I (pink) Type-II (blue) and Type-III (green) vertices are highlighted (see the text and Fig. 3.6 for a definition). Right: schematics of honeycomb spin ice showing the ice rule obeying vertices (2-in/1-out of charge +1 and 1-in/2-out of charge −1) and two excitations 3-in of charge +3 and 3-out of charge −3

(5–30) nm^3 patterned by nano-lithography on a non-magnetic Si substrate), has to be inferior to the typical magnetic domain, to provide single domains with magnetization directed along the principal axis that can be interpreted as switchable spins. Advances in lithography allows now their nano-fabrication in virtually any geometry, and, as we shall see, in these materials geometry dictates behavior.

This approach provides two advantages:

1. the low-energy dynamics, which underlies possible exotic states, is dictated by geometry, which here is open to design;
2. various characterization methods (Magnetic Force Microscopy (MFM), PhotoElectron Emission Microscopy (PEEM), Transmission Electron Microscopy (TEM), Surface Magneto-Optic Kerr Effect (MOKE), Lorentz Microscopy) allow for the direct visualization of the magnetic degrees of freedom. Indeed, even real-space, real-time characterization is possible, for unprecedented vistas of statistical mechanics in action.

These so-called Artificial Spin Ices (ASI) were employed at first to study frustration in a controllable setting, to mimic the behavior of spin ice rare earth pyrochlores, but at more useful temperature and field ranges and with direct characterization, and to provide practical implementation to celebrated, exactly solvable models of statistical mechanics, previously devised to gain an understanding of degenerate ensembles. Soon, a growing number of groups have extended the use of ASI [26], to investigate topological defects, dynamics of magnetic charges, and spin fragmentation [28–36], information encoding [37, 38], in and out of equilibrium thermodynamics [39–50], avalanches [51, 52], direct realizations of the Ising system [53–57], magnetoresistance and the Hall effect [58, 59], critical slowing down [60], dislocations [61], spin wave excitations [62], and memory effects [63, 64]. Meanwhile similar strategies [65–70] have found realization in trapped colloids [71–73], vortices in nano-patterned superconductors [74, 75] (see Sect. 3.2.4) and even at the macroscale [76]. With the evolution of nano-fabrication and of experimental protocols it is now possible to characterize the material in real-time, real-space [77–81], and to realize virtually any geometry, for direct control over the collective dynamics. This has recently opened a path toward the deliberate design of novel, exotic states [13–17, 82] often not found in natural materials [83, 84].

Frustration is a fundamental ingredient in design: it controls the interplay of length and energy scales, dictating the emergent dynamical properties that lie at the boundaries between order and disorder, and leading to a lively, quasi-disordered ensemble called *ice manifold*, to be exploited in the design of exotic behaviors.

Strongly correlated, classical spin systems have of course a long history in Physics. In classical statistical mechanics, the Ising model [85] paved the way to our understanding of long-range order from symmetry breaking as a second order phase transition, universality classes and scaling [86], and finally the renormalization group [87] with implications reaching well beyond condensed mater systems [88]. However, frustrated spin systems often do not order, generally resulting in quasi-disordered manifolds governed by some geometric or topological rule. Often their collective dynamics lends itself to emergent descriptions that are only partially reminiscent of the constitutive spin structure.

The situation is somehow similar to everyday life, where frustration results from a set of constraints that cannot be all satisfied at the same time, leading to a manifold of compromises among which the choice is most often equivalent and can be influenced by a small bias. Thus, obstructed optimization provides high susceptibility, much as in the complex social dynamics we witness everyday, and which emerges from the (perhaps eventually beneficial?) frustration of our everyday life. These analogies between social settings and frustrated materials are not merely philosophical: ideas borrowed from the frustrated spin ice physics might be exportable to the context of social networks [89].

3.2.3.1 Honeycomb Spin Ice

To familiarize with artificial spin ice we begin with the honeycomb structure. As we describe its properties, we will also introduce the characterization and annealing methods generally employed in the study of these materials.

Even before artificial spin ice realizations [40, 90, 91] (Figs. 3.4 and 3.5) honeycomb structures have been extensively studied theoretically for various reasons. In particular, as they describe the two-dimensional behavior of the three-dimensional spin ice pyrochlores under a magnetic field aligned along a particular crystalline axis. A honeycomb ice is often called the Kagome spin ice, as the spins reside on the edges of a honeycomb lattice, which is a Kagome lattice (the dual lattice of the honeycomb). In the context of artificial spin ice, honeycomb ice was initially the only simple geometry with a degenerate ice manifold, and therefore for a long time the only disordered artificial spin ice. Indeed, as we will see in the next subsection, square ice has a frustrated yet perfectly ordered, antiferromagnetic ground state.

As we have seen above, in general magnetic, elongated nano-islands can be described as nano-spins, binary degrees of freedom describing their magnetization along their principal axis. This is, however, already an approximation of the magnetic texture of the nano-structure: both direct characterization and micromagnetic simulations show significant relaxation of the magnetization field at the tips of the islands, due to the local field of the surrounding islands. That such relaxation might imply hidden variables not taken into account by a simple modeling of the nanoisland as an Ising spin has been shown in the case of square ice [94] and it would seem that a more general investigation might be required.

A further approximation, which seems to work surprisingly well, implies modeling the inter-island interaction at the nearest neighbor level, via a vertex model. There, one assigns energies to the various vertex configurations as in Fig. 3.4. As we can see, the lowest energy is ascribed to the six vertices obeying a generalization of the ice rule: one spin pointing in and two out, or two pointing in and one out.

The nano-islands being (non-ideal) magnetic dipoles, one expects the nearest neighbor approach to eventually break down. It does indeed, and in enticing ways,

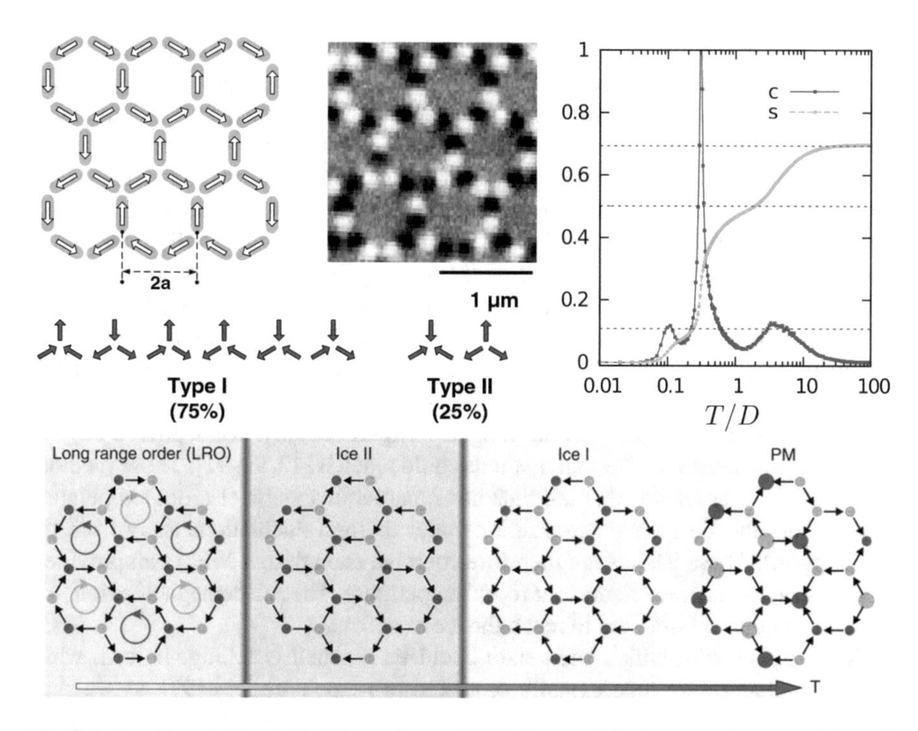

Fig. 3.4 Top, from left to right: Schematics and MFM image of the hexagonal arrays with the 8 vertices of the honeycomb/Kagome artificial spin ice. White arrows show the vertex Ice I state, and the percentages indicate the vertex multiplicity. Type-I vertices have lower energy than Type-II and correspond to the generalized ice rule. Temperature dependence (top right) of the specific heat c and entropy per spin s of the Kagome spin ice obtained by Ref. [95]. The dashed lines show values of entropy per spin $s = 0.693$ (Ising paramagnet), 0.501 (Ice I), and 0.108 (charge-ordered spin ice, or Ice II). Bottom: the four phases of Kagome ice ordered by increasing temperature. Figures are adapted from Refs. [14, 40, 60]

revealing low entropy phases within the ice manifold. If we consider each moment as a dumbbell of positive and negative magnetic charges (much as one would do with electric charges in an electric dipole), we see that, unlike in pyrochlore spin ice (or in the square ice of the following subsection) the low energy vertices are magnetically charged, because of their odd coordination (see Sect. 3.3.1). One thus expects that the next nearest neighbor interaction can be described as magnetic charge interaction between vertices. That is indeed the case.

These equilibrium phases of the system have been investigated numerically [95, 96] via Metropolis Monte-Carlo simulations with full dipolar interaction. Figure 3.4 shows that at high temperature the system is paramagnetic. As temperature is reduced, it crosses over toward a disordered ice-manifold, called Ice I, where each vertex has charge ± 1 and thus represents a neutral, disordered plasma of charges. This is as much as a vertex model approximation would explain, because the ice-rule minimizes the energy of the vertices.

However, if we further lower the temperature, we see a transition toward charge ordering. There, the disordered plasma of the magnetic charges residing on the vertices orders within an ionic crystal. The transition appears to be of the Ising class and is due to the Coulomb interaction among magnetic charged vertices. It can be replicated within a vertex-model approximation, but only if one adds further interaction via Coulomb coupling between the charges of the vertices [95, 96]. Note that such state, often called Ice II, while being charge-ordered, is still disordered in the spin structure: there is an exponentially growing (in the number of spins) number of possible spin configurations that correspond to the charge ordered state. And yet, finally, by further lowering the temperature, another transition is theoretically predicted to lead to a long-range ordered state (LRO in Fig. 3.4), where order is brought in by the long range effects of the dipolar interaction.

These states were variously investigated experimentally. Ice I proved easy to reach. Indeed, even non-thermal methods could reach it [40, 90, 91]. Those methods pertain to thicker islands that are thus not superparamagnetic at room temperature (that is, do not flip their magnetization under thermal fluctuations away from the Curie point). These islands are therefore coercive enough that MFM can provide a non-destructive characterization at room temperature. The AC demagnetization [41] of such samples is sufficient to reach the ice manifold.

The facility with which such state could be reached is telling. Indeed, while magnetic charges are topologically protected in pyrochlore ices [97], as we shall see in Sect. 3.3.1, they are not bona fide topological numbers in the ice manifold of Kagome ice. There, vertices of odd coordination can gain and lose charge freely from the surrounding, disordered, and overall neutral plasma of magnetic charges. Consequently, the ice-manifold can be explored *from within* by consecutive single-spin flips, without any need for collective moves of entire loops of spins. Note that instead, the charge-ordered state, or Ice II, cannot be explored by individual spin flips. A glimpse of the Ice II phase shown in Fig. 3.4 should convince that any spin flip within the manifold will locally destroy the charge order [98].

Signatures of the Ice II state were first suspected after AC demagnetization [35]. They were subsequently investigated via thermal methods capable of providing a bona fide thermal spin ensemble [92, 93]. These methods are of three kinds: annealing from above the Curie point of the nanoislands, typically at higher than room temperature followed by characterization of the static ensemble at room temperature where the islands are more coercive and characterization is non-destructive [80, 92, 99]; thermalization with real-time, real-space characterization [16, 79, 81]; and thermalization without real-space characterization [60, 77]. In the first, the material is not superparamagnetic at room temperature, but it is heated slightly above the Curie temperature of the nano-islands (which can vary, depending on the size and chemical composition of the nano-structure, from about 600 °C for permalloy down to about 100 °C for Fe-Pd alloys) and then annealed down into a frozen state, which is subsequently characterized, for instance via MFM. In the second method the nano-islands are chosen to be thin enough (usually thickness of 2–3 nm) to be superparamagnetic at room temperature or below, and thus need to be characterized via PEEM at a proper beam source. In the third, various averaged

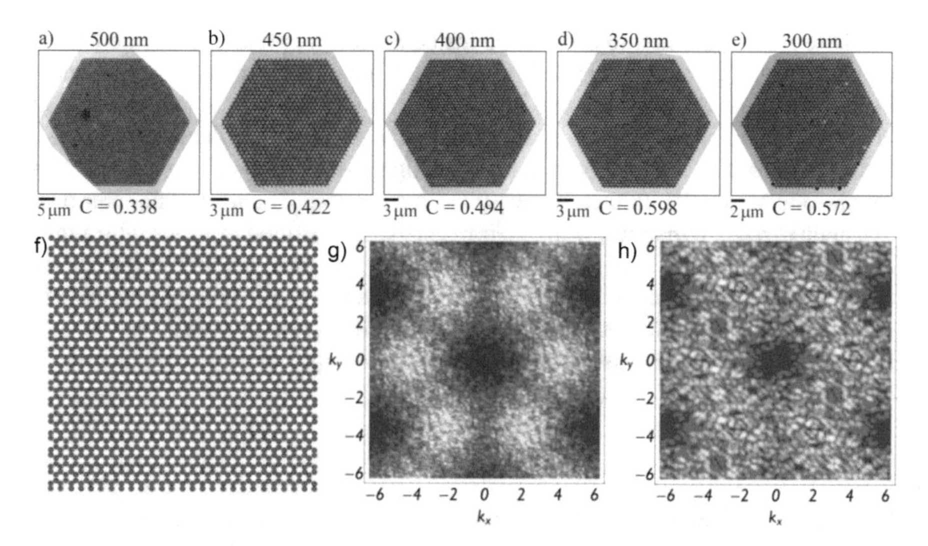

Fig. 3.5 Magnetic charge ordering in Kagome ice. (**a**)–(**e**) Charge domain maps obtained via Lorenz TEM relative to the annealing of Fe-Pd alloy artificial Kagome ices of different edge length (from 500 to 300 nm) show increasing size of the "ionic" crystallites of charges as the lattice constant decreases, and thus the mutual interaction among magnetic charges increases. C is the charge–charge correlation parameter ($C = 1$ for a fully charge-ordered state). Images adapted from Ref. [92]. (**f**) Charge map obtained via MFM after annealing of permalloy Kagome ice of lattice constant 260 nm showing incipient domains of charge-ordering and (**g**) its static structure factor showing incipient peaks corresponding to crystalline order. (**h**) Static structure factor for lattice constant 490 nm, showing no incipient peaks (images adapted from Ref. [93])

quantities are extracted, such as the average flip rate of spins, e.g. through muon spectroscopy [60]—while spin noise spectroscopy [100, 101] should also provide an interesting method.

Figure 3.5 shows the results of thermal annealing on artificial hexagonal ice made of permalloy [99], which demonstrate formation of crystallites of magnetic charges, due to the Coulomb interaction between the charges themselves. More control over the size of those ionic crystallites has also been obtained by employing an alloy of iron and palladium, of much lower Curie temperature [92]. However, nobody has yet reported any direct evidence of complete charge order in such a material, nor of the zero entropy phase of long range spin order.

Indirect indications that such low-entropy phases within the Kagome ice manifold—or at least some kind of phases, possibly close to the ones theoretically predicted—can be reached were obtained via muon spectroscopy studies. These studies concerned islands that were too small and therefore too active to be imaged directly, but whose rate of magnetic flipping could be deduced from the relaxation time of muons implanted on a gold cap over the two-dimensional array. There, the critical slowing down of the spins was measured and found to correspond to that of the numerically predicted transitions, where parameters for the numerical simulations were taken from the material [60].

These results, the first to probe deep inside the ice manifold of Kagome lattice, represent a strong corroboration of the existence of a complex phase diagram, most likely the theoretical predicted one. However, we should not forget that topological or ordered states can be hard to reach via spin dynamics of the Glauber kind [102], and that indeed the actual spin dynamics might be even more complex than a simple Glauber model. In fact, these "spins" are nanoscopic objects with their own magnetic reversal dynamics, while their magnetic state exhibits various relaxations effects coupled to the local fields generated by other spins, leading to previously neglected symmetry breakings [94]. Such specificity might bias certain kinetic pathways, leading to non-equilibration or ergodicity breaking even in ice models that are not theoretically susceptible to these phenomena.

3.2.3.2 Square Ice

With the exception of the work of Tanaka et al. [90], early works on artificial spin ice concentrated on the square geometry (Fig. 3.3) [25, 39, 40]. Square ice also represented the benchmark on which to test demagnetization and annealing methods which lead to experimental protocols for thermalized ensembles [44, 77, 80].

It is important to understand immediately that square artificial spin ice is not the square ice of Lieb [8]. Firstly, because it admits topological defects in the form of magnetic monopoles absent in the six-vertex model—in other words it admits 10 vertex-configurations above the ice rule, for a total of 16 vertices. But most importantly because it is not degenerate. In this sense it shares similarities with the Rys F-model [12]. Yet, such similarities should not be overstated, as the (physically unnatural) absence of monopoles in the Rys F-model leads to an infinitely continuous transition to antiferromagnetic ordering [9], whereas in artificial square ice the transition is of second order.

Figure 3.6 shows the energetic hierarchy of vertices with 90° angles (including those of coordination $z = 3, 2$ to be discussed later), from lower (left) to higher (right) energy. Because of the anisotropy of the dipolar interaction, nearest neighbor perpendicular islands interact more strongly than collinear ones, leading to the lifting of the degeneracy within the ice manifold. The system, if modeled at the vertex level, can be mapped into a J_1, J_2 antiferromagnetic Ising model on a square lattice [10], with a transition to antiferromagnetic ordering, ordering which indeed has been obtained experimentally via thermal annealing protocols [80, 93], as shown in Fig. 3.6.

Within the ordered state of square ice, potentially interesting transitions have been proposed [103–105]. Because the system is not degenerate, creating and separating a couple of monopoles requires energy proportional to the number of Type-II vertices in the Dirac string (see Fig. 3.6). Much like quarks or Nambu monopoles [106] these pairs are linearly confined, and the tensile strength of their Dirac string drives the ordering as the temperature is reduced. There, however, one can imagine that a topological transition corresponding to monopole deconfinement

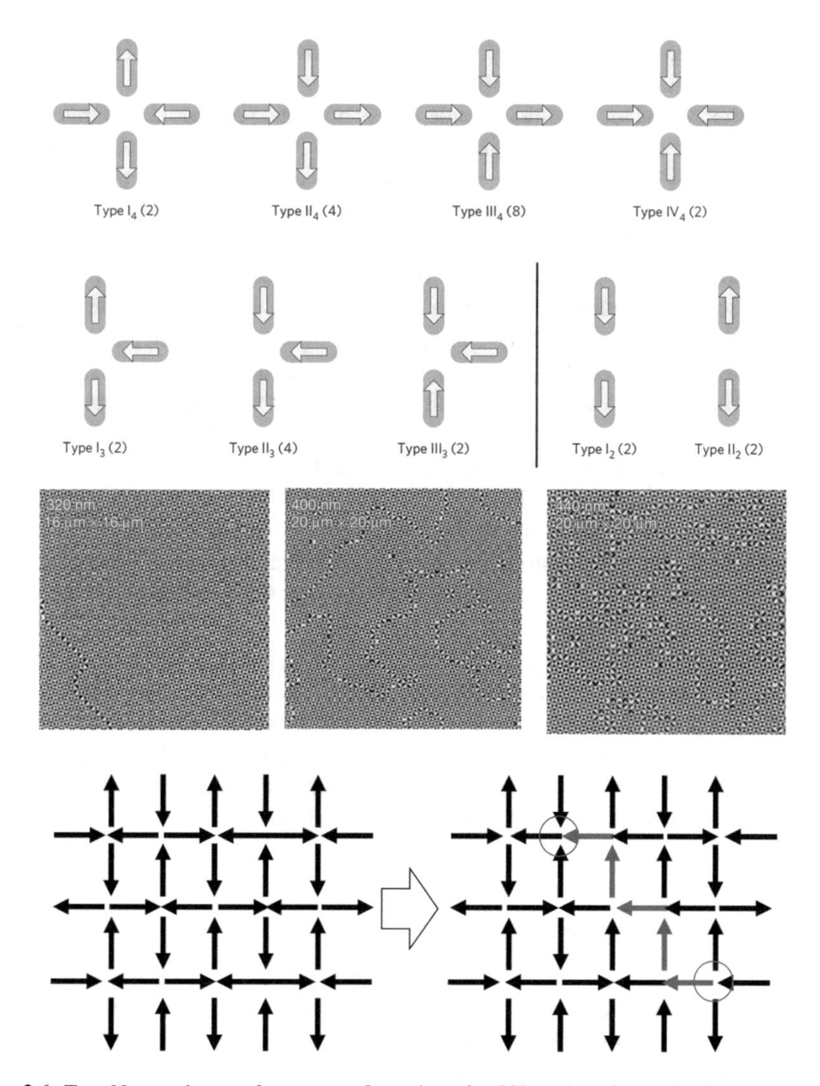

Fig. 3.6 Top: Nomenclature of vertex-configurations for 90° angles of coordination $z = 4, 3, 2$ (degeneracy in brackets) listed in order of increasing energy. Middle: MFM images of thermally annealed square ice at different lattice constants showing an ordered domain crossed by a Dirac string (for the specimen at 320 nm) and a multi-domain ensemble separated by domain walls of monopoles and dirac strings (at 400 and 440 nm); note also the frozen in monopole pairs (figures adapted from Ref. [99]). Bottom: the ordered lowest energy state of square ice as an antiferromagnetic tiling of Type-I$_4$ vertices; creating and separating a monopole pair entails a Dirac string (red) of Type-II$_4$ vertices, that are energetically more costly than the Type-I$_4$, leading to tensile strength of the string and this to the linear confinement of the pair. This should not be confused with the disordered manifold of truly degenerate square ice (Fig. 3.6). There, in a setting of constrained disorder, the Dirac strings have no tensile strength, as the system is not energetically reminiscent of them (though they control an entropic interaction between monopoles)

might take place under proper conditions when the energy of the Dirac string is offset by its fluctuating entropy [103–105].

These theoretical ideas could become timely now that properly degenerate square ice can be realized by finely gauging such degeneracy. Indeed, one way to achieve such degeneracy is to raise the vertical islands with respect to the horizontal ones [96] as to weaken the interaction between islands converging perpendicularly into a vertex. This method has recently been pursued experimentally [107] demonstrating a degenerate manifold whose static structure factor coincides with the numerically computed one for a six-vertex model, and thus providing the first artificial realization of a two-dimensional Coulomb phase (see Sect. 3.3). It was also proposed to iterate such design on the axis perpendicular to the array, leading to layered structures that are geometrically different but topologically equivalent to three-dimensional spin ice pyrochlores [108]. Those have not found realization yet. The only three-dimensional realization of artificial spin ice was obtained by filling the voids of an artificial opal film with Cobalt [109, 110], a promising approach to bring to room temperature some of the features of spin ice pyrochlores. Of course, as always with three-dimensional realizations, the challenge there lies not only in nano-fabrication, but also in characterization, as real-space methods are generally surface methods in these materials.

Another way to produce a Coulomb phase in square ice has been presented recently, and involves "rectangular ice" where vertical and horizontal islands differ in length, and degeneracy is obtained for a proper critical value of their ratio [104, 111]. Finally, a latest method, realized experimentally, corresponds to mixing nanoislands with nanodots placed at the center of the vertex, to gauge the relative energy difference in vertex-configurations [112].

3.2.4 Particle-Based Ice

Proposals for realizations of frustrated analogues of spin-ice materials are not, however, limited to magnetic systems. In a series of numerical works based on brownian dynamics, Libal et al. have proposed systems of so-called particle-based ices. These are two-dimensional arrays of traps arranged along the edges of a lattice (typically square [65], or hexagonal [68]). Each trap contains one particle, that can move along the trap. The traps attract the particle via a double well potential, for which the extreme ends of the trap are the two preferential positions. This forces the particle to be in the proximity of one of the two vertices connected by the trap (Fig. 3.7). The particles repel each other, the interaction depending on their specific nature. When brownian dynamics is performed on these systems they obey the ice rule in the strong interaction regime—that is when the local interaction energy exceeds the potential barrier of the double well potential.

The model is general and can be realized via different kinds of "particles." Some of these theoretical proposals [66, 68] pertained to pinned quantum vortices in properly nano-structured superconductors, which were recently realized experimen-

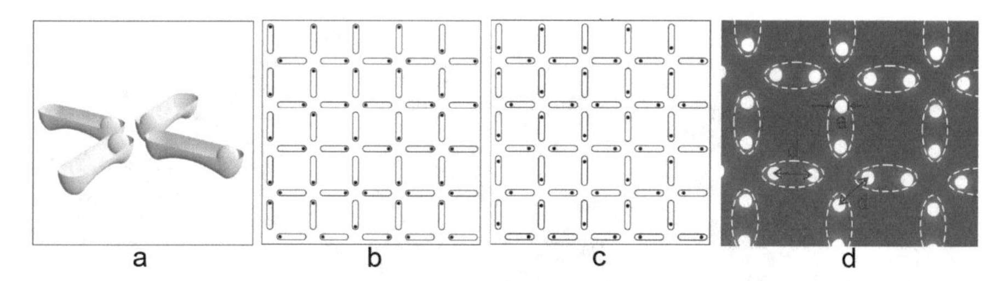

Fig. 3.7 Particle-Based Ice. (**a**) Schematics of the basic unit cell with four double well traps each capturing one colloid. (**b**) Random vertex distribution. (**c**) Long-range ordered square ice ground state, obeying the ice rule. (**d**) Scanning electron microscopy of a nano-patterned substrate for pinning of superconductive vortices in MoGe thin films ($a = 102\,\text{nm}$, $d = 300\,\text{nm}$). Images (**a**)–(**c**) from Ref. [65]. Image (**d**) from Ref. [74]

tally [74, 75]. Of course, research in pinning superconductive vortices to a substrate nano-patterned with holes had been conducted since the late seventies [113] with the goal of increasing critical currents. Latimer et al., however, arranged the pinning with the intention of reproducing the frustration of an ice-like material [74]. They fabricated superconducting thin films of MoGe containing pairs of circular holes arranged as in Fig. 3.7d: a square lattice whose vertices can accommodate in principle four vortices each. Then, when the applied magnetic field is at half the matching value, at each vertex only two holes are occupied, and two are not, following the ice rule, as predicted numerically [66]. Clearly the particle-ice models of Libal, Olson, Reichhardt, and collaborators are over-constrained compared with the experimental realization of Latimer et al. In the latter there are no semi-occupied link-shaped traps and the vortices can simply pin to in the ratio of vortices to hole dictated by the field.

More recently, faithful realizations of those particle-based models were achieved at the microscale [71, 73]. Magnetic colloids are confined by gravity in photolithographically patterned double wells, as shown in Fig. 3.8 and preferentially sit in one of the two wells in each bistable microtrap. When a magnetic field $\vec{B}$ is applied perpendicularly to the system, the colloids magnetize and repel mutually with law $\sim B^2 r^{-3}$. As the field is ramped up, a random distribution of colloids evolves into an ice-rule obeying configuration (Fig. 3.8).

While the same ice rule has also been observed in natural and artificial magnetic spin ice materials, its origin in particle-based ices is different, as we shall see in Sect. 3.3.2 when discussing the different levels of frustration in these systems.

3.3 Theoretical Themes

We have so far given definiteness to our material by introducing a set of rather different physical systems and indulging more to their phenomenology than to their underlying mathematical structures. In doing so, however, we have presaged certain

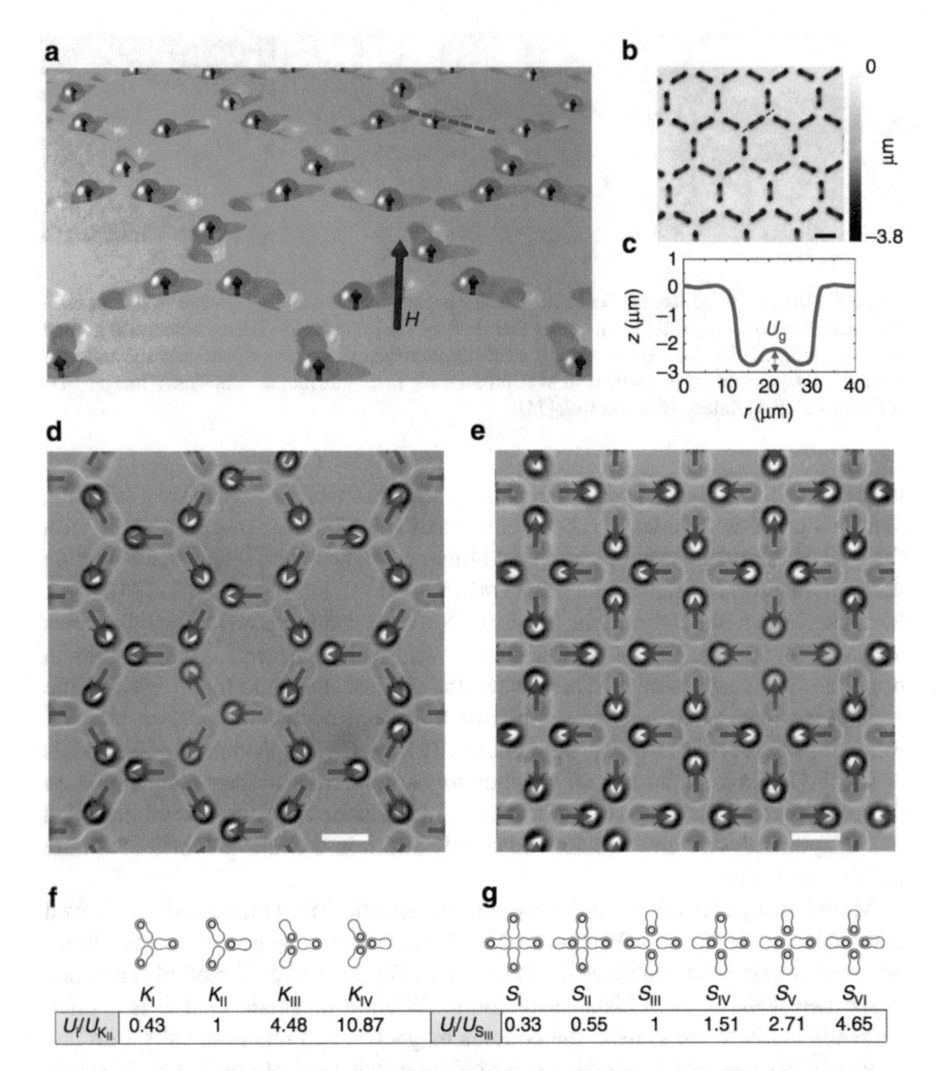

$U_i/U_{K_{II}}$	0.43	1	4.48	10.87

$U_i/U_{S_{III}}$	0.33	0.55	1	1.51	2.71	4.65

Fig. 3.8 (**a**) Schematic view of the colloidal spin ice made by a honeycomb lattice of double-well islands filled with paramagnetic colloids. The applied field B perpendicular to the plane induces repulsive dipolar interactions between the particles. (**b**) Optical profilometer image of the honeycomb spin ice, and (**c**) the cross-section of a double well with a small central hill, giving a gravitational potential U_g. (**d**, **e**) Equilibrium state of a honeycomb ice (**d**) (lattice constant $a = 44\,\mu\text{m}$) and a square ice (**e**) (lattice constant $a = 33\,\mu\text{m}$). Blue arrows denote spin direction, while green circles highlight vertices of type $K_I I$ (in **d**) and $S_I II$ (in **e**). Scale bars, 20 µm for all images. (**f** and **g**) Vertex configurations for honeycomb (**f**) and square (**g**) ices. The lowest panel shows the normalized magnetostatic energy for each type of vertex. Image and caption from Ref. [71]

notions, such as charges, topological order, the different role of frustration, and different kinds of frustration, which we will try now to put on more solid grounds. This will also help us introduce new approaches and geometries based on a deeper understanding of the interplay between frustration and the ice rule, to produce different emergent exotic behaviors.

We can consider Sect. 3.3.1 in a broader sense as "kinematic," as it provides a description of the ice rule and ice manifolds, one that can be topological. The Sect. 3.3.2 discusses instead the *origin* of such description and explores its relationship with frustration—which, as we shall see, can be of different types.

3.3.1 Ice Rule, Topological Charges, and Topological Order

We have seen that in pyrochlore spin ice and water ice, as well as in square artificial spin ice, the ice rule corresponds to two spins pointing into a vertex, and two pointing out. Instead, for honeycomb ice, it corresponds to one spin pointing in, and two out, or vice versa.

In view of the more complex geometries that we will discuss later, let us generalize the notion of ice manifold and ice rule for a general lattice, or graph, or network [89], whose edges are spins impinging in vertices of various coordination z. Then we say that a vertex of coordination z with n spins pointing toward it has *topological charge*

$$q = 2n - z, \tag{3.1}$$

corresponding to the difference between spins pointing in and out. In general, the ice rule can be considered a local minimization for $|q|$ at each vertex (typically, but not necessarily, by nearest neighbor spin–spin interaction).

For a lattice of even coordination, such as the square ice or pyrochlore ice introduced before, the ice manifold is characterized by zero charge, $q = 0$ on each vertex. However, for lattices of odd coordination there cannot be any charge cancellation, and thus in the ice manifold each vertex will have charge $q = \pm 1$. These appear in equal fraction, as the total charge of a system of dipoles must always be zero. That is the case of Kagome ice, but also of ladder ice which we have not discussed [114].

The definition in (3.1) can apply to lattices of mixed coordination, and in fact to general graphs. This can lead to more complex geometries, where the frustration is of a different kind, and which we will discuss later.

When the spins represent magnetic moments, these topological charges are also magnetic charges. Indeed a multipole expansion [5] shows that charged vertices interact via Coulomb law. We have seen in honeycomb ice how their interaction drives a transition toward charge ordering.

Even in absence of a magnetic interaction, when the ice-manifold is degenerate one expects that excitations above the manifold must interact entropically, owing

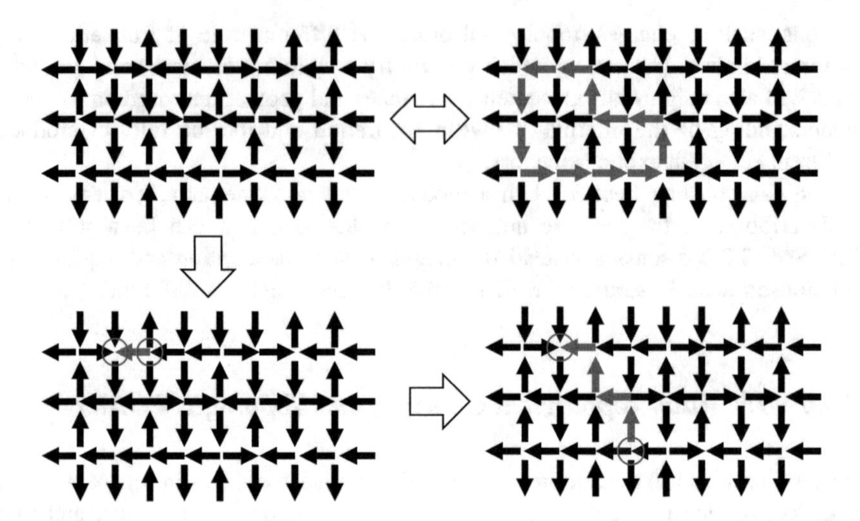

Fig. 3.9 The ice manifold (top left), an ensemble of spins obeying the ice rule (in each vertex two arrows point in, two out). One can obtain another realization of the ensemble only by flipping a proper loop of spins. Flipping a single spin creates a couple of magnetic monopoles of opposite charge (positive is red, negative is blue). The monopoles can be separated by further spin flips (creating a "Dirac string," shown in red), interact via Coulomb interaction, but are topologically protected, as they can only be created and annihilated in pairs

to the different ways to arrange the underlying spins to obtain a given charge distribution. One is also not surprise to learn that in a three-dimensional material such entropic interaction follows a Coulomb law, given that r^{-1} is the green function of the Poisson problem in three dimension. And thus the Coulomb interaction between magnetic monopoles in rare earth spin ice is expected to be renormalized by a term proportional to temperature, due to these entropic effects [115]. Similarly one expects the entropic interaction among topological charges to be logarithmic in two dimensions.

Another interesting aspect is that these charges are topologically protected monopoles in square or pyrochlore ice: charges can only be created and annihilated in opposite pairs. This is tied to an underlying topological structure. To facilitate understanding consider the convenient two-dimensional schematics of Fig. 3.9, which represent a disordered ice manifold, an ensemble of spins where all the vertices obey the ice rule. The reader will notice that it is impossible to explore the manifold by single spin flips, without breaking the ice-rule. Only by flipping proper loops of spins we can obtain a new configuration within the ice-manifold.

If we flip one spin only, we create two defects (3-in/1-out and vice versa). We can separate those defects by further flips, and we have two deconfined magnetic monopoles [5]. We can also annihilate these monopoles by bringing them together via a different path: at the end the overall spin flips must amount to a loop. Of course these monopoles are in effect simply the opposite ends of a long, floppy dipole, in red in figure, called the Dirac string; however, owing to the disorder of the manifold,

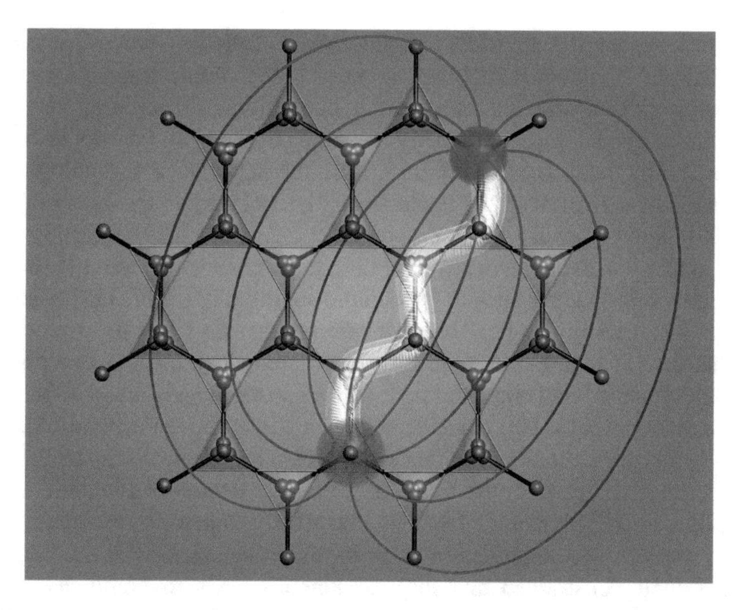

Fig. 3.10 Two magnetic monopoles (red and blue) and a Dirac string (highlighted in white) in a three-dimensional spin ice material, from Ref. [5]

the system is no longer reminiscent of the Dirac string connecting the monopoles. Thus, excitations over the ice manifold can be described by a fractionalization of the spins into individual, separable magnetic charges which interact via a Coulomb law.

One could indeed create a single monopole in an open system but that would simply imply pushing the second one at the boundaries. If we, however, placed the ensemble of Fig. 3.9 on a torus, thus without boundaries, then clearly there cannot be a net monopole charge. Even in an open system, the total net charge will be proportional to the flux of the magnetic moment through the boundaries, and as the latter is bound by the net magnetization of the spins, one finds that the density of net charge must scale at least with the reciprocal length of the system. These two-dimensional considerations extend to the three-dimensional spin ice. Figure 3.10 shows monopoles and diract strings in three-dimensional spin ice materials, where they were initially introduced [5].

Much as one labels the disorder of a dimer cover model via a height function, one can also label the disorder of a pyrochlore spin ice via a field $\vec{M}$, a proper coarse graining of the magnetization, such that the pure ice-manifold is characterized by

$$\vec{\nabla} \cdot \vec{M} = 0, \tag{3.2}$$

and of course $\vec{\nabla} \cdot \vec{M} = q\delta(x - x_0)$ for a monopole excitation in x_0. From this approach one can derive the correlations between spins in the ice-manifold and find that they are algebraic, and indeed dipolar [116, 117].

This set of properties are taken to define what is called a Coulomb phase, which is an example of classical topological order [118]. While quantum topological order has provided a valuable framework to conceptualize disordered states of spin liquids that escape a Landau symmetry breaking paradigm and cannot be obviously characterized by local correlations [119, 120], the importance of topological states had been recognized even earlier in classical physics [121]: in the theory of dislocations [122], liquid crystals [123], or topological transitions [124]. Recently, whether in direct analogy with quantum physics [125], in purely abstract terms [118, 126], or motivated by real systems such as pyrochlore spin ices [97, 117, 127], a consistent notion of *classical* topological order in *discrete* systems has been proposed, to conceptualize (1) a degenerate, locally disordered manifold (2) described by a topologically non-trivial, emergent field (3) whose topological defects (in spin ice, magnetic monopoles [5, 19]) coincide with excitations above the manifold.

Topological protection implies that states *within* the manifold can be linked only via collective changes of entire loops of a discrete degree of freedom. Thus any realistic low-energy dynamics happens necessarily *above the manifold*, through creation, motion, and annihilation of pairs of protected topological excitations. Typically, their constrained and discrete kinetics leads to ergodicity breaking, fractionalization and thus various forms of glassy behaviors [118].

The fact that monopoles are topologically protected has consequences on the kinetics: unusual forms of glassiness can result, for instance from quenches, where the monopoles are trapped in and cannot easily annihilate by simple diffusion [128]. This is often a feature of topologically protected systems with a discrete kinetics.

Note also that such protection for magnetic charges depends on the coordination of the lattice. In a lattice of even coordination, the ice manifold is characterized by a local zero magnetic charge on every vertex and any individual spin flip produces two $q = \pm 2$ excitations. There, magnetic charges are protected topological charges, and can be created and annihilated only in couples (Fig. 3.9). However, in a lattice of odd coordination, such as the honeycomb lattice of Sect. 3.2.3.1, it is possible to explore the ice manifold by single, consecutive spin flips if they do not alter the constraint that each vertex must host a charge $q = \pm 1$. There, excitations $q = \pm 3$ are clearly not protected and therefore are not proper topological charges: the manifold is a plasma of charges, and an excitation can lose its charge to the surrounding plasma.

The same is true for lattices of mixed coordination involving oddly coordinated vertices, of the kind that we will discuss later. Note also, however, that the fact that magnetic charges are not proper topological charges does not imply necessarily that the ice manifold is not a topological phase. We will see later that topological order can also be found in novel, non-trivial geometries of artificial spin ice characterized by vertex-frustration, such as Shakti spin ice [13–15]. There, higher level topological charges can be identified in an emergent description of the ice-manifold which is not reminiscent of the underlying spin or magnetic charge structure [129].

3.3.2 *Ice Rule and Frustration(s)*

In the previous subsection we have provided a more general description of the ice-rule. It is time to investigate its origin from frustration. In common folklore, "frustration" is used interchangeably for "degeneracy," and in particular the ice rule is generally taken to come from frustration, and to lead to degeneracy. None of that is strictly correct.

A frustrated system can be perfectly ordered, as we saw in artificial square ice, which also obeys the ice rule. Water ice is degenerate, obeys the ice rule but its degeneracy is not a consequence of any discernible frustration of its interaction, but rather of the freedom afforded by the stoichiometry of the water molecule. Moreover, frustration can be of different kinds. It can be "local," associated with pairwise interactions, but also more global and collective in nature, when it is only present in a system of a large number of elementary degrees of freedom. That is the case of vertex-frustration, and also of frustration in particle-based ice, as we will see later.

3.3.2.1 Frustration of Pairwise Interaction

The concept of geometric frustration in its broader mathematical form involves a geometric system describing a manifold of degrees of freedom and a set of prescriptions on how they should arrange with respect to each other. The system is frustrated if there are loops along which not all the prescriptions can be satisfied (Fig. 3.11). Clearly the concept is very general and extends beyond Physics. One immediately recognizes topology in the nature of such definition: at least in theory, any homotopy, that is any continuous transformation that does not tear those loops, will lead to a system of the same frustration. In practice, in Physics that is often

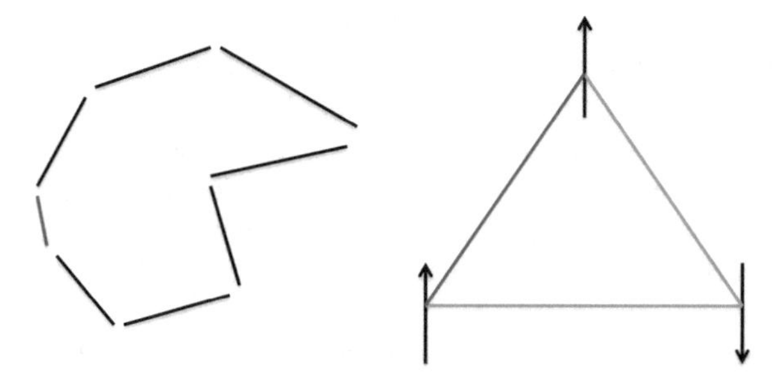

Fig. 3.11 Geometric frustration can be understood schematically as a set of prescriptions that cannot be satisfied simultaneously around certain loops. The red link on the figure on the left represents an "unhappy link" in a generally frustrated system. More specifically, for an Ising antiferromagnet (right) the loop in question is a loop of interactions among nearest neighbors and the prescription is the minimization of a pairwise interaction. On a triangular lattice, triangular loops are frustrated, as one of the three links (red) must be unhappy

only partially true when frustration is frustration of interactions, as the latter tend to depend on geometry.

Generally speaking, in Physics: (a) these "prescriptions" correspond to the optimization of a certain energy, and (b) that energy is usually a pairwise interaction, typically between binary degrees of freedom. We will see later how weakening the request (b) can be useful to define new kinds of frustration in spin ice materials.

An early and classical example of these two points is the famous antiferromagnetic Ising model on a triangular lattice [130], a system of binary Ising spins interacting antiferromagnetically on a triangular lattice (Fig. 3.11). There the antiferromagnetic configuration among nearest neighboring spins cannot be satisfied simultaneously on a triangular plaquette, leading to a disordered manifold. The disorder is, however, non-trivial, it is constrained disorder and its entropy per spin is of course not merely $s = k_B \ln(2) \simeq 0.6931 k_B$. Because rules apply, due to frustration, of all the energy links only one per plaquette is frustrated in the lowest energy configuration, leading to an entropy per spin $s \simeq 0.3383 k_B$, different from zero, and about half of the entropy of a completely random configuration.

This kind of frustration of the pairwise interaction is the most commonly studied in physics. It is however not the only form of frustration.

3.3.2.2 Vertex Frustration

As most realistic interactions in Physics are geometric, they break the topological nature of frustration in a real system. In the case of spin ice systems, for instance, the dipolar interaction between in-plane, shape anisotropic nanoislands depend on their relative orientation. This is seen, for instance, in square ice of Fig. 3.6, where the degeneracy between Type I_4 and Type II_4, both ice rule vertices, is lifted by the fact that mutually perpendicular islands interact more strongly than parallel ones. This difference in strength among the interactions does not kill frustration, but it does remove degeneracy, as the weakest energy link (between collinear, nearest neighboring islands) will always be frustrated, leading to a frustrated yet ordered antiferromagnetic ensemble. The antiferromagnetically ordered artificial square ice of Fig. 3.6 is an example of frustrated yet ordered system.

Another example of real interactions breaking the topological structure of frustration is provided by a comparison of the hexagonal lattice with the brickwork lattice, in Fig. 3.12. The two are topologically equivalent, yet their magnetic ensemble is different. The former is degenerate, the latter is ordered, at least when captured by a vertex model.

This issue might appear trivial, in the Hyperuranion of pure ethereal, theoretical concepts, yet it has truly limiting consequences for the design of new magnetic spin ice materials. And indeed, until 2014, the only degenerate artificial spin ice was the honeykomb/kagome geometry. As both nano-fabrication and characterization protocols evolved, it became clear that the initial inspiration of the entire artificial spin ice project—to design exotic behaviors in the geometry of interacting, binary degrees of freedom—could become viable, if not for one problem: in real systems,

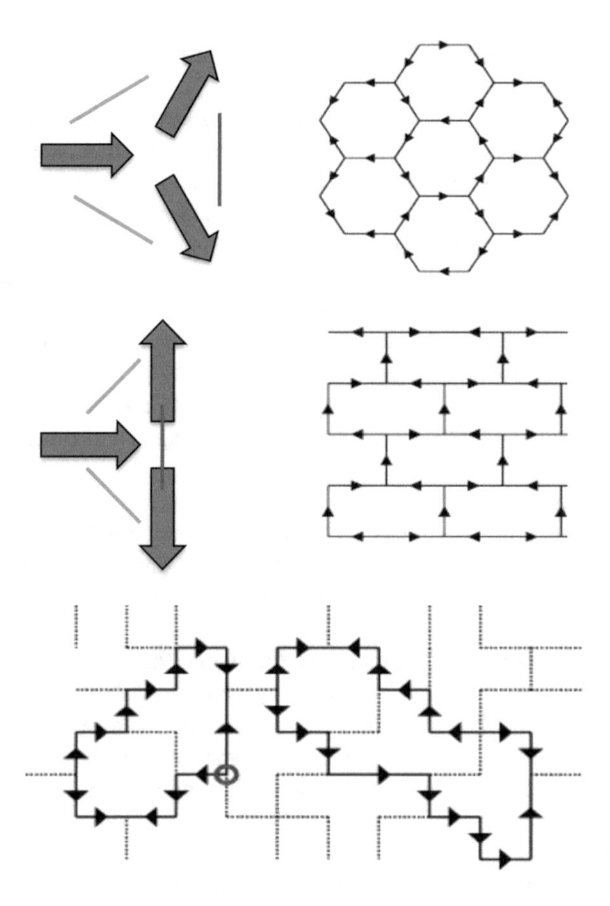

Fig. 3.12 Geometry vs. Topology. Topologically equivalent geometries leads to completely different spin ensembles at low energy, due to the anisotropy of the dipolar interaction. The honeycomb spin ice (top right) is topologically equivalent to the ladder spin ice (middle, right) yet the nearest neighbor interactions lead to an ordered ground state in the latter (see also Fig. 3.6 for the energetic hierarchy of the vertices) and a disordered manifold in the former. Pairwise interactions are frustrated in both systems, however in the honeycomb lattice all the spins interacting in the vertex have the same mutual angle (top left) and thus any of the three interactions can be frustrated, whereas it is energetically favorable to frustrate the interaction between parallel spins in the ladder lattice (middle left). At the bottom is an example of vertex-frustration, where the allocation of vertices of lowest energy is frustrated, leading to "unhappy vertices" (blue circles) on certain loops, instead of unhappy energy links (red lines above)

the frustration of the pairwise interaction is wedded to the geometry, because the dipolar interaction is not topologically invariant, but instead depends very much on the mutual arrangements of the dipoles.

To overcome this limitation and gain freedom in the design of new materials capable of various states and unusual behaviors, the first step is to decouple frustration from geometry. As the pairwise interaction is anisotropic, something else will have to be frustrated. Perhaps a cluster of islands with its own set of energy levels. A natural candidate is the vertex itself.

Consider a geometry made of 90° vertices of coordination $z = 4, 3, 2$ (Figs. 3.6 and 3.12). Each vertex has a unique configuration of minimal energy (up to a flip of all the spins). Imagine now arranging them in such a way that, however, around certain loops not all vertices can be assigned to the lowest energy configuration [13]. This will lead to "unhappy vertices" (UV), that is, topologically protected local excitations (Fig. 3.12) that cannot be removed from the ground state. Note that these are not real excitations of the system, as they in fact belong to its low energy state. We have called them local excitations as locally each vertex would admit lower energy configurations, which are however prevented by the collective structure. In proper geometries, the degeneracy of the allocation of such vertices grows exponentially with the size of the system, leading to a degenerate low-energy manifold [13, 17].

Crucial here is that within this manifold the system is best described not by the disordered spin texture, rather it is captured by an emergent description that labels the possible allocation of these protected local excitations. As a consequence, other emergent properties appear that are in general not obvious nor indeed apparent in the local spin structure.

It is of some importance to understand that while vertex models [11] had been introduced to describe frustrated systems, they were themselves not frustrated. They could simply subsume the degeneracy of a frustrated system within the degenerate energetics of the vertices. Vertex-frustrated geometries can thus be considered the first frustrated vertex models, a fact which might elicit some theoretical interest beyond the usefulness in the design of artificial magnets with properties not found in natural ones.

Vertex-frustration is of course a nearest-neighbor level concept, although it can induce topological states that are collective. However, the real materials being made of dipoles, other phases are present within their vertex-frustrated low energy manifold, much like inner phases are present in the diagram of Kagome above. We will see in the next section how this comes about in three such geometries: Shakti, Tetris, and Santa Fe.

3.3.2.3 Collective Frustration

A yet different case of frustration is provided by particle based ices, already discussed in Sect. 3.2.4. We saw there that they have been studied numerically in square and hexagonal geometries and then realized experimentally using magnetic colloids gravitationally trapped into microgrooves and also vortices pinned in nanopatterned superconductors. In all these cases it was shown that they obey the ice rule at low energy. Because of that, ideas and results have been exchanged among the two materials, and they have been often considered as equivalent.

However, despite similarities, the two systems differ essentially in energetics and frustration. A nearest neighbor analysis will readily convince the reader. The nearest neighbor energy of a magnetic spin ice vertex with n spin converging in the vertex is typically proportional to the square of its charge, or

$$E_n \propto q_n^2, \tag{3.3}$$

thus favoring the ice rule. Indeed we have shown in this section that the ice rule can be seen as a precept for minimization of the topological (or also magnetic) charge of a vertex. However, for particle ice, the energy of a vertex scales instead as

$$E_n \propto n(n-1), \tag{3.4}$$

thus favoring large negative charges ($n = 0$ and $n = 1$) that violate the ice-rule. This is seen in the energy hierarchy of vertices reported in Fig. 3.8.

Thus, we encounter the essential difference: while local energetics promotes the ice rule in magnetic spin ices, it opposes it in particle-based ices. There, its origin is instead collective, [70] as it was recently verified numerically and experimentally [131].

While its vertex reaches the lowest energy for large negative charges, obviously the total charge of particle ice must be zero. It is not possible for all vertices to be negatively charged at the same time, and the ice rule emerges as a collective compromise among vertices. This form of collective frustration is in a way similar to an extreme form of the vertex frustration described previously, as here each vertex is in a locally excited state. It is, however, also collective in as much as it depends on the size of the system: it is fully obtained only in the thermodynamic limit. Indeed, in a finite realization individual vertices can push their charge to the boundaries. The resulting charge accumulation, however, is limited by the size of the boundaries, and thus the *density* of topological charge in the bulk must scale as the reciprocal length of the boundaries, leading to the emergence of the ice rule in the thermodynamic limit.

The reason why these two different systems behave similarly can be investigated quantitatively in a mean field approximation [70], of which we summarize below the main results. If we constrain the total charge to be zero, then the thermodynamic ensemble at equilibrium is controlled by *effective* vertex energies

$$\tilde{E}_n = E_n - q_n \phi, \tag{3.5}$$

where ϕ is a constant, a Lagrange multiplier determined by the requirement of total zero charge. Thus, for a lattice of coordination z the choice $\phi \sim (z-1)$ returns a spin-ice-like effective energetics, or $\tilde{E}_n \sim q_n^2$, which explains the ice rule of colloidal ice in simple lattices. The reader will notice that the formula in Eq. (3.5) is suggestive. The *collective* effect of the particle-sharing vertices can be subsumed into a field ϕ, which modifies the energetics of the *individual* vertex. One can indeed introduce a less crude approximation, where the mean field is allowed to fluctuate in the material, and ϕ is in fact an entropic, emergent field which conveys local correlations [70].

When the lattice has multiple coordinations, however, there is no value of ϕ that produces an effective ice-like energetics for more than one coordination. For $z = 4, 3$, charge conservation imposes $\phi = (3-1)/2 = 1/2$ and thus the effective energetics maintains the ice rule on $z = 3$ vertices. On $z = 4$ vertices, however, it ascribes the same effective energy to the negative ($q = -2$) monopoles and to the ice rule ($q = 0$) vertices [70].

This spontaneous emergence of negative magnetic charges leads to *ice rule fragility* in geometries of different and mixed coordination. This phenomenon is typical of particle spin ices and was investigated recently, both numerically and experimentally, and we will encounter it at the end of the next section.

While this approach based on topological charges allows for quantitative predictions validated by numerical simulations and experiments [131] an exact mapping can be established between magnetic spin ices and particle based ices. In particle ices, particles in positions $\{\mathbf{y}\}$ repel with isotropic interaction ϕ. Their total energy is given by

$$H = \sum_{\mathbf{y} \neq \mathbf{y}'} \phi\left(|\mathbf{y} - \mathbf{y}'|\right) \tag{3.6}$$

which does not look much similar to a magnetic spin ice Hamiltonian. Yet, both systems are described by binary variables, at least at equilibrium. We can represent the position $\mathbf{y}^{+}$ of a particle in a trap by —● or ●—. Then we ascribe a *positive* charge to the real ● particles and we consider the empty locations $\mathbf{y}^{-}$ of the traps as virtual *negative* charges ○, which repel (attract) other negative (positive) charges. With this definition, we can then fractionalize a trap on an edge $\mathbf{x}$ as

$$-\!\!\bullet = \frac{1}{2}\bullet\!\!-\!\!\bullet + \frac{1}{2}\circ\!\!-\!\!\bullet, \tag{3.7}$$

i.e. a *positive dumbbell* ●—● (a trap doubly occupied by positive charges), plus a *dipole* of negative and positive charges represented by a spin $\vec{\sigma}$ = ○—● located in $\mathbf{x}$, the center of the trap so that $\mathbf{y}^{\pm} = \mathbf{x} \pm \vec{\sigma}/2$. Then, because spins are binary variables, it is easy to see that the energy in (3.6) can always be rewritten as

$$H = \frac{1}{2} \sum_{\mathbf{x} \neq \mathbf{x}'} \sigma^{i}_{\mathbf{x}} J_{ii'}\left(\mathbf{x} - \mathbf{x}'\right) \sigma^{i'}_{\mathbf{x}'} - \sum_{\mathbf{x}} \vec{\sigma}_{\mathbf{x}} \cdot \vec{B}(\mathbf{x}). \tag{3.8}$$

The first term expresses the spin ice part of the hamiltonian and $J_{ii'}(\mathbf{x})$ is a tensor field that can be reconstructed from ϕ. The second term represents the interaction between dipoles and the positive dumbbells which generate a background field $\vec{B}$. Thus a particle-based ice is equivalent to a magnetic spin ice when $\vec{B} = 0$. Clearly that is true if a lattice has point reflection symmetry in the middle points $\{\mathbf{x}\}$ of each edge. This explains why the hexagonal and square PI follow the ice rule, as found previously numerically and experimentally [65, 71, 73]. However, for more complex geometries that is not necessarily true, as we will see in the next section, leading to many phenomena specific to particle ice, from ice rule fragility, to inner phases, to order to disorder transitions, as described in Ref. [132].

3.4 Ice Manifolds and Emergent States by Artificial Design

In Sect. 3.2 we have seen simple realizations of ice rule via the more elementary kind of frustration: frustration of the pairwise interaction among binary degrees of freedom. In Sect. 3.3 we have then indulged more theoretically on the origin of the ice rule from frustration and on its extensions, and we have attempted to go beyond the generally accepted understanding of such concepts. We will now show how a deeper understanding of frustration and the ice rule can lead to the design of novel phenomena in artificial realizations.

We will start with vertex frustration in artificial magnetic ices. As we explained previously, their degeneracy comes from (relative) freedom in allocation of unhappy vertices, those topologically protected local excitations. Typically such allocation leads to an emergent description that is not directly reminiscent of the underlying spin structure, and that can be exploited for a variety of new phenomena, not seen in previous systems, from topological protection to dimensional reduction.

3.4.1 Emergent Ice Rule, Charge Screening, and Topological Protection: Shakti Ice

Consider the Shakti geometry in Fig. 3.13 [14]. Each minimal, rectangular loop of Shakti is frustrated: trying to arrange spins on it we realize that at least one unhappy vertex is needed, according to the energy hierarchy of Fig. 3.6. Indeed each minimal loop must contain an odd number of unhappy vertices [13, 14]. Because each unhappy vertex always affects two nearby loops and costs energy, the lowest energy configuration is realized when nearby loops are dimerized by a single unhappy vertex as in Fig. 3.13b [13]. If one considers the geometry, one finds (Fig. 3.13c) that each plaquette made by two rectangular loops will host two unhappy vertices in 4 possible locations, much like the ice rule in water ice prescribes that 2 hydrogen atoms are within the tetrahedron containing each oxygen atom (Fig. 3.1), in 2 of the 4 possible allocations. In both cases the same ice-rule applies, but here in emergent form: not in terms of the original spins, but in terms of allocation of unhappy vertices. Thus, the lowest energy manifold, at the nearest neighbor vertex description employed here, corresponds then to an emergent six-vertex model. This has been shown experimentally (Fig. 3.14) [15].

Nonetheless, as we had cautioned before, this nearest neighbor description defines the ice-manifold, within which other, non-trivial phenomena intervene, due to the long range nature of the interaction. A particularly interesting one regards the

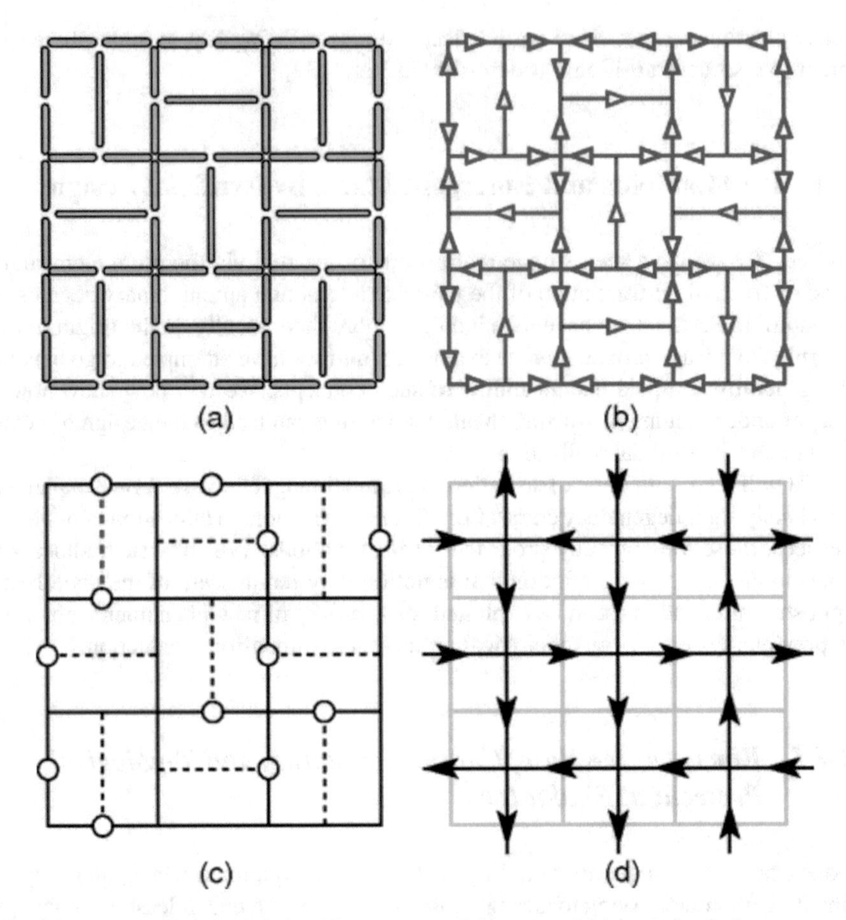

Fig. 3.13 Theory of Shakti spin ice. The structure of the system (**a**) is such that its lowest energy spin ensemble (**b**) is disordered. A look at the spin structure in (**b**) does not seem particularly insightful. However, if we translate that spin map unto a picture of the allocation of locally excited vertices, denoted by circles in (**c**) we then see that each plaquette will host two and only two unhappy vertices in four possible positions. This is equivalent to a six-vertex model (**d**) where pseudo-spins are assigned to each plaquettes and point toward (away from) the unhappy vertices in plaquette of vertical (horizontal) long island. Figures adapted from Ref. [14]

screening of magnetic charges. Shakti has multiple coordination, therefore while in its low-energy state all the vertices of coordination $z = 4$ are in the ice rule, they are surrounded by vertices of coordination $z = 3$ which always have a magnetic charge ± 1 (in natural units, previously defined), and are disordered. When a vertex of coordination $z = 4$ hosts a magnetic monopole, the overall neutral plasma of charge around it rearranges to screen it, as shown in Fig. 3.14 [15].

It is important to understand that magnetic monopoles of the kind described in the previous section are *not* proper topological charges for Shakti, as they are not protected. Each $z = 4$ vertex is surrounded by $z = 3$ vertices, which as we know

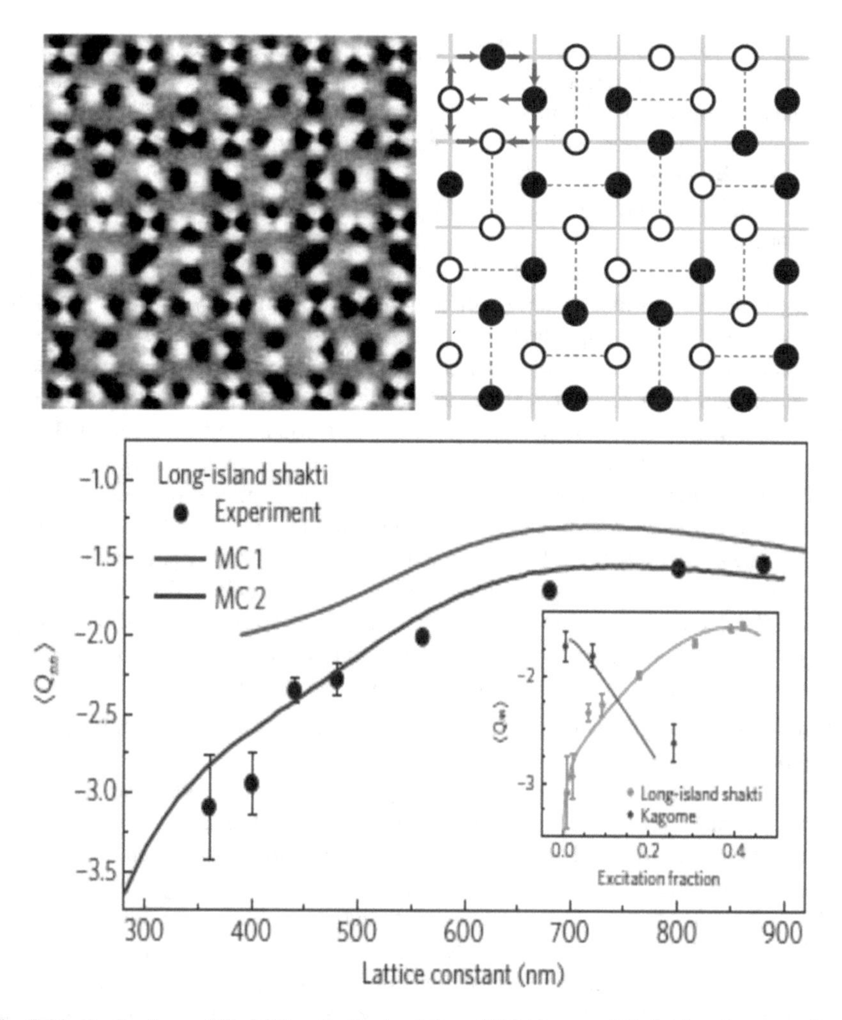

Fig. 3.14 Realizations of Shakti Ice. On the top left, an MFM image of Shakti ice after annealing. On the top right, the experimental data of the left part is translated in terms of allocation of the unhappy vertices, on plaquettes (black dots). One can see how an emergent ice rule describes the system, as each plaquette can have only two of four slots for unhappy vertices occupied. Bottom: screening of monopoles from magnetic charges $\langle Q_{nn} \rangle$ denotes the average magnetic charge surrounding a magnetic monopole on a $z = 4$ vertex, at the nearest neighbor level. Figures adapted from Ref. [15]

are always charged. Surrounded by a sea of charges, a monopole can gain or lose charge to and from it. Thus monopoles can pop up spontaneously or be reabsorbed, and do not need an anti-monopole to annihilate, as they do in pyrochlore spin ice, or square ice.

Despite that, however, the Shakti state is a bona fide topological phase, as we will explain now. In fact some other topological charge can be identified in it.

That the low-energy manifold has topological protection can be immediately suspected by noting that a single spin flip takes the system out of the manifold, and only a proper loop of collective spin flips realizes change *within* the manifold. This can be understood easily from Fig. 3.13, as all spins impinging in a $z = 3$ vertex also impinge into $z = 4, 2$ vertices, which are in their lowest energy in the ice manifold (Type I_4, I_2 of Fig. 3.6). Flipping any spin will thus necessarily cause excitations.

To identify the topological structure, we go back to the properties of the low-energy state. We saw that because each unhappy vertex affects two nearby plaquettes (Fig. 3.15) and costs energy, the lowest energy configuration is realized when nearby plaquettes are "dimerized" by a single UV [13]. The ice manifold of Shakti is thus described by a dimer cover model on the lattice connecting the rectangular plaquettes, which is topologically equivalent to a square lattice (Fig. 3.15) (from now on called "dimer lattice"), and which can be solved exactly [133].

The following is then standard: a discrete, emergent vector field $\vec{E}$ can be introduced, perpendicular to each edge, of length 1 (o 3) if the edge is unoccupied (or occupied) by a dimer, and direction entering (exiting) a gray square of Fig. 3.15 from top or bottom, and exiting (entering) it from the sides. The "line integral" $\int_\gamma \vec{E} \cdot d\vec{l}$ for such a discrete vector field along a directed line γ crossing the edges is the sum of the vectors along the line with sign taken along the line's direction. For a complete cover the emergent field is irrotational ($\oint_\gamma \vec{E} \cdot d\vec{l} = 0$) leading to the definition of a "height function" [118] h such that $\vec{E} = \vec{\nabla} h$ and thus demonstrating the topological state.

Beyond the standard dimer model, this picture can incorporate the low-energy excitations of Shakti ice as scramblings of the cover. As Fig. 3.15 shows, above the ground state a frustrated plaquette (i.e. a node of the dimer lattice) can be dimerized three times instead of one (over-dimerization) by UVs, or also diagonally by a Type-II_4 or a Type-II_2 vertex. In the presence of such scramblings the emergent vector field $\vec{E}$ is no longer irrotational. Indeed its circulation around any topologically equivalent loop encircling a defect defines the quantized topological charge of the defect as $q = \frac{1}{4} \oint_\gamma \vec{E} \cdot d\vec{l}$ (Fig. 3.15). Thus, the excitations of the Shakti ice manifold are topological charges, turning the discrete scalar field h that defines its order into a multivalued phase.

We have now the full picture: a topological phase, which cannot be explored from within, but only via a discrete kinetics of excitations whose topological charge is conserved. This picture is emergent, and not at all evident from, or indeed reminiscent of, the original spin structure. It also has consequences for the kinetics, in terms of ergodicity breaking, non-equilibration, and glassiness, as it is typical of a topological state with topologically protected excitations, that cannot be reabsorbed into the manifold individually, and which evolves via a discrete kinetics. All these issues have been recently investigated numerically and experimentally [129], showing that Shakti ice might provide the first artificial, controllable, modifiable and fully characterizable magnetic system which provides non-topographic vistas of ergodicity breaking and non-equilibration as consequences of a classical topological order.

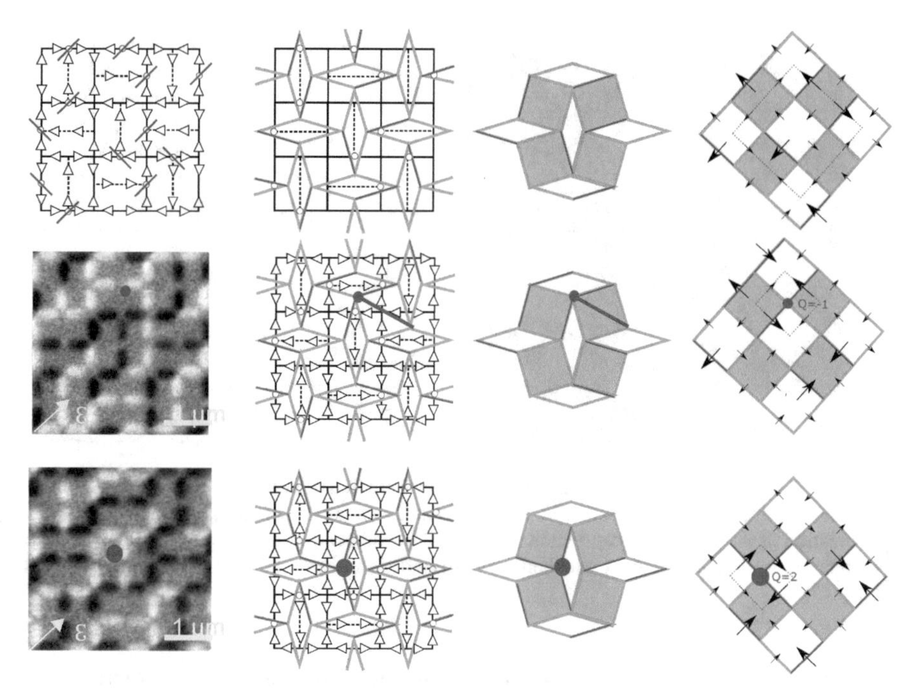

Fig. 3.15 Top: Shakti manifold as a dimer cover model. From left to right: Disordered spin ensemble for the ground state of Shakti ice manifold. The manifold is completely described by the allocation of the UVs (circles) which affect two nearby rectangular plaquettes (connected by the blue segments). Thus, an unhappy vertex is a dimer (blue segments) connecting frustrated plaquettes, and the ground state is a complete dimer-cover model on the (Ochre color) lattice with nodes in the center of rectangular plaquettes, topologically equivalent to a square lattice. There we introduce the emergent vector field $\vec{E}$, as in the text. The circulation of the vector field along any closed loop is zero. Middle and Bottom: The Shakti's low-energy manifold. XMCD image of Shakti spin ice, for a spin ensemble with one excitation (red and blue dots) and the corresponding emergent dimer cover representation. Now excitations appear as multiple occupancy and/or diagonal dimers (Type II_2s). $\vec{E}$ is no longer irrotational and its circulation defines the topological charge as $q = \frac{1}{4}\oint_\gamma \vec{E} \cdot d\vec{l}$ (image adapted from Ref [129])

3.4.2 *Dimensionality Reduction: Tetris Ice*

While Shakti spin ice provides a topologically protected low-energy manifold, no such protection is present in the ground state of Tetris ice, which can be explored by consecutive spin flips. As Fig. 3.16 shows, the lattice can be decomposed into T-shaped “tetris” pieces and it has a principal axis of symmetry. The geometry represents a layered one-dimensional systems. On the blue islands in Fig. 3.16 there cannot be any Type-II_3 unhappy vertex [13], and therefore the blue portion of the lattice, which we call backbone, must be ordered at the lowest energy. The unhappy vertices must reside on the red portions, which we call staircases, and which therefore remain disordered at low temperature. As temperature is lowered,

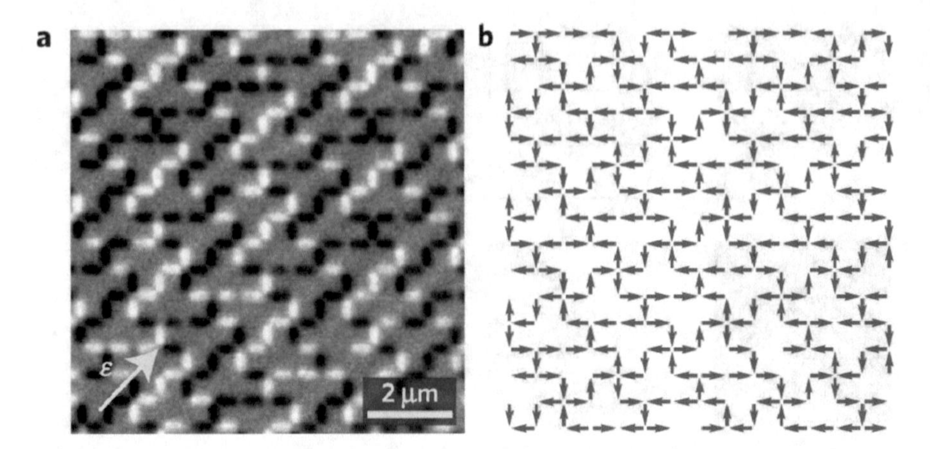

Fig. 3.16 Tetris Ice. (**a**) XMCD-PEEM image of a 600 nm Tetris lattice. The black/white contrast indicates whether the magnetization of an island has a component parallel or antiparallel to the polarization of the incident X-ray, which is indicated by the yellow arrow. (**b**) Map of the moment configurations showing ordered backbones (blue) and disordered staircases (red) (images from Ref. [16])

we have thus a dimensional reduction of an alternating ordered-disordered one-dimensional system, which was indeed confirmed experimentally [16].

This dimensional reduction is also apparent in the kinetics. Tetris was the first of the new geometries to be characterized in real-time, real-space, and from the supplementary information of Ref. [16] it is possible to watch animations of Tetris kinetics as the temperature is lowered or raised. Starting at high temperature, all the spins flip at about the same rate. As the temperature is lowered, ordered domains begin to form in correspondence with the backbones, where eventually the spins become static, while the spins on the staircases continue to fluctuate.

While the low-energy state described above has been confirmed experimentally, it follows from a nearest neighbor approximation. The profile of low-energy excitations, however, has not been yet studied in any systematic way, and promises interesting new effects. For instance, as one-dimensional systems, one expects that the backbones can never order completely, and will always host excitations above the low-energy manifold. Of course Tetris is in fact a two-dimensional system, which decomposes into one-dimensional ones only in the lowest energy configuration. Slightly above such a manifold, one expects correlations among excitations that belong to different backbones. Such correlations must be controlled both by the magnetic interaction between these defects—as Tetris is, after all, a system of dipoles that can interact at long-range—but also through entropic interactions. Indeed, the backbones are separated by disordered staircases of non-zero density of entropy, and whose entropy is affected by defects and excitations on the nearby backbones.

None of the above issues has yet been studied theoretically, and they might indeed provide a useful setting to explore the onset of phase decoupling into lower-dimensional states, a broader problem relevant to liquid crystal phases [134] or weakly coupled sliding phases [135, 136].

3.4.3 Polymers of Topologically Protected Excitations: Santa Fe Ice

We end this vista on how novel and unusual spin ice geometries influence topology with Santa Fe [14, 84] of Fig. 3.17, which was inspired by a terra cotta floor in the homonymous New Mexican capital—incidentally, the oldest in the United States. While Shakti and Tetris are maximally frustrated, which means that any minimal loop inside the geometry needs to be affected by an unhappy vertex, in Santa Fe only the dashed loops in the figure are frustrated and they are surrounded by unfrustrated ones. It is an inviolable topological constraint that at any energy frustrated loops can be affected by only an odd number of excitations, and unfrustrated ones by only an even number (or none).

An unhappy vertex on a frustrated loop of the Santa Fe lattice affects a nearby unfrustrated one. However, an unfrustrated loop can only be affected by an even number of defects, and thus there will be a second unhappy vertex on it, affecting in turn a nearby unfrustrated loop, et cetera. It follows that the low-energy state must consist of magnetic "polymers", whose "monomers" are local excitations. These polymers must begin from and end into frustrated loops.

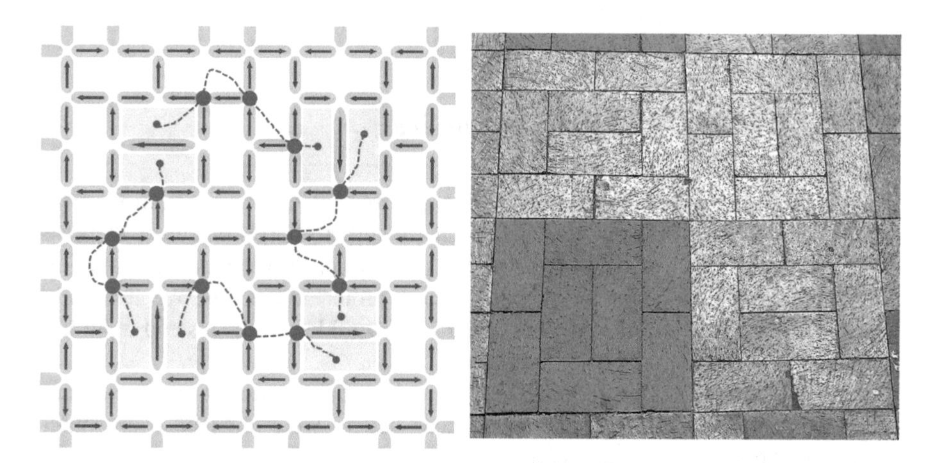

Fig. 3.17 The Santa Fe Ice can support both frustrated (shaded, green) and unfrustrated loops. There, "polymers" of unhappy vertices (blue dots) thread through unfrustrated loops to connect frustrated ones. On the right, brick floor in Santa Fe, New Mexico, USA

As each monomer costs energy, the lowest energy configuration of the magnetic ensemble will correspond to the shortest possible polymers, which are made of three monomers, each connecting nearby frustrated loops as in Fig. 3.17. The entropy of such a state can easily be computed exactly. As each polymer dimerizes two frustrated loops, the degeneracy is given by the dimer cover model on the square lattice whose node is made of nearby frustrated loops, times the number of ways in which polymers can be chosen once their pinned ends are fixed. Thus the ice manifold decomposes into the direct product of two states: the dimer-cover manifold, which classifies which loops are joined by which polymers, and the degeneracy of the polymers themselves.

At low temperature the kinetics must reduce to the fluctuations of the magnetic polymers without changing the pinning location of their ends, and thus without changing the dimer cover picture. Thus, the low-energy manifold can be explored from within, but only in part: only the polymers' configurations can change, not their pinning sites. To change the latter, excitations above the low-energy manifold are needed.

The kinetics within the manifold remains local and the polymer's fluctuations are uncorrelated, at least in a nearest neighbor energy approximation. As the temperature rises the polymers lengthen to include more than three monomers. At that point they can bump into each other, fuse in a cross, and then separate in different ways. This transition can lead to a different dimerization of the frustrated loops, as the new polymers emerging from "collisions" of old ones are now pinned to different ending point. Thus the dimer-cover ensemble is explored via this mechanism of polymer colliding, fusing together and then breaking again into different ones. This of course involves excitations over the ice manifold, further demonstrating the partial topological protection that pertains only to the dimer-cover sector of the low-energy manifold.

3.4.4 Ice Rule Fragility in Particle Ices

We have shown above how extensions of the notion of frustration from pairwise interactions to frustration in the allocation of vertex energy can lead to novel emergent phenomena, such as topological protection, dimensionality reduction, emergent ice rule, or polymers of topologically protected charges. We show now how the collective frustration typical of particle-based ices leads instead to an ice-rule fragility absent in magnetic realizations [70].

A recent paper [131] has reported experiments and simulations of particle-based spin ice in mixed-coordination lattices. These lattices were obtained from square lattices, decimated in such a way as to obtain only $z = 4, 3$ vertices, and η is the ratio of $z = 3$ vertices over $z = 4$ ones. At zero decimation, $\eta = 0$.

Experimentally, the system is the same as the gravitationally confined colloids described in Sect. 3.2.4, where paramagnetic colloids are set into microgrooves with two preferential orientations. The difference is that the lattice is now decimated by

Fig. 3.18 Experimental results for ice rule fragility. (**a**) An experimental image of the undecimated system shows the expected antiferromagnetic ordered configuration. (**b**)–(**e**) Experimental images of systems at increasing decimation $\eta = 0.1892$, 1.3158, 2.3846, and 5.2857, respectively for ice rule fragility. As decimation increases, more negative charge (blue glows) forms on $z = 4$ vertices, in violation of the ice rule. At low decimation most of the charge on $z = 3$ vertices is positive (red glows), whereas at high decimation the ratio between positive and negative charges on $z = 3$ vertices tends to one, as there is less negative charge to be cancelled on $z = 4$ vertices (images from Ref. [70])

leaving some of the microgrooves empty of their colloid. As a field perpendicular to the system is ramped, the paramagnetic colloids become magnetized and thus mutually repulsive. They evolve into a collective low-energy configuration, mapped via video microscopy and particle tracking. Figure 3.18 shows experimental results for different decimations. The images suggest that the ice rule is broken in the $z = 4$ sublattice, where $q = -2$ charges appear spontaneously. At the same time, the ice rule is still obeyed on the $z = 3$ sublattice, where only charges $q = \pm 1$ appear.

The authors accompanied their experimental results with numerical analysis on larger arrays. Figure 3.19a, b reports the experimentally and numerically obtained vertex statistics $n_{z4,q}$ and $n_{z3,q}$ grouped by charge q versus η along with theoretical predictions, which are based on the arguments of Sect. 3.3.2.3. There, one sees more precisely that in the $z = 4$ sector, negative monopoles of charge $q = -2$ appear along ice rule vertices and increase in density with increasing decimation. Figure 3.19c plots the total density of negative charge $q_{z4} = \sum_q n_{z4,q} q$ appearing on the $z = 4$ sublattice, as a function of decimation expressed via η.

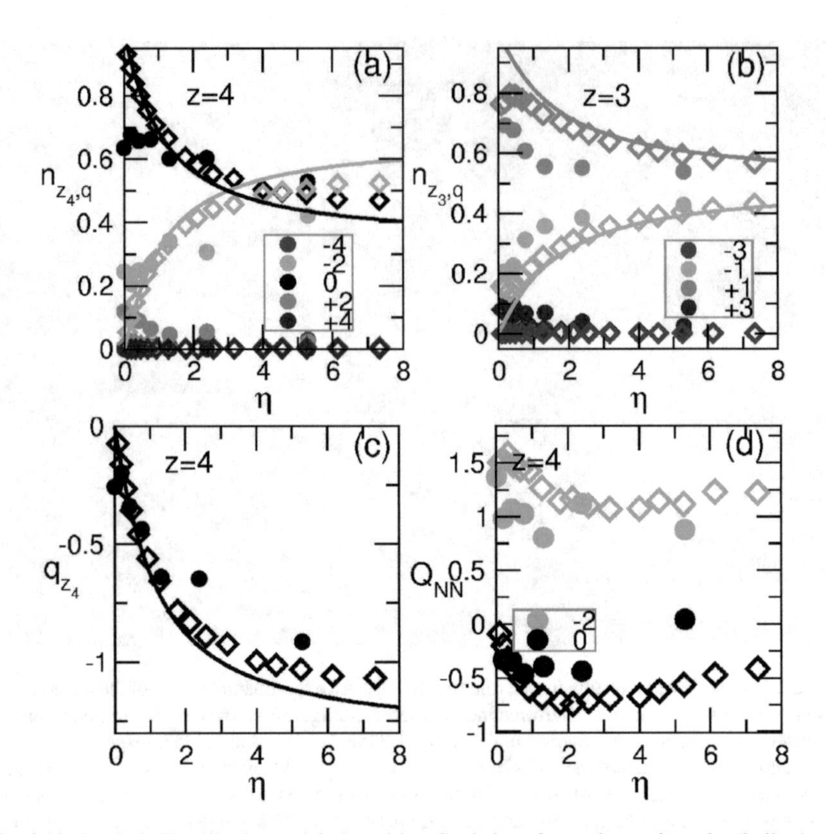

Fig. 3.19 Ice Rule Fragility in particle-based ice. Statistics of experimental results (bullets) and numerical results (diamonds) compared to theoretical predictions (solid lines). (**a**) Vertex statistics $n_{z_4,q}$ at equilibrium vs $\eta = N_{z_3}/N_{z_4}$ for $z = 4$ vertices grouped by topological charge q. Dark blue: $q = -4$; light blue: $q = -2$; black: $q = 0$; pink: $q = +2$; red: $q = +4$. All the non-ice-rule vertices are suppressed except $q = -2$ monopoles. (**b**) Vertex statistics $n_{z_3,q}$ vs η for $z = 3$ vertices. Dark blue: $q = -3$; light blue: $q = -1$; pink: $q = +1$; red: $q = +3$. Only ice rule vertices are present ($q = \pm 1$), but positive $q = +1$ charges exceed negative ones. As $\eta \to \infty$, the $z = 4$ sector disappears and thus $n_{z_3,q=1}$ and $n_{z_3,q=-1}$ tend to the same value of 1/2. (**c**) Net density of charge q_{z_4} forming on $z = 4$ vertices vs η as a measure of ice rule violation. (**d**) Charge screening Q_{NN} of $q = -2$ monopoles (blue) and "screening" of $q = 0$ ice rule vertices (black) on $z = 4$ vertices vs η (images from Ref. [131])

The $z = 3$ vertices (Fig. 3.19b) remain in the pseudo-ice manifold, however, the $z = 3$ vertices are positively charged overall. Indeed, as explained in Sect. 3.2.3.1, ice rule vertices of odd coordination are always charged with $q = \pm 1$. Therefore, they can therefore adsorb the negative charge of the $z = 4$ vertices without leaving the ice-manifold, simply by changing their relative ratio in favor of vertices of charge $q = 1$, as explained also in Sect. 3.3.2.3. Finally, Fig. 3.19d shows that the $q = 1$ charges on $z = 3$ vertices also rearrange locally to screen the $z = 4$ monopoles. This suggests that charge screening is not unique to magnetic charges that interact via a Coulomb law in magnetic ices [15, 93, 96], and in fact charge ordering was also recently found in kagome colloidal ice [137].

These results unambiguously demonstrate the inherent fragility of the ice rule in particle based ice as theoretically postulated in Ref. [70, 132]. As explained in Sect. 3.3.2.3, this fragility is due to the collective origin of the frustration in particle ice, where the ice rule is not enforced by the local energetics, unlike in magnetic ices. Rather, it emerges as a collective compromise among vertices in the thermodynamic limit. This effect is typical of these particular ice-rule materials. Such a breakdown is not possible in magnetic spin ices, where indeed the ice manifold had been found completely robust against decimation [13], mixed coordination [15, 16], and dislocations [61], and where even isolated clusters of magnetic vertices obey the ice rule at low energy [114].

3.5 Conclusion

We have elaborated on how the ice-rule emerges as a unifying topological concept to describe a large variety of material systems. It often originates in frustration, yet frustration can be of different kinds. While different forms of frustration might lead to similar manifolds at equilibrium, at least in simple geometries, a deeper understanding of the relationship between ice rule and frustration can unearth much more complex phenomena, both in the equilibrium thermodynamics and in the kinetics. These phenomena can now be explored in depth through deliberate design of artificial frustrated realization at the nano- or micro-scale.

References

1. W.F. Giauque, M.F. Ashley, Molecular rotation in ice at 10 k. free energy of formation and entropy of water. Phys. Rev. **43**(1), 81 (1933)
2. W.F. Giauque, J.W. Stout, The entropy of water and the third law of thermodynamics. the heat capacity of ice from 15 to 273 k. J. Am. Chem. Soc. **58**(7), 1144–1150 (1936)
3. L. Pauling, The structure and entropy of ice and of other crystals with some randomness of atomic arrangement. J. Am. Chem. Soc. **57**(12), 2680–2684 (1935)
4. V.F. Petrenko, R.W. Whitworth, *Physics of Ice* (Oxford University Press, Oxford, 1999)
5. C. Castelnovo, R. Moessner, S.L. Sondhi, Magnetic monopoles in spin ice. Nature **451**(7174), 42–5 (2008)
6. J.D. Bernal, R.H. Fowler, A theory of water and ionic solution, with particular reference to hydrogen and hydroxyl ions. J. Chem. Phys. **1**(8), 515–548 (1933)
7. J.F. Nagle, Lattice statistics of hydrogen bonded crystals. I. the residual entropy of ice. J. Math. Phys. **7**(8), 1484–1491 (1966)
8. E.H. Lieb, Residual entropy of square ice Phys. Rev. **162**(1), 162 (1967)
9. E.H. Lieb, Exact solution of the f model of an antiferroelectric. Phys. Rev. Lett. **18**(24), 1046 (1967)
10. F.Y. Wu, Critical behavior of two-dimensional hydrogen-bonded antiferroelectrics. Phys. Rev. Lett. **22**, 1174–1176 (1969)

11. R.J. Baxter, *Exactly Solved Models in Statistical Mechanics* (Academic, New York, 1982)
12. F. Rys, Ueber ein zweidimensionales klassisches Konfigurationsmodell. Ph.D. thesis, 1963
13. M.J. Morrison, T.R Nelson, C. Nisoli, Unhappy vertices in artificial spin ice: new degeneracies from vertex frustration. New J. Phys. **15**(4), 045009 (2013)
14. G.-W. Chern, M.J. Morrison, C. Nisoli, Degeneracy and criticality from emergent frustration in artificial spin ice. Phys. Rev. Lett. **111**, 177201 (2013)
15. I. Gilbert, G.-W. Chern, S. Zhang, L. O'Brien, B. Fore, C. Nisoli, P. Schiffer, Emergent ice rule and magnetic charge screening from vertex frustration in artificial spin ice. Nat. Phys. **10**(9), 670–675 (2014)
16. I. Gilbert, Y. Lao, I. Carrasquillo, L. O'Brien, J.D Watts, M. Manno, C. Leighton, A. Scholl, C. Nisoli, P. Schiffer, Emergent reduced dimensionality by vertex frustration in artificial spin ice. Nat. Phys. **12**(2), 162–165 (2016)
17. R.L. Stamps, Artificial spin ice: the unhappy wanderer. Nat. Phys. **10**(9), 623–624 (2014)
18. N. Bjerrum, Structure and properties of ice. Science **115**(2989), 385–390 (1952)
19. I.A. Ryzhkin, On magnetic relaxation in rare earth metal perchlorate metals. Zhurnal Ehksperimental'noj i Teoreticheskoj Fiziki **128**(3), 559–566 (2005)
20. A.P. Ramirez, A. Hayashi, R.J. Cava, R. Siddharthan, B.S. Shastry, Zero-point entropy in 'spin ice'. Nature **399**, 333–335 (1999)
21. B.C. den Hertog, M.J.P. Gingras, Dipolar interactions and origin of spin ice in ising pyrochlore magnets. Phys. Rev. Lett. **84**(15), 3430 (2000)
22. S.T. Bramwell, M.J. Gingras, Spin ice state in frustrated magnetic pyrochlore materials. Science **294**(5546), 1495–1501 (2001)
23. M.J. Harris, S.T. Bramwell, D.F. McMorrow, T.H. Zeiske, K.W. Godfrey, Geometrical frustration in the ferromagnetic pyrochlore ho 2 ti 2 o 7. Phys. Rev. Lett. **79**(13), 2554 (1997)
24. S.D. Bader, Colloquium: opportunities in nanomagnetism. Rev. Mod. Phys. **78**(1), 1 (2006)
25. R.F. Wang, C. Nisoli, R.S. Freitas, J. Li, W. McConville, B.J. Cooley, M.S. Lund, N. Samarth, C. Leighton, V.H. Crespi, P. Schiffer, Artificial 'spin ice' in a geometrically frustrated lattice of nanoscale ferromagnetic islands. Nature **439**(7074), 303–306 (2006)
26. C. Nisoli, R. Moessner, P. Schiffer, Colloquium: artificial spin ice: designing and imaging magnetic frustration. Rev. Mod. Phys. **85**(4), 1473 (2013)
27. L.J. Heyderman, R.L. Stamps, Artificial ferroic systems: novel functionality from structure, interactions and dynamics. J. Phys.: Condens. Matter **25**(36), 363201 (2013)
28. E. Mengotti, L.J. Heyderman, A.F. Rodríguez, F. Nolting, R.V. Hügli, H.-B. Braun, Real-space observation of emergent magnetic monopoles and associated dirac strings in artificial kagome spin ice. Nat. Phys. **7**(1), 68–74 (2010)
29. S. Ladak, D.E. Read, G.K. Perkins, L.F. Cohen, W.R. Branford, Direct observation of magnetic monopole defects in an artificial spin-ice system. Nat. Phys. **6**, 359–363 (2010)
30. S. Ladak, D. Read, T. Tyliszczak, W.R. Branford, L.F. Cohen, Monopole defects and magnetic coulomb blockade. New J. Phys. **13**(2), 023023 (2011)
31. K. Zeissler, S.K. Walton, S. Ladak, D.E. Read, T. Tyliszczak, L.F. Cohen, W.R. Branford, The non-random walk of chiral magnetic charge carriers in artificial spin ice. Sci. Rep. **3**, 1252 (2013)
32. C. Phatak, A.K. Petford-Long, O. Heinonen, M. Tanase, M. De Graef, Nanoscale structure of the magnetic induction at monopole defects in artificial spin-ice lattices. Phys. Rev. B **83**(17), 174431 (2011)
33. S. Ladak, D.E. Read, W.R. Branford, L.F. Cohen, Direct observation and control of magnetic monopole defects in an artificial spin-ice material. New J. Phys. **13**(6), 063032 (2011)
34. S.D. Pollard, V. Volkov, Y. Zhu, Propagation of magnetic charge monopoles and dirac flux strings in an artificial spin-ice lattice. Phys. Rev. B **85**(18), 180402 (2012)
35. N. Rougemaille, F. Montaigne, B. Canals, A. Duluard, D. Lacour, M. Hehn, R. Belkhou, O. Fruchart, S. El Moussaoui, A. Bendounan et al. Artificial kagome arrays of nanomagnets: a frozen dipolar spin ice. Phys. Rev. Lett. **106**(5), 057209 (2011)
36. B. Canals, I.-A. Chioar, V.-D. Nguyen, M. Hehn, D. Lacour, F. Montaigne, A. Locatelli, T. Onur Menteş, B. Santos Burgos, N. Rougemaille, Fragmentation of magnetism in artificial kagome dipolar spin ice. Nat. Commun. **7**, 11446 (2016)

37. P.E. Lammert, X. Ke, J, Li, C. Nisoli, D.M. Garand, V.H. Crespi, P. Schiffer, Direct entropy determination and application to artificial spin ice. Nat. Phys. **6**(10), 786–789 (2010)
38. Y.-L. Wang, Z.-L. Xiao, A. Snezhko, J. Xu, L.E. Ocola, R. Divan, J.E. Pearson, G.W. Crabtree, W.-K. Kwok, Rewritable artificial magnetic charge ice. Science **352**(6288), 962–966 (2016)
39. C. Nisoli, R. Wang, J. Li, W. McConville, P. Lammert, P. Schiffer, V. Crespi, Ground state lost but degeneracy found: the effective thermodynamics of artificial spin ice. Phys. Rev. Lett. **98**(21), 217203 (2007)
40. C. Nisoli, J. Li, X. Ke, D. Garand, P. Schiffer, V.H. Crespi, Effective temperature in an interacting vertex system: theory and experiment on artificial spin ice. Phys. Rev. Lett. **105**(4), 047205 (2010)
41. X. Ke, J. Li, C. Nisoli, P.E. Lammert, W. McConville, R. Wang, V.H. Crespi, P. Schiffer, Energy minimization and ac demagnetization in a nanomagnet array. Phys. Rev. Lett. **101**(3), 037205 (2008)
42. L.F. Cugliandolo, Artificial spin-ice and vertex models. J. Stat. Phys. **167**(3–4), 499–514 (2017)
43. D. Levis, L.F. Cugliandolo, L. Foini, M. Tarzia, Thermal phase transitions in artificial spin ice. Phys. Rev. Lett. **110**(20), 207206 (2013)
44. J.P. Morgan, A. Stein, S. Langridge, C.H. Marrows, Thermal ground-state ordering and elementary excitations in artificial magnetic square ice. Nat. Phys. **7**(1), 75–79 (2010)
45. Z. Budrikis, J.P. Morgan, J. Akerman, A. Stein, P. Politi, S. Langridge, C.H. Marrows, R.L. Stamps, Disorder strength and field-driven ground state domain formation in artificial spin ice: experiment, simulation, and theory. Phys. Rev. Lett. **109**(3), 037203 (2012)
46. Z. Budrikis, P. Politi, R.L. Stamps, Diversity enabling equilibration: disorder and the ground state in artificial spin ice. Phys. Rev. Lett. **107**(21), 217204 (2011)
47. P.E. Lammert, V.H. Crespi, C. Nisoli, Gibbsianizing nonequilibrium dynamics of artificial spin ice and other spin systems. New J. Phys. **14**(4), 045009 (2012)
48. C. Nisoli, On thermalization of magnetic nano-arrays at fabrication. New J. Phys. **14**(3), 035017 (2012)
49. I.A. Chioar, B. Canals, D. Lacour, M. Hehn, B. Santos Burgos, T.O. Menteş, A. Locatelli, F. Montaigne, N. Rougemaille, Kinetic pathways to the magnetic charge crystal in artificial dipolar spin ice. Phys. Rev. B **90**(22), 220407 (2014)
50. J.P. Morgan et al., Real and effective thermal equilibrium in artificial square spin ices. Phys. Rev. B **87**(2), 024405 (2013)
51. R.V. Hügli, G. Duff, B. O'Conchuir, E. Mengotti, L.J. Heyderman, A.F. Rodrı́guez, F. Nolting, H.B. Braun, Emergent magnetic monopoles, disorder, and avalanches in artificial kagome spin ice. J. Appl. Phys. **111**(7), 07E103 (2012)
52. P. Mellado, O. Petrova, Y. Shen, O. Tchernyshyov, Dynamics of magnetic charges in artificial spin ice. Phys. Rev. Lett. **105**(18), 187206 (2010)
53. S. Zhang, J. Li, I. Gilbert, J. Bartell, M.J. Erickson, Y. Pan, P.E. Lammert, C. Nisoli, K.K. Kohli, R. Misra et al., Perpendicular magnetization and generic realization of the Ising model in artificial spin ice. Phys. Rev. Lett. **109**(8), 087201 (2012)
54. U.B. Arnalds, J. Chico, H. Stopfel, V. Kapaklis, O. Bärenbold, M.A. Verschuuren, U. Wolff, V. Neu, A. Bergman, B. Hjörvarsson, A new look on the two-dimensional ising model: thermal artificial spins. New J. Phys. **18**(2), 023008 (2016)
55. C. Nisoli, Nano-ising. New J. Phys. **18**(2), 021007 (2016)
56. I.A. Chioar, N. Rougemaille, A. Grimm, O. Fruchart, E. Wagner, M. Hehn, D. Lacour, F. Montaigne, B. Canals, Nonuniversality of artificial frustrated spin systems. Phys. Rev. B **90**(6), 064411 (2014)
57. I.A. Chioar, N. Rougemaille, B. Canals, Ground-state candidate for the classical dipolar kagome Ising antiferromagnet. Phys. Rev. B **93**(21), 214410 (2016)
58. W.R. Branford, S. Ladak, D.E. Read, K. Zeissler, L.F. Cohen, Emerging chirality in artificial spin ice. Science **335**(6076), 1597–1600 (2012)

59. B.L. Le, J. Park, J. Sklenar, G.-W. Chern, C. Nisoli, J.D. Watts, M. Manno, D.W. Rench, N. Samarth, C. Leighton, P. Schiffer, Understanding magnetotransport signatures in networks of connected permalloy nanowires. Phys. Rev. B **95**, 060405 (2017)
60. L. Anghinolfi, H. Luetkens, J. Perron, M.G. Flokstra, O. Sendetskyi, A. Suter, T. Prokscha, P.M. Derlet, S.L. Lee, L.J. Heyderman, Thermodynamic phase transitions in a frustrated magnetic metamaterial. Nat. Commun. **6**, 8278 (2015)
61. J. Drisko, T. Marsh, J. Cumings, Topological frustration of artificial spin ice. Nat. Commun. **8**, 14009 (2017)
62. S. Gliga, A. Kákay, R. Hertel, O.G. Heinonen, Spectral analysis of topological defects in an artificial spin-ice lattice. Phys. Rev. Lett. **110**(11), 117205 (2013)
63. I. Gilbert, G.-W. Chern, B. Fore, Y. Lao, S. Zhang, C. Nisoli, P. Schiffer, Direct visualization of memory effects in artificial spin ice. Phys. Rev. B **92**(10), 104417 (2015)
64. A. Libál, C. Reichhardt, C.J. Olson Reichhardt, Hysteresis and return-point memory in colloidal artificial spin ice systems. Phys. Rev. E **86**(2), 021406 (2012)
65. A. Libál, C. Reichhardt, C.J. Olson Reichhardt, Realizing colloidal artificial ice on arrays of optical traps. Phys. Rev. Lett. **97**(22), 228302 (2006)
66. A. Libál, C.J. Olson Reichhardt, C. Reichhardt, Creating artificial ice states using vortices in nanostructured superconductors. Phys. Rev. Lett. **102**(23), 237004 (2009)
67. A. Libal, C. Nisoli, C. Reichhardt, C.J. Reichhardt, Dynamic control of topological defects in artificial colloidal ice (2016). Arxiv preprint arXiv:1609.02129
68. C.J. Olson Reichhardt, A. Libal, C. Reichhardt, Multi-step ordering in kagome and square artificial spin ice. New J. Phys. **14**(2), 025006 (2012)
69. D. Ray, C.J. Olson Reichhardt, B. Jankó, C. Reichhardt, Strongly enhanced pinning of magnetic vortices in type-ii superconductors by conformal crystal arrays. Phys. Rev. Lett. **110**(26), 267001 (2013)
70. C. Nisoli, Dumping topological charges on neighbors: ice manifolds for colloids and vortices. New J. Phys. **16**(11), 113049 (2014)
71. A. Ortiz-Ambriz, P. Tierno, Engineering of frustration in colloidal artificial ices realized on microfeatured grooved lattices. Nat. Commun. **7**, 10575 (2016)
72. P. Tierno, Geometric frustration of colloidal dimers on a honeycomb magnetic lattice. Phys. Rev. Lett. **116**(3), 038303 (2016)
73. J. Loehr, A. Ortiz-Ambriz, P. Tierno, Defect dynamics in artificial colloidal ice: real-time observation, manipulation, and logic gate. Phys. Rev. Lett. **117**(16), 168001 (2016)
74. M.L. Latimer, G.R. Berdiyorov, Z.L. Xiao, F.M. Peeters, W.K. Kwok, Realization of artificial ice systems for magnetic vortices in a superconducting moge thin film with patterned nanostructures. Phys. Rev. Lett. **111**, 067001 (2013)
75. J. Trastoy, M. Malnou, C. Ulysse, R. Bernard, N. Bergeal, G. Faini, J. Lesueur, J. Briatico, J.E. Villegas, Freezing and melting of vortex ice (2013). Arxiv preprint arXiv:1307.2881
76. P. Mellado, A. Concha, L. Mahadevan, Macroscopic magnetic frustration. Phys. Rev. Lett. **109**(25), 257203 (2012)
77. V. Kapaklis, U.B. Arnalds, A. Harman-Clarke, E. Th. Papaioannou, M. Karimipour, P. Korelis, A. Taroni, P.C.W. Holdsworth, S.T. Bramwell, B. Hjörvarsson, Melting artificial spin ice. New J. Phys. **14**(3), 035009 (2012)
78. U.B. Arnalds, A. Farhan, R.V. Chopdekar, V. Kapaklis, A. Balan, E.Th. Papaioannou, M. Ahlberg, F. Nolting, L.J. Heyderman, B. Hjörvarsson, Thermalized ground state of artificial kagome spin ice building blocks. Appl. Phys. Lett. **101**(11), 112404 (2012)
79. A. Farhan, P.M. Derlet, A. Kleibert, A. Balan, R.V. Chopdekar, M. Wyss, L. Anghinolfi, F. Nolting, L.J. Heyderman, Exploring hyper-cubic energy landscapes in thermally active finite artificial spin-ice systems. Nat. Phys. **9**, 375–382 (2013)
80. J.M. Porro, A. Bedoya-Pinto, A. Berger, P. Vavassori, Exploring thermally induced states in square artificial spin-ice arrays. New J. Phys. **15**(5), 055012 (2013)
81. V. Kapaklis, U.B. Arnalds, A. Farhan, R.V. Chopdekar, A. Balan, A. Scholl, L.J. Heyderman, B. Hjörvarsson, Thermal fluctuations in artificial spin ice. Nat. Nanotechnol. **9**(7), 514–519 (2014)

82. D. Shi, Z. Budrikis, A. Stein, S.A. Morley, P.D. Olmsted, G. Burnell, C.H. Marrows, Frustration and thermalization in an artificial magnetic quasicrystal. Nat. Phys. **14**(3), 309 (2018)
83. I. Gilbert, C. Nisoli, P. Schiffer, Frustration by design. Phys. Today **69**(7), 54–59 (2016)
84. C. Nisoli, V. Kapaklis, P. Schiffer, Deliberate exotic magnetism via frustration and topology. Nat. Phys. **13**(3), 200–203 (2017)
85. E. Ising, Beitrag zur theorie des ferromagnetismus. Zeitschrift für Physik A Hadrons and Nuclei **31**(1), 253–258 (1925)
86. L.P. Kadanoff, Scaling laws for ising models near tc, in *From Order To Chaos: Essays: Critical, Chaotic and Otherwise* (World Scientific, Singapore, 1993), pp. 165–174
87. K.G. Wilson, Renormalization group and critical phenomena. I. Renormalization group and the kadanoff scaling picture. Phys. Rev. B **4**(9), 3174 (1971)
88. K.G. Wilson, J. Kogut, The renormalization group and the ϵ expansion. Phys. Rep. **12**(2), 75–199 (1974)
89. B. Mahault, A. Saxena, C. Nisoli, Emergent inequality and self-organized social classes in a network of power and frustration. PloS One **12**(2), e0171832 (2017)
90. M. Tanaka, E. Saitoh, H. Miyajima, T. Yamaoka, Y. Iye, Magnetic interactions in a ferromagnetic honeycomb nanoscale network. Phys. Rev. B **73**(5), 052411 (2006)
91. Y. Qi, T. Brintlinger, J. Cumings, Direct observation of the ice rule in an artificial kagome spin ice. Phys. Rev. B **77**(9), 094418 (2008)
92. J. Drisko, S. Daunheimer, J. Cumings, Fepd 3 as a material for studying thermally active artificial spin ice systems. Phys. Rev. B **91**(22), 224406 (2015)
93. S. Zhang, I. Gilbert, C. Nisoli, G.-W. Chern, M.J. Erickson, L. O'Brien, C. Leighton, P.E. Lammert, V.H. Crespi, P. Schiffer, Crystallites of magnetic charges in artificial spin ice. Nature **500**(7464), 553–557 (2013)
94. S. Gliga, A. Kákay, L.J. Heyderman, R. Hertel, O.G. Heinonen, Broken vertex symmetry and finite zero-point entropy in the artificial square ice ground state. Phys. Rev. B **92**(6), 060413 (2015)
95. G.-W. Chern, P. Mellado, O. Tchernyshyov, Two-stage ordering of spins in dipolar spin ice on the kagome lattice. Phys. Rev. Lett. **106**, 207202 (2011)
96. G. Möller, R. Moessner, Magnetic multipole analysis of kagome and artificial spin-ice dipolar arrays. Phys. Rev. B **80**(14), 140409 (2009)
97. C. Castelnovo, R. Moessner, S.L. Sondhi, Spin ice, fractionalization, and topological order. Annu. Rev. Condens. Matter Phys. **3**(1), 35–55 (2012)
98. A.J. Macdonald, P.C.W. Holdsworth, R.G. Melko, Classical topological order in kagome ice. J. Phys.: Condens. Matter **23**(16), 164208 (2011)
99. S. Zhang, I. Gilbert, C. Nisoli, G.-W. Chern, M.J. Erickson, L. O'Brien, C. Leighton, P.E. Lammert, V.H. Crespi, P. Schiffer, Crystallites of magnetic charges in artificial spin ice. Nature **500**(7464), 553–557 (2013)
100. N.A. Sinitsyn, Y.V Pershin, The theory of spin noise spectroscopy: a review. Rep. Prog. Phys. **79**(10), 106501 (2016)
101. S.A. Crooker, D.G. Rickel, A.V. Balatsky, D.L. Smith, Spectroscopy of spontaneous spin noise as a probe of spin dynamics and magnetic resonance. Nature **431**(7004), 49–52 (2004)
102. R.J. Glauber, Time-dependent statistics of the ising model. J. Math. Phys. **4**(2), 294–307 (1963)
103. L.A.S. Mól, W.A. Moura-Melo, A.R. Pereira, Conditions for free magnetic monopoles in nanoscale square arrays of dipolar spin ice. Phys. Rev. B **82**(5), 054434 (2010)
104. F.S. Nascimento, L.A.S. Mól, W.A. Moura-Melo, A.R. Pereira, From confinement to deconfinement of magnetic monopoles in artificial rectangular spin ices. New J. Phys. **14**(11), 115019 (2012)
105. L.A. Mól, R.L. Silva, R.C. Silva, A.R. Pereira, W.A. Moura-Melo, B.V. Costa, Magnetic monopole and string excitations in two-dimensional spin ice. J. Appl. Phys. **106**(6), 063913 (2009)
106. Y. Nambu, Strings, monopoles, and gauge fields. Phys. Rev. D **10**(12), 4262 (1974)

107. Y. Perrin, B. Canals, N. Rougemaille, Extensive degeneracy, coulomb phase and magnetic monopoles in artificial square ice. Nature **540**(7633), 410–413 (2016)
108. G.-W. Chern, C. Reichhardt, C. Nisoli, Realizing three-dimensional artificial spin ice by stacking planar nano-arrays. Appl. Phys. Lett. **104**(1), 013101 (2014)
109. A.A. Mistonov, N.A. Grigoryeva, A.V. Chumakova, H. Eckerlebe, N.A. Sapoletova, K.S. Napolskii, A.A. Eliseev, D. Menzel, S.V. Grigoriev, Three-dimensional artificial spin ice in nanostructured co on an inverse opal-like lattice. Phys. Rev. B **87**(22), 220408 (2013)
110. A.A. Mistonov, I.S. Shishkin, I.S. Dubitskiy, N.A. Grigoryeva, H. Eckerlebe, S.V. Grigoriev. Ice rule for a ferromagnetic nanosite network on the face-centered cubic lattice. J. Exp. Theor. Phys. **120**(5), 844–850 (2015)
111. I.R.B. Ribeiro, F.S. Nascimento, S.O. Ferreira, W.A. Moura-Melo, C.A.R. Costa, J. Borme, P.P. Freitas, G.M. Wysin, C.I.L. de Araujo, A.R. Pereira, Realization of rectangular artificial spin ice and direct observation of high energy topology (2017). Arxiv preprint arXiv:1704.07373
112. E. Östman, H. Stopfel, I.-A. Chioar, U.B. Arnalds, A. Stein, V. Kapaklis, B. Hjörvarsson, Interaction modifiers in artificial spin ices. Nat. Phys. **14**(4), 375 (2018)
113. A.T. Fiory, A.F. Hebard, S. Somekh, Critical currents associated with the interaction of commensurate flux-line sublattices in a perforated al film. Appl. Phys. Lett. **32**(1), 73–75 (1978)
114. J. Li, S. Zhang, J. Bartell, Cristiano Nisoli, X. Ke, P. Lammert, V. Crespi, P. Schiffer, Comparing frustrated and unfrustrated clusters of single-domain ferromagnetic islands. Phys. Rev. B **82**(13), 134407 (2010)
115. C. Castelnovo, R. Moessner, S.L. Sondhi, Debye-hückel theory for spin ice at low temperature. Phys. Rev. B **84**(14), 144435 (2011)
116. S.V. Isakov, K. Gregor, R. Moessner, S.L. Sondhi, Dipolar spin correlations in classical pyrochlore magnets. Phys. Rev. Lett. **93**(16), 167204 (2004)
117. C.L. Henley, The 'coulomb phase' in frustrated systems. Annu. Rev. Condens. Matter Phys. **1**(1), 179–210 (2010)
118. C.L. Henley, Classical height models with topological order. J. Phys.: Condens. Matter **23**(16), 164212 (2011)
119. X.-G. Wen, Vacuum degeneracy of chiral spin states in compactified space. Phys. Rev. B **40**(10), 7387 (1989)
120. X.-G. Wen, Quantum orders and symmetric spin liquids. Phys. Rev. B **65**(16), 165113 (2002)
121. P.M. Chaikin, T.C. Lubensky, *Principles of Condensed Matter Physics* (Cambridge University Press, Cambridge, 2000)
122. V. Volterra, Sur l'équilibre des corps élastiques multiplement connexes, in *Annales Scientifiques de l'École Normale Supérieure*, vol. 24 (Elsevier, New York, 1907), pp. 401–517
123. M. Vasil'evich Kurik, O.D. Lavrentovich, Defects in liquid crystals: homotopy theory and experimental studies. Phys.-Usp. **31**(3), 196–224 (1988)
124. J. Michael Kosterlitz, D. James Thouless, Ordering, metastability and phase transitions in two-dimensional systems. J. Phys. C: Solid State Phys. **6**(7), 1181 (1973)
125. C. Castelnovo, C. Chamon, Topological order and topological entropy in classical systems. Phys. Rev. B **76**(17), 174416 (2007)
126. R. Zachary Lamberty, S. Papanikolaou, C.L. Henley, Classical topological order in abelian and non-abelian generalized height models. Phys. Rev. Lett. **111**(24), 245701 (2013)
127. L.D.C. Jaubert, M.J. Harris, T. Fennell, R.G. Melko, S.T. Bramwell, P.C.W. Holdsworth, Topological-sector fluctuations and curie-law crossover in spin ice. Phys. Rev. X **3**(1), 011014 (2013)
128. C. Castelnovo, R. Moessner, S.L. Sondhi, Thermal quenches in spin ice. Phys. Rev. Lett. **104**(10), 107201 (2010)
129. Y. Lao, F. Caravelli, M. Sheikh, J. Sklenar, D. Gardeazabal, J.D. Watts, A.M. Albrecht, A. Scholl, K. Dahmen, C. Nisoli, P. Schiffer, Classical topological order in the kinetics of artificial spin ice. Nat. Phys. **14**, 723–727 (2018)
130. G.H. Wannier, Antiferromagnetism. the triangular ising net. Phys. Rev. **79**(2), 357 (1950)

131. A. Libál, D.Y. Lee, A. Ortiz-Ambriz, C. Reichhardt, C.J.O. Reichhardt, P. Tierno, C. Nisoli, Ice rule fragility via topological charge transfer in artificial colloidal ice. Nat. Commun. **9**(1), 4146 (2018). https://doi.org/10.1038/s41467-018-06631-1
132. C. Nisoli, Unexpected phenomenology in particle-based ice absent in magnetic spin ice. Phys. Rev. Lett. **120**(16), 167205 (2018)
133. P.W. Kasteleyn, The statistics of dimers on a lattice: I. the number of dimer arrangements on a quadratic lattice. Physica **27**(12), 1209–1225 (1961)
134. P.G. De Gennes, G. Sarma, Tentative model for the smectic B phase. Phys. Lett. A **38**(4), 219–220 (1972)
135. C.S. O'Hern, T.C. Lubensky, J. Toner, Sliding phases in xy models, crystals, and cationic lipid-dna complexes. Phys. Rev. Lett. **83**(14), 2745 (1999)
136. S.L. Sondhi, K. Yang, Sliding phases via magnetic fields. Phys. Rev. B **63**(5), 054430 (2001)
137. A. Libál, C. Nisoli, C.J.O. Reichhardt, C. Reichhardt, Inner phases of colloidal hexagonal spin ice. Phys. Rev. Lett. **120**, 027204 (2018)

Chapter 4
Glassy Phenomena and Precursors in the Lattice Dynamics

M. E. Manley

Abstract Broad classes of functional materials exhibit glass-like phenomena originating with the frustration of a soft phonon driven phase transition, including relaxor ferroelectrics and shape memory strain glasses. While the soft phonon mechanism is mostly understood, how this mechanism becomes frustrated in the presence of disorder remains intensely debated. A common structural feature of the frustrated state is nanoscale regions of local ferroic displacements that form well above the ordering temperature; these are called polar nanoregions (PNRs) in relaxor ferroelectrics and ferroelastic nanodomains (FND) in the strain glasses. The existence of these small regions provides a basis to explain glass-like slow relaxation phenomena, which can manifest in the lattice dynamics as phonon over damping. However, this does not explain why the long-range order becomes localized into PNRs or FNDs, or why this happens specifically at the nanoscale. Recent scattering experiments and theories suggest an exciting new way to think about these problems in terms of the physics of lattice vibrations in chemically disordered crystals. More generally, probing the lattice dynamics of these systems sheds new light on the microscopic origin of the nanoregions, glassy behavior, and enhanced functional properties.

4.1 Introduction

The frustration of ferroic phase transformations by disorder is at the heart of many technologically important functional materials. Some of highest performing piezoelectric materials used in industry, PMN-xPT ($(Pb(Mg_{1/3}Nb_{2/3})O_3)_{1-x}$-$(PbTiO_3)_x$) and PZN-$x$PT ($(Pb(Zn_{1/3}Nb_{2/3})O_3)_{1-x}$-$(PbTiO_3)_x$) [1, 2], are relaxor-based ferroelectrics—meaning they exhibit a ferroelectric ordering temperature but also retain many of the disordered-induced relaxor characteristics, including

M. E. Manley (✉)
Material Science and Technology Division, Oak Ridge National Lab, Oak Ridge, TN, USA
e-mail: manleyme@ornl.gov

T. Lookman, X. Ren (eds.), *Frustrated Materials and Ferroic Glasses*, Springer Series in Materials Science 275, https://doi.org/10.1007/978-3-319-96914-5_4

polar nanoregions (PNRs) and a frequency-dependent dielectric response [3]. The ultrahigh piezoelectric response in these materials has been attributed to both a shear instability associated with being close to a morphotropic phase boundary [4] and an additional shear softening caused by PNRs hybridizing with transverse acoustic phonons [5]. Phase-field modeling [6] and experiments [5, 7] also show that alignment of the PNRs in a field can further enhance the macroscopic piezoelectric response. Shape memory strain glasses exhibit glass-like behavior, including frequency-dependent mechanical damping, ferroelastic nanodomains (FNDs), and superelasticity, and also retaining the useful property of shape memory, only with a technologically interesting gentle shape recovery [8].

In both classes of materials the nanoscale regions (PNRs or FNDs) begin as dynamically fluctuating regions at high temperatures before "freezing" in at some lower temperature. These dynamic nanoregion precursors are reminiscent of the soft phonon precursors to the un-frustrated ferroic transitions. Figure 4.1 illustrates the prototypical soft phonon mechanisms for both the un-frustrated ferroelectric and shape memory martensitic displacive phase transitions. In the ferroelectric case, Fig. 4.1a, the soft phonon is a transvers optic (TO) phonon at the zone center that corresponds to the ferroelectric displacements. Above the transition temperature (T_C) the TO phonon is stable, but as the material is cooled towards the transition temperature the phonon gradually decreases in frequency as the forces soften until the phonon displacements freeze in and ferroelectric domains form. This description is for the ideal second-order transition. Typically the unit cell distorts slightly in the direction of polarization. In practice there may also be considerable mode damping and coupling between the TO and transverse acoustic (TA) phonons. In the case of strong damped-mode coupling, a so-called "waterfall" effect can occur where the TO intensity cascades vertically into the TA phonon. This effect appears different depending on the zone measured in a neutron scattering experiment and can be understood in terms of a simple damped harmonic oscillator model [9], although the underlying causes for mode damping can be from anharmonicity in the interatomic

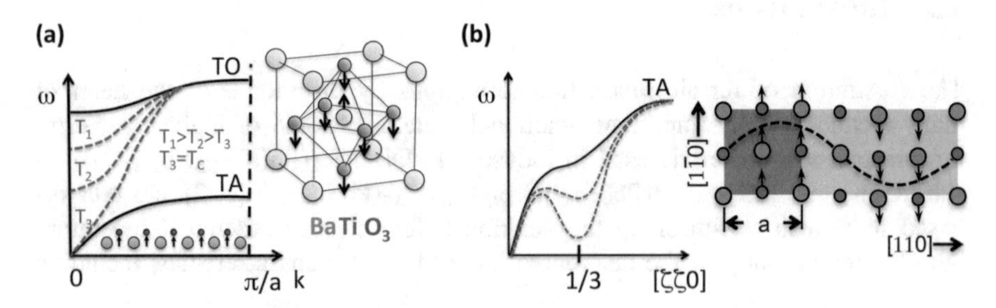

Fig. 4.1 Soft phonon precursors to two types of ferroic phase transitions. (**a**) Ferroelectric soft phonon in the transverse optic (TO) phonon. Out-of-phase motion of blue and red site atoms at $k = 0$ (infinite wavelength) separates the electrically charged atoms to induce ferroelectric order as the TO mode slows. (**b**) Soft transverse acoustic (TA) phonon that is typical of shape memory alloys. In-phase motion of atoms results in a shuffling of the crystal planes across three unit cells, resulting in a tripling of the unit cell in martensitic phase

potentials or disorder in the case of imperfect crystals. For the shape memory alloys the transition displacements correspond, in part, to a transverse acoustic (TA) phonon and typically occur at around $\boldsymbol{q} \cong \left[{}^{1}/_{3}, {}^{1}/_{3}, 0\right]$, c.f. Fig. 4.1b. This soft phonon corresponds to a shuffling of the $[1\bar{1}0]$ planes along $[1, -1, 0]$ across three unit cells. The transition is also accompanied by a shear strain and has a weakly first-order character. The first-order character, that there is an energy barrier to the transition (or latent heat), means that the soft phonon does not actually go all the way to zero frequency before the transition occurs, but in the case of a very weakly first-order transition it can soften considerably [10].

This basic soft phonon mechanism was first proposed to explain the onset of the ferroelectric behavior independently by William Cochran in 1959 [11] and Philip Anderson in 1960 [12]. Interestingly, Cochran concluded that the onset of ferroelectric properties is probably always a lattice dynamics problem, but that at the time there was no prospect of a detailed application of the theory to disordered crystals [11]. This conclusion makes sense since the electric polarization underlying ferroelectric behavior is inexorably tied to the atomic displacements of the TO phonon. If we take this as a starting point for understanding relaxor ferroelectrics, the problem reduces to understanding how the presence of disorder changes the lattice vibrations that drive ferroelectric behavior. In this sense we can see that measurements of the lattice dynamics is crucial to understanding the onset of relaxor ferroelectric behavior. A key difference between a normal ferroelectric and a relaxor ferroelectric is that in the latter case polar displacements become spatially localized in polar nanoregions, like small islands of ferroelectricity. Therefore, the problem is further reduced to understanding how TO-phonon-like displacements become spatially localized in the presence of disorder.

4.2 Phonon Localization in Relaxor Ferroelectrics

In the time since the soft mode theory was first developed, new theories have emerged in fields unrelated to ferroelectrics showing that lattice vibrations can spontaneously localize in crystals under certain conditions. One way of localizing lattice vibrations is through interplay of nonlinearity and lattice discreteness, in modes called discrete breathers or intrinsic localized modes (ILMs) [13–15]. The essence of this mechanism is that of a non-resonance condition [14]. A feature of nonlinearity (anharmonicity) is that a local dynamical fluctuation can have a shifted frequency because the frequency depends on amplitude. Furthermore, because of discreteness, the spectrum of plane waves have a cutoff frequency and can have gaps, meaning that there are frequencies where no plane waves reside. Under right circumstances these local fluctuations can develop out of resonance with the plane wave spectrum. In this case, the local dynamic fluctuation does not interact with the plane waves and an ILM forms as a new mode. This mechanism has been proposed as a possible explanation for PNRs and relaxor ferroelectric behavior

[16–18]. In this model the ILMs form preferentially at impurity sites and gradually slow to form the PNRs [16]. Another way lattice vibrations may localize is in the presence of disorder by Anderson localization. Anderson localization is in essence a wave interference effect where waves scattered by disorder constructively interfere in small regions and destructively interfere elsewhere [19]. Anderson localization was first developed by Philip Anderson in 1960 [20] to explain why some metals become insulators in the presence of disorder from impurities. In this case it is the interferences in the electron wave functions that prevent the diffusion of electrons. Anderson won the 1977 Nobel Prize in physics for this contribution. The idea has since been expanded to many other types of waves, ranging from light waves to ultrasound and phonons [19]. Only recently, however, has the idea been applied to relaxor ferroelectrics. Akbarzadeh et al. [21] first made the suggestion in a 2012 publication on ab initio-based simulations of a relaxor ferroelectric. In the analysis of their simulations they found that the only factor that mattered for the formation of PNRs was the presence disorder, which led them to suggest Anderson localization as a plausible mechanism for PNR formation. Sherrington [22] followed up on this suggestion and showed that Anderson localization could also be derived from equations used to describe relaxor ferroelectric behavior. Sherrington also noted that this idea resolved several issues with the phase diagram not explained by the random electric field model alone [23]. For example, the fact that the dilution of random electric fields does not decrease the relaxor ferroelectric onset temperature [22] suggests that there is more to PNR formation than the quenching of random fields [23]. Of course, from the point of view of neutron scattering the electric fields are not detected directly, only the resulting arrangements and motions of the atomic nuclei are probed.

The contribution of neutron scattering experiments has been to show that atomic scale resonant local modes (in band) occur first at high temperatures and that this leads to nanoscale phonon localization in PMN-PT and PZN-PT relaxor ferroelectrics when cooled into the relaxor state [24–26]. It was also found that the temperature of nanoscale phonon localization corresponded to the temperature that the PNRs "freeze" [26]. The dynamic fluctuating PNRs that are known to preempt the static PNRs can be understood in terms of weak localization [26], a known precursor to strong (Anderson) localization [19]. These experiments were originally designed as an attempt to confirm the ILM theory of PNR formation [16–18]. The observations themselves pointed towards an effect more consistent with the Anderson localization idea [21, 22] rather than ILMs [16–18]. Chief among these is the fact that localization occurs within the bands, which is counter to the non-resonant condition of ILM formation [13–15]. The localization coherence length is also matched to the wavelength of the localized mode, which is expected for a wave interference effect such as that driving Anderson localization. Perhaps the most interesting conclusion derived from this analysis is that the PNRs get their size ($\sim$2 nm) and shape from the wavelength of the localized phonons, and this is determined from a single resonance frequency [24] set by atomic scale local resonant modes (scattering centers). In this section, the rational behind the idea that localization is basically a wave interference effect and ultimately Anderson localization is described.

Figure 4.2 illustrates the basic features observed in the dynamical structure of PMN-30%PT shown in reference [24]. At temperatures above the PNR freezing temperature, T_f, and near the Burns temperature (T_d) (where dynamically fluctuating PNRs first appear) a dispersionless mode is observed in resonance with the TO phonon with a flat intensity profile (Fig. 4.2a). A lack of dispersion indicates a mode that is stationary since it has zero group velocity, $v_g = dE/dk$. A flat intensity profile (constant) in reciprocal space indicates a mode that is fully localized spatially in real space, since the Fourier transform of a constant is a delta function. On cooling below T_f, while the mode remains dispersionless (stationary), the intensity

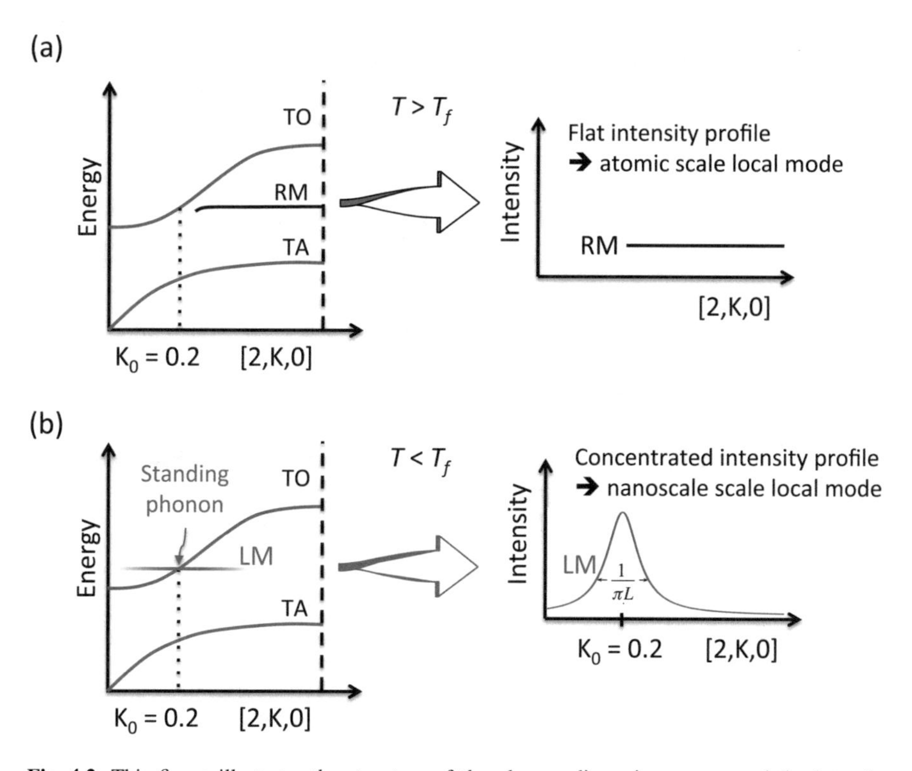

Fig. 4.2 This figure illustrates the structure of the phonon dispersion curves and the intensity profile of the resonance mode (RM) and localized mode (LM) above and below the PNR freezing temperature, T_f, respectively, after [24]. (**a**) Resonance mode associated with the off-centering of Pb atoms appears in resonance with the transverse optic (TO) phonon. On the right, the intensity profile in momentum (along $\mathbf{Q} = [2,K,0]$) appears flat, which indicates an atomic scale local mode (a constant in reciprocal space corresponds to a delta function in real space). (**b**) Below T_f the dynamics transition to a state where the local mode intensity becomes concentrated near the TO phonon. On the right, the intensity profile becomes peaked in a Lorentzian shape, with a coherence length, L, of 2 nm. The 2 nm coherence length is equal to the wavelength of the TO phonon at the crossing (standing phonon) and the size of the PNRs

of the localized phonons changes to a state where it is concentrated at the crossing with the TO phonon (Fig. 4.2b). The localized mode now has a coherence length (size) that equals a single TO wavelength at the crossing, and this size matches the PNRs [24]. As described below, these results are explained in terms of an Anderson-type localization mechanism, where constructive interference of the TO phonons interacting with randomly distributed localized resonance modes results in the localization of the ferroelectric TO phonons.

To make sense of these results, it is informative to start by considering the apparently paradoxical result that the localized mode becomes longer lived (shaper in energy) as it forms resonance crossings with the TO phonon [24]. It is also useful to first imagine the problem in one dimension, and then to build on this intuition. As depicted in Fig. 4.3a, a localized mode in resonance with traveling phonons, in this case the TO phonons, is expected to radiate these phonons, in the same way that a disturbed water surface send out ripples. Now the radiation of energy through the TO phonons is expected to quickly dissipate the resonance mode energy,

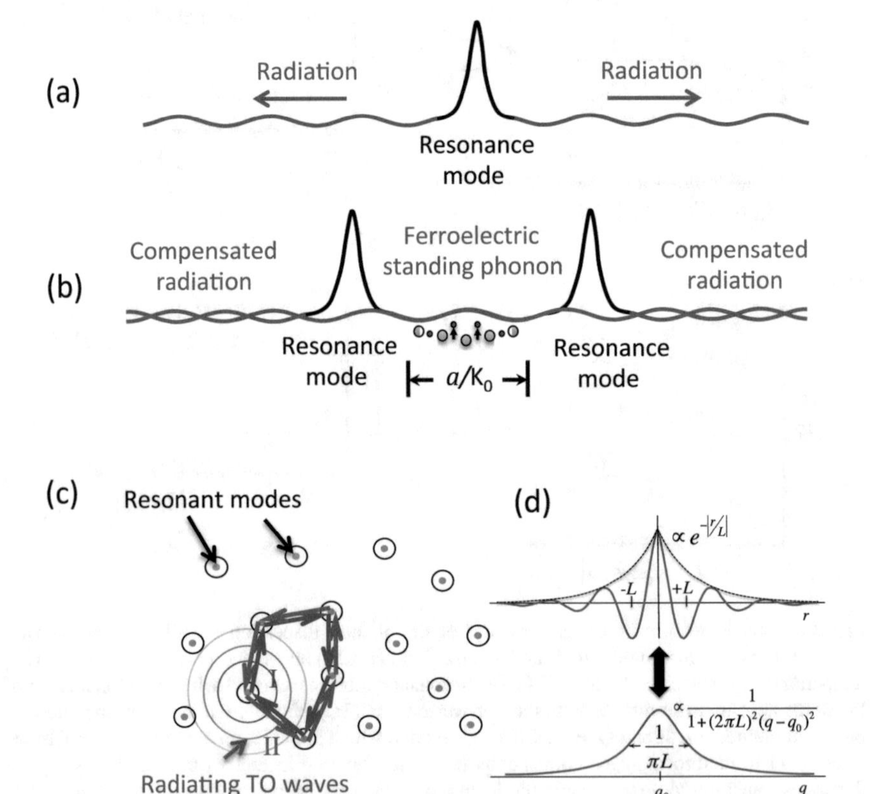

Fig. 4.3 Trapping ferroelectric modes with resonance modes. (**a**) Resonance mode radiating transverse optic (TO) phonons. (**b**) Trapped TO phonon in 1D. (**c**) Anderson localization scheme. (**d**) Exponential localization transformation from real space to reciprocal space

thereby reducing it to a short-lived fluctuation rather than a long-lived excitation. As depicted in Fig. 4.3b, however, it is possible to "trap" (localize) the radiating phonons between two resonance modes in 1D. The radiating phonons constructively interfere a standing wave between the pair while destructively interfering outside of the pair, resulting in a long-lived bound state. This long-lived state will then appear sharp in energy, yet dispersionless since it is a standing wave. The size, or coherence length, of the standing wave trapped between the resonant modes will be matched to the wavelength of the trapped mode, consistent with the observations illustrated in Fig. 4.2b. Furthermore, since it is a spatial localization of the TO phonon, it is expect that the intensity should be spread out from the position on the dispersion curve where TO phonon appears. The reason for this can be understood from the convolution theorem. In real space the bare TO phonon is a wave extended throughout the crystal with a periodicity that places it at a particular **q** point in the Brillouin zone. Localization can be viewed as an envelope function that attenuates the mode outside of the local standing wave region. The Fourier transform of the wave multiplied by this envelope function will be the convolution of the **q** point with the transform of the envelope function. If the localization is exponential, this will produce a Lorentzian broadening function. That local mode intensity profile actually appears Lorentzian, Fig. 4.2b, indicates exponential localization [24], which is strong localization [19]. As intuitively appealing as this simple model is, the real problem involves randomly distributed resonance modes in three dimensions.

The problem of how a radiating wave becomes trapped (localized) in a random three-dimensional distribution of resonance modes is essentially the Anderson localization problem [19, 20]. The basic argument is illustrated in Fig. 4.3c: waves emanating (scattered) from a single resonance mode and traveling in equal by opposite directions around the same random scattering path constructively interfere on returning. Anderson localization is exponential in real space, which transforms to Lorentzian in reciprocal space (Fig. 4.3d), and the characteristic length scale, L, is the wavelength of the trapped phonon. Invoking the convolution theorem, the localized TO phonon is as a **q**-broadened TO mode centered at the crossing and with a coherence length matching the wavelength. This explains the observed dynamical structure [24]. The trapped wavelength of 2 nm matches the coherence length of the localized phonon and the size of the PNRs. The picture emerging here naturally introduces a new concept, the trapping wave vector. This is the wave vector where the localization occurs and indicates the size of the PNRs that form. This has implications beyond the size of the PNRs as viewed along the $\mathbf{Q} = [2,K,0]$ direction since the points where TO phonon crosses the resonance mode form a higher dimensional surface with a predictable shape.

The relationship between the TO phonon dispersion and the resonance mode explains the evolution of the size, shape, and positions in reciprocal space of the PNRs, as well as antiferroelectric nanoregions [24] that occur in PMN-xPT for $x = 0$–30% [27]. Interestingly, the relationship between the TO-phonon anisotropy and the PNR anisotropy was noted before the concept of Anderson localization was first suggested. Matsuura et al. [28] first recognized the fact that the anisotropy in the TO phonon dispersion, which depends on PT content, closely follows the anisotropy of the PNR diffuse scattering. In fact, they argued that the PNRs may be related to a “trapping” of the TO phonons, but did not recognize that Anderson localization could provide the mechanism [28]. In view of how the trapping wavevector maps with direction, Fig. 4.4a, it can be seen why the anisotropy of the PNR diffuse scattering follows directly from the anisotropy of the TO phonon [24]. The intersection of the TO phonon dispersion surface with the resonance mode frequency carves out the size and shape of the PNR diffuse scattering. This is expected if the PNRs are driven by the phonon localization. The diffuse

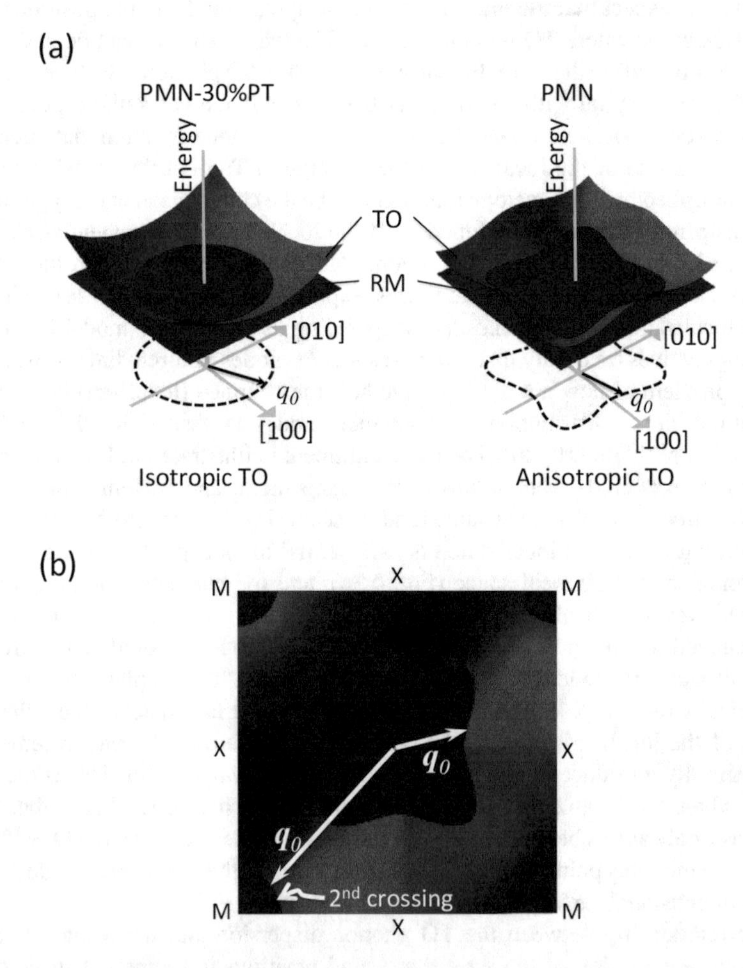

Fig. 4.4 Trapping wavevectors expected for a single resonance mode (RM) frequency crossing the dispersion surface of the transverse optic (TO) phonon. (**a**) A comparison of the trapping wavevectors, q_0, expected near the zone center for PMN-30%PT and PMN. The trapping wavevector loci map to the observed diffuse scattering anisotropy, after [24]. (**b**) Locations in reciprocal space where the second crossing introduces trapping wavevectors near the M points. These are the wavevectors where antiferroelectric distortions were observed [27]

scattering represents the static, or average, structural relaxation around the Anderson localized phonons, which have sizes and shapes determined by the trapped phonon wavevector.

Following the TO phonon throughout the Brillouin zone it also crosses the resonance mode frequency near the high symmetry M point on the zone edge [24], illustrated in Fig. 4.4b. Inelastic neutron scattering measurements reveal that a similar phonon localization effect occurs at these crossing points [24]. Specifically, the intensity of the local mode concentrates near the second crossing near the M point on cooling below the same PNR freezing temperature T_f. This part of momentum space corresponds to TO phonons with antiferroelectric atomic displacements, meaning the polar displacements alternate polarization from one unit cell to the next. Based on the above arguments about phonon localization this result suggests that antiferroelectric nanoregions should also form at the M points. Indeed, Swainson et al. [27] reported in 2009 that diffuse elastic scattering at the M points occurs in PMN. They attribute the diffuse scattering to antiferroelectric distortions. Hence, knowing the TO phonon dispersion surface and the resonance mode frequency, it is possible to explain the size and shape of the PNRs as well as the existence of antiferroelectric distortions at the zone edge M points [24].

The resonance modes can be attributed to the off-centering of Pb atoms in the lead-based relaxor ferroelectrics (PMN-PT and PZN-PT) [5, 29, 30]. The Pb atomic displacements occur along the [100] crystallographic directions [5, 29, 30] and the application of an electric field to a single crystal aligns these local atomic displacements and the local vibrational modes [5]. As illustrated in Fig. 4.5, above the Burns temperature the resonance modes (RM) and the associated Pb atom off-centering occur at a density where the RMs are about 5% of the intensity of the TO phonon [5]. Recent measurements on PZN-5%PT [26] show that on cooling below the Burns temperature the intensity of the RMs quickly increases from about 5% to about 10% of the TO intensity just above the PNR freezing temperature. From neutron pair distribution function analysis we know that there is a corresponding increase in the density of Pb atoms off-centering [29].

In the context of phonon localization, the increase in the density of RMs, or Pb atom off-centering, on cooling is important because a critical number is needed to establish localization [26]. The condition for localization is met when the distance between the RMs becomes comparable to the wavelength of the phonon that becomes localized. This is known as the Ioffe-Regel criterion [19] and can be expressed as $ql \lesssim 1$, where q is the phonon wavevector and l is the distance between RMs (Fig. 4.5). The exact value of the criterion at the transition to Anderson localization is not known, but can be determined empirically. The closest analogous experimental system where this can be observed directly is with the localization of transverse ultrasound waves in brazed aluminum beads, where the transition occurs at $ql \approx 1.8$ [31]. In the case of PZN-5%PT the value of 10% just above the PNR freezing temperature corresponds to $ql \approx 1.9$ [26], based on the size of the PNRs in PZN-5%PT and the average distance between RMs assuming a 10% concentration. Hence, the transition to localization in PZN-5%PT occurs at a temperature that could be reasonably expected for the number of resonance

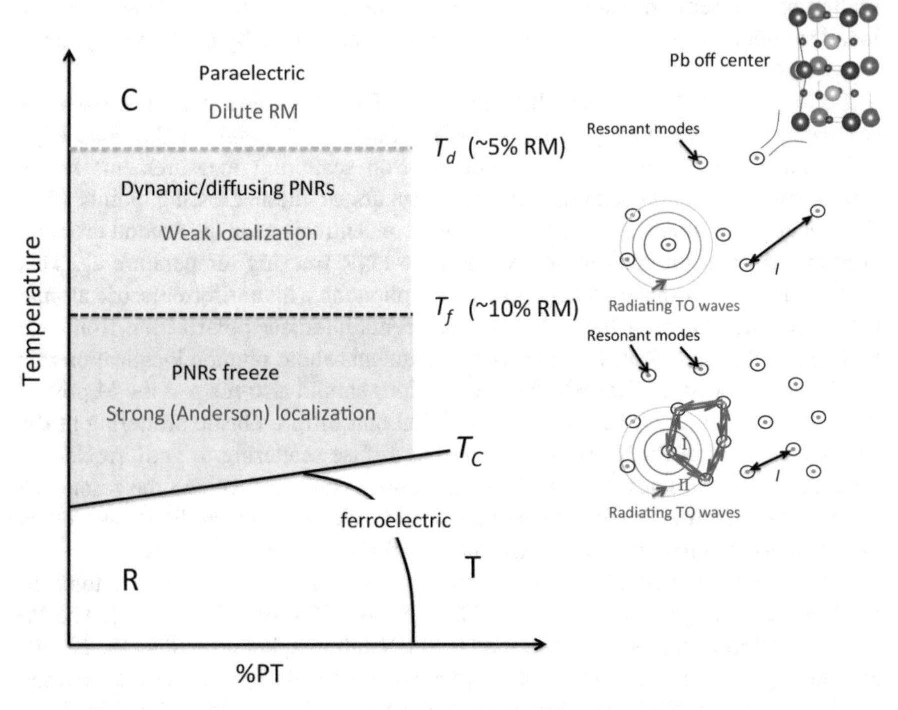

Fig. 4.5 This figure illustrates the relationship between the temperature dependence of the resonance modes (RMs), weak and strong phonon localization, and dynamic and static PNR formation in PMN-xPT ($(Pb(Mg_{1/3}Nb_{2/3})O_3)_{1-x}$-$(PbTiO_3)_x$) and PZN-$x$PT ($(Pb(Zn_{1/3}Nb_{2/3})O_3)_{1-x}$-$(PbTiO_3)_x$) relaxor ferroelectrics. The general behavior, where the onset of strong localization occurs on cooling when a critical number of scattering centers (resonance modes) form, is consistent with Anderson-type localization

modes. Note that this localization temperature corresponds to the PNR freezing temperature, Fig. 4.5. Above this temperature there is a state where the PNRs behave diffusively, where they fluctuate and move around in the lattice [32]. This is also expected in the localization picture. A precursor to Anderson or strong localization is weak localization. In the case of weak localization the diffusive motion slows down severely but it does not halt [19]. This provides a simple explanation for the existence of dynamic/diffusing PNRs occurring as a precursor to the freezing in of the PNRs (Fig. 4.5). When strong phonon localization occurs the phonons become standing waves, fixed in place. The PNR structural distortions are then caused by the structural relaxations that occur around these localized TO phonons. In the precursor state of weak localization the trapped standing phonons are not long lived and energy is able to slowly diffuse around. Consequently, the PNR structural distortions also diffuse around slowly.

Underlying the Pb atom off-centering and all the interatomic forces are the presence of random electric fields and the forces that polar interactions create. There do exist theories that deal with the effects of random fields specifically and how they manifest in the dielectric response of relaxor ferroelectrics [23], but they do not treat the underlying atomic displacements explicitly. From the point of view of the lattice dynamics, the atoms are moved by the resulting energy landscape and the phonons are scattering by the resulting atomic rearrangements. The observed PNR sizes and shapes then follow from the way the lattice vibrations interfere (Figs. 4.3 and 4.4). For this reason the localization mechanism, like the soft phonon mechanism described in Fig. 4.1, is more general and may be applied to a broader class of problems where disorder and phase instability meet. On the other hand, a more complete theory of relaxor ferroelectrics will surely require a complete account of both the polar interactions and the physics of lattice vibrations in a disordered lattice.

4.3 Coupling of PNRs to Phonons and the Ultrahigh Piezoelectricity in Relaxor-Based Ferroelectrics

That the mechanism driving nanoregion formation may relate to disorder is hardly surprising given that disorder is a key feature of all of these systems. However, nonlinear or anharmonic effects also play an important role, particularly as related to the bulk response of these materials near the structural instabilities. After all, being near a nearly second-order soft-mode phase transition indicates that a system is going to be inherently anharmonic in the lattice dynamics. So we cannot ignore the interplay of the soft phonon instability and disorder in understanding these materials from the perspective of the lattice dynamics. Unfortunately, there is no good theoretical model that treats both disorder localization (Anderson) and nonlinear localization (ILMs) simultaneously, yet there have been computer simulations that show that it is possible to analytically continue from one type of localization to the other by trading between nonlinearity and disorder [33, 34]. Nevertheless, it is possible to detect the instabilities by observing electric-field-driven changes in the structure and dynamics using neutron scattering. The instabilities can be observed on multiple length scales. At the atomic scale locally off-centered Pb atoms in lead-based relaxors [29, 30] can be aligned in an electric field [5], and along with this comes an alignment of the atomic local modes [5], the same random local modes that drive the formation of the PNRs by trapping the phonons [24]. The PNRs themselves can also be aligned in a field [7] and this comes with a softening of the shear mode of the crystal [35]. The shear softening has been explained as a manifestation of coupling between the transverse acoustic phonon and the PNRs [5, 36]. This shear softening is in addition to what is expected from being close to a morphotropic phase boundary [4] and together they enhance the piezoelectric response. Hence, the key to understanding the giant electromechanical response of relaxor-based ferroelectrics is in the cooperative effect of instabilities distributed across multiple length scales.

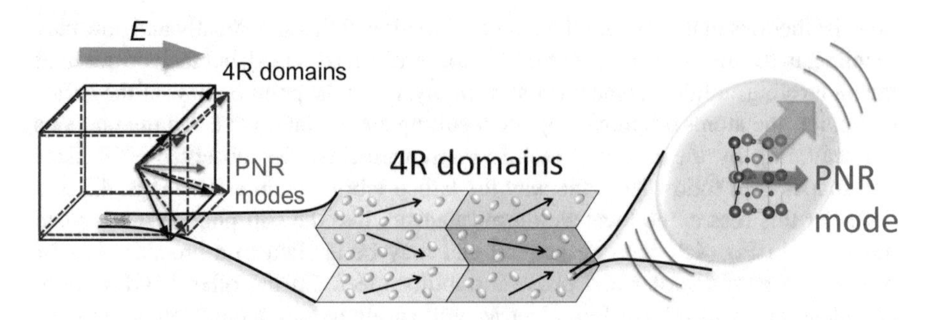

Fig. 4.6 Multiple length scales involved in high performance ferroelectric relaxor-$PbTiO_3$ single crystals

Figure 4.6 illustrates the multiple length scales involved in the macroscopic response of relaxor-based ferroelectrics. In applications, the bulk crystals are typically poled along the [100] axis [3]. In the un-poled crystal the ferroelectric domains are oriented along the eight equivalent cubic [111] directions possible for the rhombohedral (R) distortions. With poling the eight orientations are reduced to the 4R domains illustrated in Fig. 4.6. These orientations are those with polarizations closest to the [100] poling direction. The bulk electromechanical response is a result of the 4R poled domains rotating towards the [100] poling direction as an electric field is applied. The rotation of the domains requires a mechanical shearing of the individual domains, which is resisted by the elasticity of the material. Embedded within these "macro" domains (micron scale) are the much smaller PNRs, which exhibit local polar displacements along the [110] directions. The polarization of these PNRs causes them to couple to the ferroelectric polarization in the micron scale domains. The effect of this coupling is to hybridize the PNRs low energy dynamics with the transverse acoustic phonon propagating along [110] with displacements along [1,−1,0]. This hybridization results in a lower shear stiffness for the domains, which enable the polarization rotations underlying the giant electromechanical response [5]. Additionally, embedded within the PNRs are the off-centered Pb atoms, Fig. 4.6. These displacements occur along the [100] directions, and the local modes associated with these also couple to the phonons in the macro domains, although in this case to the transverse optic phonon [5].

Figure 4.7 shows inelastic neutron scattering intensity maps of the dispersion of transverse phonons in [100]-poled PMN-30%PT, after Ref. [5]. Several effects of the local dynamics interacting with the average dynamics can be observed in this data. First, comparing the transverse scans along $\mathbf{Q} = [2,K,0]$ and $\mathbf{Q} = [H,-2,0]$ (Fig. 4.7a, b), which are equivalent in the un-poled crystal, the PNR mode around 11.5 meV only appears in the direction that probes displacements parallel to the [100]-poling direction (the $\mathbf{Q} = [2,K,0]$ direction). This shows that the PNR modes align with the field and this matches an alignment of the Pb local off-centering [5]. Next, notice that the TO phonon dispersion is also a little different between the two directions. In the direction with the PNR mode (Fig. 4.7a) the TO phonon is shifted

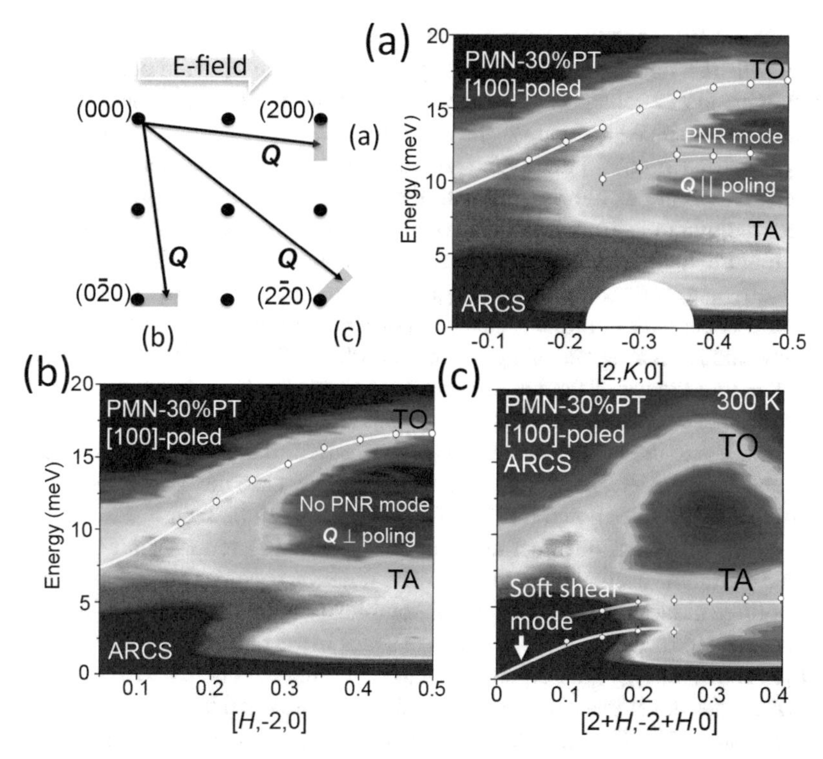

Fig. 4.7 This figure shows inelastic neutron scattering spectrum of the transverse phonons at different directions with respect to the [100] poling direction. (**a**) Polar nanoregion (PNR) modes appear between the transverse optic (TO) and transverse acoustic (TA) along poling direction. (**b**) PNR modes are absent for Q perpendicular to poling. (**c**) Alignment of PNRs enhances softening in the [110]-shear mode, which enables ultrahigh piezoelectricity. This data, which was measured on the ARCS instrument at the Spallation Neutron Source, is after Ref. [5]

to higher energy towards the zone center compared to the direction without a PNR mode (Fig. 4.7b). The shift in the TO phonon is an expected consequence of mode repulsion, also called mode anticrossing or avoided crossing [37], between the PNR mode and the TO phonon. This only happens in the ferroelectric phase because in the relaxor state at higher temperatures the phonon is localized, which means that is uncoupled from the lattice. Mode repulsion is a consequence of coupling between the modes. The reason for the uncoupling at high temperatures is a wave interference effect according to the Anderson localization picture [24]. The reason for coupling in the ferroelectric state is the interaction of the local polarization of the PNRs and off-centered Pb atoms with the ferroelectric macro domains. The more technologically interesting interaction is between the PNR distortions at low energies near the elastic line and the transverse acoustic (TA) phonon along [110] in Fig. 4.7c. These are the nanoscale displacements that occur along the [110] direction and appearing in pink in Fig. 4.6. The TA phonon hybridizes with the PNR displacements along [110] directions to form effectively two modes. The lower mode, the soft shear mode, is a hybrid mode where the PNRs move together in phase with the normal TA phonon displacements. The upper mode, which gets stiffer with the coupling, is from the other hybrid mode where the PNRs move together out of phase to oppose the motion of the TA phonon. These are called the symmetric (in phase) and antisymmetric (out of phase) modes. The net effect of the PNR vibrations hybridizing with the TA phonon is a shear softening, which enables the giant electromechanical coupling in these relaxor-based ferroelectrics. What's more, the application of an electric field along [100] tends to increase this shear softening, further aiding in the giant electromechanical response. This shear softening effect with poling has also been observed in ultrasound measurements of the same mode (at long wavelengths) in both PMN-33%PT [37] and PZN-4.5%PT [38]. The importance of the inelastic neutron scattering measurements is that they show that shear softening extends to the nanoscale [5].

In addition to the bending of the dispersion curves there are coupling effects on the inelastic intensity distributions. In the two mode case the TO phonon exhibits intensity that cascades down into the TA phonon, as can be seen in Fig. 4.7b. This is the so-called waterfall effect and can be explained in terms of a coupled two-damped-harmonic oscillators model [9]. The situation becomes more complex in the case of three modes [25]. To see the full complexity of this effect it is useful to take the data used to make Fig. 4.7a, b and take slices in the out-of-plane direction along [00*L*] as shown in Fig. 4.8. Perpendicular to the poling direction, Fig. 4.8a–c, the TO and TA modes develop a column of intensity between them that increases intensity on decreasing H in $\mathbf{Q} = [H,\text{-}2,L]$. This is the standard waterfall extended into an extra dimension. Parallel to the poling direction, Fig. 4.8d–f, the TO, TA, and PNR modes all couple to produce some interesting features. In addition to the standard TO-TA waterfall there is a mini waterfall at $\mathbf{Q} = [2,-0.3,L]$ between the TO and PNR modes. There are also pockets of intensity minimums at $\mathbf{Q} = [2,-0.3,0.1]$ as indicated in Fig. 4.8e. All of these features can be described using a three-coupled damped harmonic oscillators model [25], the results of which are shown in Fig. 4.8g–l. The strong damping in the lattice dynamics is also a glass-like symptom since it is indicative of dynamical relaxation at the nanoscale.

4.4 Summary

Recent computational [21], theoretical [22], and experimental work [24] are leading to an emerging new idea about how the nanoregions of a frustrated ferroic phase transition may result from disorder by Anderson localization. From the perspective of the lattice dynamics described here, the problem can be cast in terms of the behavior of lattice dynamics in disorder, and at the most elementary level Anderson localization is a wave interference effect. In this view we can understand that the size of the nanoregions is determined by the wavelength of coherently trapped phonons [24]. The onset of polar nanoregions is associated with the development of phonon localization at a critical density of random scattering centers [26], which

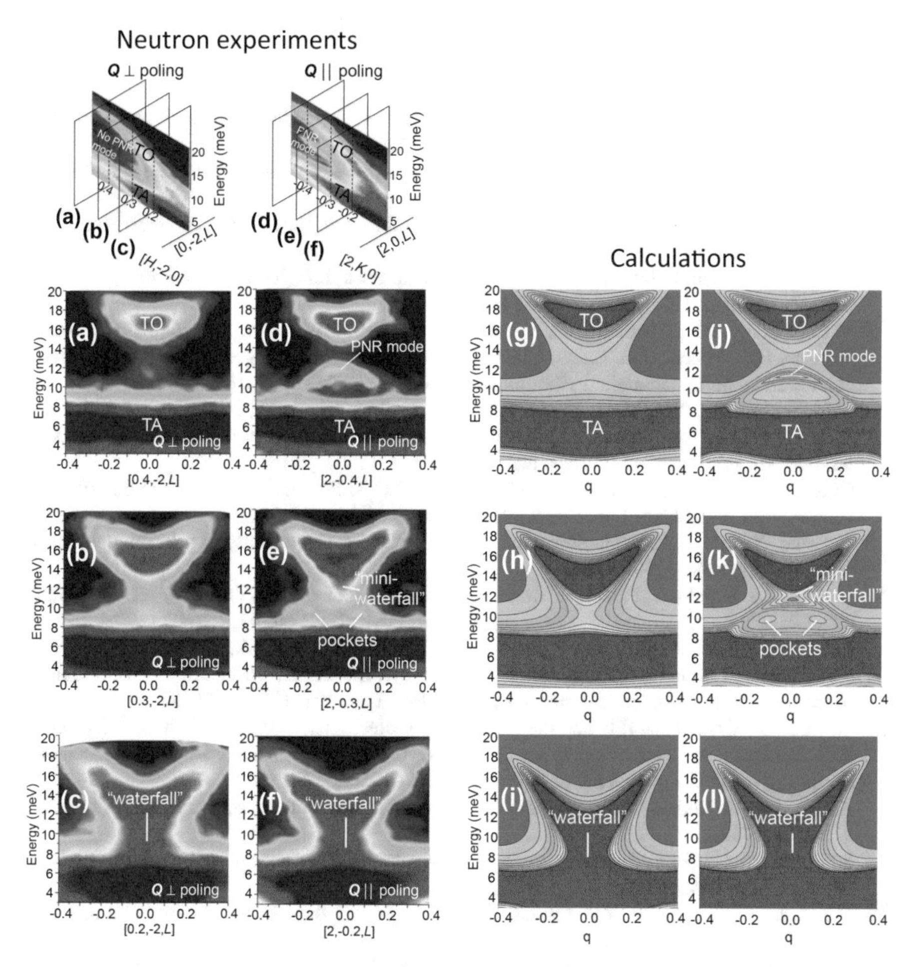

Fig. 4.8 Inelastic neutron scattering and model calculations of the "waterfall" feature in PMN-30%PT sliced into out-of-plane sections in energy-momentum space, after Ref. [25]. (**a–c**) Perpendicular to the poling direction and (**d–f**) along the poling direction. (**g–l**) Corresponding mode coupling calculations

are caused by a local off-centering of atoms. The local off-centering of atoms is itself a consequence of local lattice instability associated with the ferroic instability in the presence of chemical disorder. Hence, the formation of the nanoregions is a direct consequence of both instability and disorder via the Anderson localization wave interference mechanism.

Acknowledgements Research sponsored by the U.S. Department of Energy, Office of Basic Energy Sciences, Materials Sciences and Engineering Division.

References

1. S.-E. Park, T.R. Shrout, Ultrahigh strain and piezoelectric behavior in relaxor based ferroelectric single crystals. J. Appl. Phys. **82**, 1804–1811 (1997)
2. K. Uchino, *Piezoelectric Actuators and Ultrasonic Motors* (Kluwer Academic Publishers, Dordrecht, 1996)
3. S. Zhang, F. Li, High performance ferroelectric relaxor-$PbTiO_3$ single crystals: Status and perspective. J. Appl. Phys. **111**, 031301 (2012)
4. H. Fu, R.E. Cohen, Polarization rotation mechanism for ultrahigh electromechanical response in single-crystal piezoelectrics. Nature **403**, 281–283 (2000)
5. M.E. Manley, D.L. Abernathy, R. Sahul, D.E. Parshall, J.W. Lynn, A.D. Christianson, P.J. Stonaha, E.D. Specht, J.D. Budai, Giant electromechanical coupling of relaxor ferroelectrics controlled by polar nanoregion vibrations. Sci. Adv. **2**, e1501814 (2016)
6. F. Li, S.J. Zhang, T.N. Yang, Z. Xu, N. Zhang, G. Liu, J.J. Wang, J.L. Wang, Z.X. Cheng, Z.G. Ye, J. Luo, T.R. Shrout, L.Q. Chen, The origin of ultrahigh piezoelectricity in relaxor-ferroelectric solid solution crystals. Nat. Commun. **7**, 13807 (2016)
7. G. Xu, Z. Zhong, Y. Bing, Z.-G. Ye, G. Shirane, Electric-field-induced redistribution of polar nano-regions in a relaxor ferroelectric. Nat. Mater. **5**, 134–140 (2006)
8. Y. Wang, X. Ren, K. Otsuka, Shape memory effect and superelasticity in a strain glass alloy. Phys. Rev. Lett. **97**, 225703 (2006)
9. J. Hlinka, S. Kamba, J. Petzelt, J. Kulda, C.A. Randall, S.J. Zhang, Origin of the "waterfall" effect in phonon dispersion of relaxor perovskites. Phys. Rev. Lett. **91**, 107602 (2003)
10. J.C. Lashley, S.M. Shapiro, B.L. Winn, C.P. Opeil, M.E. Manley, A. Alatas, W. Ratcliff, T. Park, R.A. Fisher, B. Mihaila, P. Riseborough, E.K.H. Salje, J.L. Smith, Observation of a continuous phase transition in a shape-memory alloy. Phys. Rev. Lett. **101**, 135703 (2008)
11. W. Cochran, Crystal stability and the theory of ferroelectricity. Phys. Rev. Lett. **3**, 412 (1959)
12. P.W. Anderson, in *Proceedings of the 2nd Union Conference, Moscow*, ed. by G. I. Skanavi. *Physics of dialectics* (Academy of Sciences, USSR, 1960), p. c. 290
13. A.J. Sievers, S. Takeno, Intrinisic localized modes in anharmonic crystals. Phys. Rev. Lett. **61**, 970–973 (1988)
14. S. Flach, A. Gorbach, *Discrete Breathers-Advances in Theory and Applications* (Elsevier Science, New York, 2008)
15. D.K. Campbell, S. Flach, Y.S. Kivshar, Localizing energy through nonlinearity and discreteness. Phys. Today **57**, 43–49 (2004)
16. A. Bussmann-Holder, A.R. Bishop, T. Egami, Relaxor ferroelectrics and intrinsic inhomogeneity. Europhys. Lett. **71**, 249–255 (2005)
17. A.R. Bishop, A. Bussmann-Holder, S. Kamba, M. Maglione, Common characteristics of displacive and relaxor ferroelectrics. Phys. Rev. B **81**, 064106 (2010)
18. J. Macutkevic, J. Banys, A. Bussmann-Holder, A.R. Bishop, Origin of polar nanoregions in relaxor ferroelectrics: Nonlinearity, discrete breather formation, and charge transfer. Phys. Rev. B **83**, 184301 (2011)
19. A. Lagendijk, B. van Tiggelen, D.S. Wiersma, Fifty years of Anderson localization. Phys. Today **62**, 24–29 (2009)
20. P.W. Anderson, Absence of diffusion in certain random lattices. Phys. Rev. **109**(5), 1492–1505 (1958)
21. A.R. Akbarzadeh, S. Prosandeev, E.J. Walter, A. Al-Barakaty, L. Bellaiche, Finite-temperature properties of $Ba(Zr, Ti)O_3$ relaxors from first principles. Phys. Rev. Lett. **108**, 257601 (2012)
22. D. Sherrington, BZT: A soft pseudospin glass. Phys. Rev. Lett. **111**, 227601 (2013)
23. V. Westphal, W. Kleemann, M.D. Glinchuk, Diffuse phase transitions and random-field-induced domain states of the "relaxor" ferroelectric $PbMg_{1/3}Nb_{2/3}O_3$. Phys. Rev. Lett. **68**, 847 (1992)
24. M.E. Manley, J.W. Lynn, D.L. Abernathy, E.D. Specht, O. Delaire, A.R. Bishop, R. Sahul, J.D. Budai, Phonon localization drives polar nanoregions in a relaxor ferroelectric. Nat. Commun. **5**, 3683 (2014)

25. M.E. Manley, D.L. Abernathy, R. Sahul, P.J. Stonaha, J.D. Budai, Three-mode coupling interference patterns in the dynamic structure factor of a relaxor ferroelectric. Phys. Rev. B **94**, 104304 (2016)
26. M.E. Manley, A.D. Christianson, D.L. Abernathy, R. Sahul, Phonon localization transition in relaxor ferroelectric PZN-5%PT. Appl. Phys. Lett. **110**, 132901 (2017)
27. I.P. Swainson, C. Stock, P.M. Gehring, G. Xu, K. Hirota, Y. Qiu, H. Luo, X. Zhao, J.-F. Li, D. Viehland, Soft phonon columns on the edge of the Brillouin zone in the relaxor $PbMg_{1/3}Nb_{2/3}O_3$. Phys. Rev. B **79**, 224301 (2009)
28. M. Matsuura, K. Hirota, P.M. Gehring, Z.-G. Ye, W. Chen, G. Shirane, Composition dependence of the diffuse scattering in the relaxor ferroelectric compound (1−x)Pb(Mg1/3Nb2/3)O3−xPbTiO3(0<x<0.40). Phys. Rev. B **74**, 144107 (2006)
29. T. Egami, S.J.L. Billinge, in *Underneath the Bragg Peaks: Structural Analysis of Complex Materials*, ed. by R. W. Cahn. (Pergamon Materials Series, Oxford, 2003), p. 343
30. B.P. Burton, E. Cockayne, Why $Pb(B,B')O_3$ perovskites disorder at lower temperatures than $Ba(B,B')O_3$ perovskites. Phys. Rev. B **60**, R12542(R) (1999)
31. H. Hu, A. Strybulevych, J.H. Page, S.E. Skipetrov, B.A. van Tiggelen, Localization of ultrasound in a three-dimensional elastic network. Nat. Phys. **4**, 945 (2008)
32. G.-M. Rotaru, S.N. Gvasaliya, B. Roessli, S. Kojima, S.G. Lushnikov, P. Gunter, Evolution of the neutron quasielastic scattering through the ferroelectric phase transition in $93\%PbZn_{1/3}Nb_{2/3}O_3$–$7\%PbTiO_3$. Appl. Phys. Lett. **93**, 032903 (2008)
33. F.R. Archilla, R.S. MacKay, J.L. Marín, Discrete breathers and Anderson modes: two faces of the same phenomena? Phys D **134**, 406–418 (1999)
34. A.R. Bishop, S. Jimenez, L. Vazquez, *Fluctuation Phenomena: Disorder and Nonlinearity* (World Scientific, Singapore, 1995)
35. G. Xu, J. Wen, C. Stock, P.M. Gehring, Phase instability induced by polar nanoregions in a relaxor ferroelectric system. Nat. Mater. **7**, 562–566 (2008)
36. R. Pirc, R. Blinc, V.S. Vikhnin, Effect of polar nanoregions on giant electrostriction and piezoelectricity in relaxor ferroelectrics. Phys. Rev. B **69**, 212105 (2004)
37. R. Zhang, B. Jiang, W.W. Cao, Elastic, piezoelectric, and dielectric properties of multidomain $0.67Pb(Mg_{1/3}Nb_{2/3})O_3$–$0.33PbTiO_3$ single crystals. J. Appl. Phys. **90**, 3471 (2001)
38. J.H. Yin, B. Jiang, W.W. Cao, Elastic, piezoelectric, and dielectric properties of $0.955Pb(Zn_{1/3}Nb_{2/3})O_3$–$0.045PbTiO_3$ single crystal with designed multidomains. IEEE Trans. Ultrason. Ferroelectr. Freq. Control **47**, 285–291 (2000)

Chapter 5
Relaxor Ferroelectrics and Related Cluster Glasses

Wolfgang Kleemann and Jan Dec

Abstract Mesoscopic ferroic glasses such as strain glass, cluster spin-glass, and nanopolar glassy relaxors are of paramount importance in the phase diagrams of complex ferroic materials. All of them are based on supercritical chemical disorder and are preceded by precursor patterns of elastic/magnetic tweed or polar nanoregions, respectively. Within this general scheme we comment on the properties of (1) superdipolar glassy relaxors $PbMg_{1/3}Nb_{2/3}O_3$ and $Sr_{0.8}Ba_{0.2}Nb_2O_3$ in some detail, (2) structural strain glass $Ti_{50-x}Ni_{50+x}$ martensite with tweed patterns, and (3) magnetic cluster glass $La_{0.7}Ca_{0.3}Mn_{0.85}Cd_{0.15}O_3$ with tweed precursor. All of these mesoscopic "*ferroic glasses*" with their complex chemical disorder must be distinguished from "*superspin glasses*," which are known as systems of widely dispersed nanoparticles with fixed magnetic moments and inert nonmagnetic environment, e.g., multilayers $(Co_{80}Fe_{20}/Al_2O_3)_{10}$. Nonetheless, at the (super)glass transitions both families of materials exhibit the same glassy dynamic criticality and non-ergodicity of their field-induced orders.

5.1 Mesoscopic Ferroic Glasses

The concept of "*ferroic glasses*" has recently entered the discussion of disordered materials, which are ferroically ordered at the nano- and microscale, but retain mesoscale glassy disorder at low temperatures [1]. Figure 5.1 shows the generic x-T phase diagram of a defect-containing ferroic system with crossover at the critical defect concentration x_C from normal ferroic transition at Curie temperatures T_C to ferroic glass transition at glass temperatures T_g. An important additional ingredient of these specific "*ferroic glasses*" is the appearance of a precursor phase

---

W. Kleemann (✉)
Angewandte Physik, Universität Duisburg-Essen, Duisburg, Germany
e-mail: wolfgang.kleemann@uni-due.de

J. Dec
Institute of Materials Science, University of Silesia, Katowice, Poland
e-mail: jan.dec@us.edu.pl

T. Lookman, X. Ren (eds.), *Frustrated Materials and Ferroic Glasses*, Springer Series in Materials Science 275, https://doi.org/10.1007/978-3-319-96914-5_5

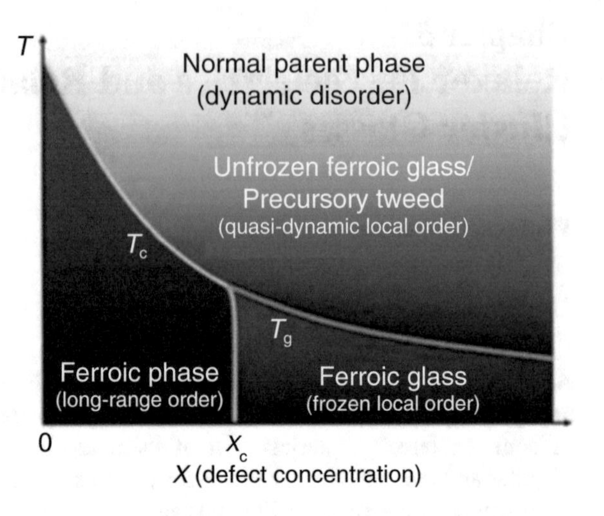

Fig. 5.1 Generic phase diagram of a defect-containing ferroic system showing the crossover at the critical defect concentration x_C from a normal ferroic transition at T_C to a ferroic glass transition at the glass temperature T_g in dependence on the defect concentration x. Reproduced with permission from [1]

at intermediate temperatures, $T > T_C/T_g$, which is called "*precursory tweed*" with quasi-dynamic local order in the case of structural "*strain glass*" (Sect. 5.4) and needs to be specified in the cases of electric and magnetic "*ferroic glasses*" (Sects. 5.2, 5.3, and 5.5, respectively).

It was conjectured that this phase diagram can be applied to ferroelastic, ferroelectric, and ferromagnetic systems. In particular three subgroups were discussed [2]: magnetic "*cluster spin glass*," structural "*strain glass*," and polar "*relaxor ferroelectrics*," all of which are finally stabilized by microscopic intrinsic defects. First, the formation of "*strain glass*" is attributed to ferroelastic systems exhibiting random nanostress fields as in the nonstoichiometric martensitic compound $Ti_{50-x}Ni_{50+x}$ [3]. Second, after the discovery of "*cluster spin glass*" in magnetic materials with nonmagnetic dopants [4] charge and geometric frustration in colossal magneto-resistance (CMR) materials $La_{0.7}Ca_{0.3}Mn_{0.7}Co_{0.3}O_3$ [5] and $La_{0.7}Ca_{0.3}Mn_{0.7}Cd_{0.3}O_3$ [6] are believed to be at the origin of "*cluster spin glass*" formation. Finally, "*relaxor ferroelectrics*" such as lanthanum-doped lead zirconate-titanate, PLZT 8/65/35 [7] and other structurally disordered ionic compounds like lead magno-niobate, $PbMg_{1/3}Nb_{2/3}O_3$ (PMN), and tungsten bronze-type $Sr_{1-x}Ba_xNb_2O_3$ (SBN) became approved as "*ferroic glass*" by virtue of their "*superdipolar glass*" ground state [8]. Similarly to the case of "*strain glass*" the action of quenched random fields—albeit electric ones due to the charge disorder—is at the origin of the mesoscopic glassy disorder. This insight has taken as long as 60 years after the discovery of PMN [9] and will be detailed in the next two sections.

5.2 Relaxor Ferroelectrics

5.2.1 *Solid Solutions of $PbMg_{1/3}Nb_{2/3}O_3$-$PbTiO_3$ (PMN-PT)*

The disordered perovskite $PbMg_{1/3}Nb_{2/3}O_3$ (PMN) exhibits typical relaxor-type ferroelectric properties, which have been studied since 1954 [9]. The relaxor state is characterized by the frustration of the local polarization, which prevents long-range ferroelectric order from developing globally. Although the local symmetry of the polar domains is rhombohedral, the macroscopic symmetry of PMN remains cubic below the temperature of maximum permittivity. A ferroelectric phase can be induced either by application of an electric field along <111>$_{cub}$ or by partial substitution of the complex $(Mg_{1/3}Nb_{2/3})^{4+}$ ions by Ti^{4+}. In both cases, the nanopolar domains transform into macrodomains with cubic-to-rhombohedral symmetry breaking. Upon substitution a complete series of solid solutions xPMN-$(1-x)$PT (PMN-PT) forms, where a morphotropic phase boundary (MPB) located at $x_c \approx 0.65$ marks the frontier between the tetragonal (ferroelectric side, T) and rhombohedral (relaxor side, R) phases, as illustrated in Fig. 5.2. Note that this presentation has been redrawn after the original publication [10] within the spirit of Fig. 5.1, where the "*defect*" ions, $(Mg_{1/3}Nb_{2/3})^{4+}$ (concentration x) are diluents of the "*pure*" ferroic phase, $PbTiO_3$. In the following we shall restrict our discussion to the extremely defective case of pure PMN, $x = 1$.

5.2.2 *Superglass Transition of PMN*

In contrast to conventional ferroelectric crystals like $BaTiO_3$ [11] the structurally disordered relaxor ferroelectrics like PMN escape the familiar soft-mode scheme

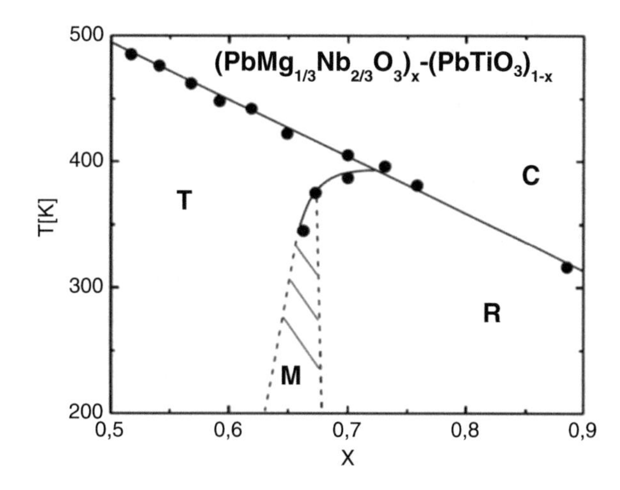

Fig. 5.2 Phase diagram of the PMN-PT solid solution system. Solid circles and related phase boundaries separate cubic (C), rhombohedral (R), and tetragonal (T) regions, respectively, while the shaded area represents the monoclinic (M) region. Redrawn after [10]

of ferroelectrics [12]. In the early 1990s the discussion of the observed polar ordering at the nanoscale culminated in two competing models, the "*dipole glass*" [13] and the "*domain state*" of polar nanoregions (PNR) under the constraint of charge disorder-induced quenched electric random fields (RFs) [14], respectively. Only gradually both ideas converged into the model of a "*non-ergodic ferroelectric cluster glass*" ground state emerging from the high-T PNR ensemble under their random electrostatic interaction via a mesoscopic random bond glass transition [8, 15–18]. In the following we shall outline the main arguments applicable to the specific case of the cubic system PMN and expand them to other systems. As remarked by Ahn et al. [19] in this field some reservation still remains and deserves extra consideration.

The name "*relaxor ferroelectrics*" was coined by Cross [20] to ferroelectric compounds fulfilling the following criteria:

(a) The temperature-dependent dielectric susceptibility, $\chi'(T)$, exhibits a broad and smeared peak.
(b) The frequency dependence of its temperature, $T_m(f)$, hints at dielectric relaxation processes.
(c) No macroscopic symmetry breaking is observed even at lowest temperatures.

Criteria (a) and (b) are evident from $\chi'(T)$ curves in Fig. 5.3 obtained on a (001) oriented single crystal of PMN at temperatures $197 \leq T \leq 297$ K and *ac* frequencies $10^{-3} \leq f \leq 10^5$ Hz of the dielectric spectrometer (Solartron 1260 impedance analyzer with 1296 dielectric interface; amplitude of the *ac* probing field: $E_{ac} \approx 500$ V/m [21]).

Fitting the asymptotic lowest frequency data of $T_m(f)$ within $10^{-3} \leq f \leq 2 \times 10^{-2}$ Hz (Fig. 5.4; left dataset) to the power law of glassy critical dynamics [22]

$$\tau\,(T_m) = \tau_0\big(T_m/T_g - 1\big)^{-z\nu} \tag{5.1}$$

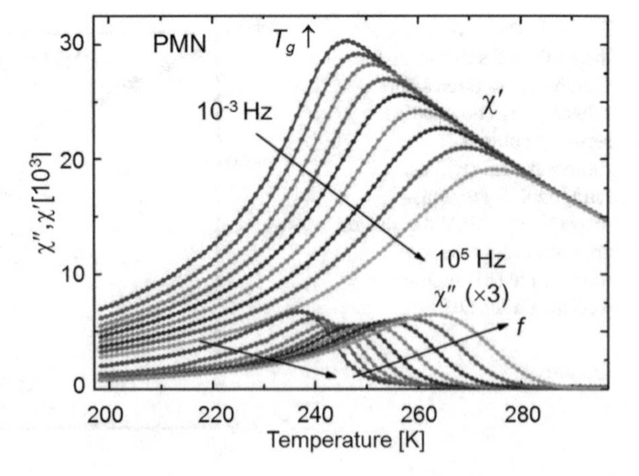

Fig. 5.3 Temperature dependences of the dielectric susceptibility components χ' and χ'' of PMN at decade stepped frequencies within $10^{-3} \leq f \leq 10^5$ Hz. The glass temperature $T_g = 238.8$ K is marked by an arrow. Reproduced with permission from [21]

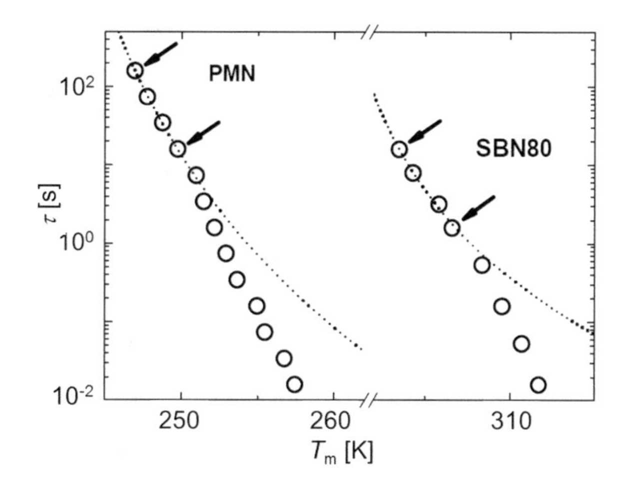

Fig. 5.4 Low-frequency relaxation times $\tau = (2\pi f)^{-1}$ vs. peak temperatures $T_m(f)$ of χ' of PMN (left) and SBN80 (right curve) best fitted to Eq. (5.1) for data points delimited by arrows (dotted lines)

yields the (static) glass temperature $T_g = (238.8 \pm 1.1)$ K (Fig. 5.3, arrow). The relatively large attempt time $\tau_0 = (4.3 \pm 0.1) \times 10^{-10}$ s is in accordance with the mesoscopic size of the elementary dipole moments being active in PMN in the critical regime. They are attributed to the above mentioned PNR, which have been proved in relaxors like PMN, e.g., by neutron pair distribution function analysis [23]. The dynamic critical exponent $z\nu = 7.9 \pm 0.3$ agrees within uncertainties with that of the 3D magnetic dipolar glass $LiHo_{0.045}Y_{0.955}F_4$, $z\nu = 7.8 \pm 0.2$ [24].

Slightly above the critical region of PMN, $T/T_g > 1.04$ (Fig. 5.4; left dataset), the dynamics of the cluster system is no longer described by Eq. (5.1). Here the "*superparaelectric*" regime [20] is entered, where virtually noninteracting electric "*supermoments*" correspond to single PNR and give rise to local polarization and giant susceptibility with large frequency dispersion. This "*Vogel-Fulcher* (VF) *regime*" is often described by a relaxation time corresponding to the cusp temperature T_m of $\chi'(f, T)$ [25],

$$\tau_{VF}(T_m) = \tau_{VF}^0 \exp\left[E_0/(T_m - T_{VF})\right]. \tag{5.2}$$

Although satisfactory fits hold for PMN at frequencies within $10^2 \le f \le 10^5$ Hz [26], the significance of the emerging parameters τ_{VF}^0, E_0 and T_{VF} is ambiguous. Unlike the glass temperature T_g in Eq. (5.1) the interpretation of T_{VF} as a "freezing temperature" is denied by theory [25]. We therefore abstain from further discussing the VF approach here and prefer the direct modeling of the interaction-free superparaelectric permittivity as proposed by Lu and Calvarin [27]. Their simulations of the dielectric response of 95PMN-5PT (see Fig. 15 in [27]) agree qualitatively with our χ' data shown in Fig. 5.1. This corroborates the absence of a phase transition within the underlying interaction-free approach.

Indispensable properties of genuine spin and dipolar glass phases at $T < T_g$ are their non-ergodicity [28, 29]. Figure 5.5 shows the standard procedure, which

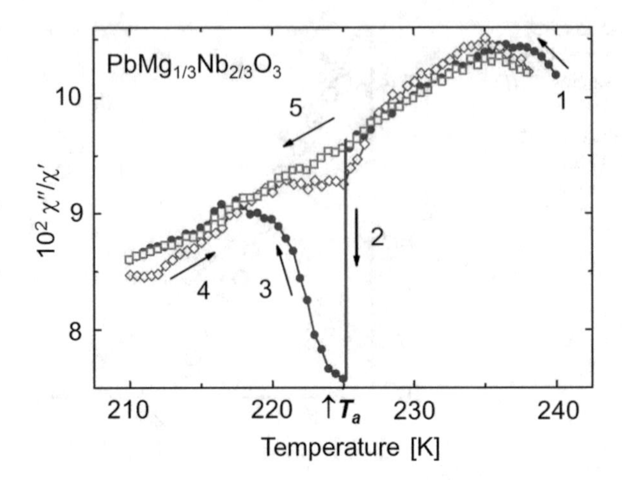

Fig. 5.5 Aging, rejuvenation, and memory of PMN reflected by the susceptibility component ratio $\chi''/\chi'(T)$ after ZFC from $T = 240$ K at rate $dT/dt = \pm 0.003$ K/s and frequency $f = 0.1$ Hz on first cooling to 210 K with intermittent halt ($\Delta t = 2.5 \times 10^3$ s) at $T_a = 225$ K (curves 1–3), continuous reheating via memorized dip at T_a to 238 K (curve 4), and continuous cooling back to 210 K (curve 5). Reproduced with permission from [21]

proves that the glass phase is thermally out-of-equilibrium except after being "*aged*" for a sufficiently long time at constant temperature T_a. This is done here after continuous cooling the PMN sample from 240 to $T_a = 225$ K $< T_g$ (curve 1) within 4.5×10^3 s followed by a halt at T_a of $\Delta t = 2.5 \times 10^5$ s (line 2). As a result the most sensitive quantity, χ''/χ' with the imaginary component χ'' of the susceptibility, is found to decrease by ≈20% when approaching equilibrium by mere waiting. By further cooling to $T = 210$ K at the previous rate a strong upward-trend indicates "*rejuvenation*" toward the unaged curve (curve 3), which is subsequently reproduced on continuously heating up to $T = 238$ K (curve 4) and recooling to $T = 210$ K (curve 5). A dip of about 20% of the initial "*dielectric hole*" is recovered around T_a during heating in curve 4. This is attributed to the "*memory*" of the near-ground state being approached at T_a during the first extensive aging procedure. Non-global aging is an essential feature of the chaotic glass state [30]. Another strong indicator for glassiness is the stretched exponential relaxation kinetics during the aging process. For example, at $T_a = 200$ K it is found [31]

$$\chi''/\chi'\,(f = 25\text{ Hz}) = \chi_0 \exp\left[-(t/\tau)^\beta\right] + \chi_1 \tag{5.3}$$

with amplitudes $\chi_0 = 5630 \pm 35$ and $\chi_1 = 2208 \pm 25$, basic relaxation time $\tau = 88 \pm 3$ s and extremely small stretching exponent $\beta = 0.011 \pm 0.003$ as shown in Fig. 5.6.

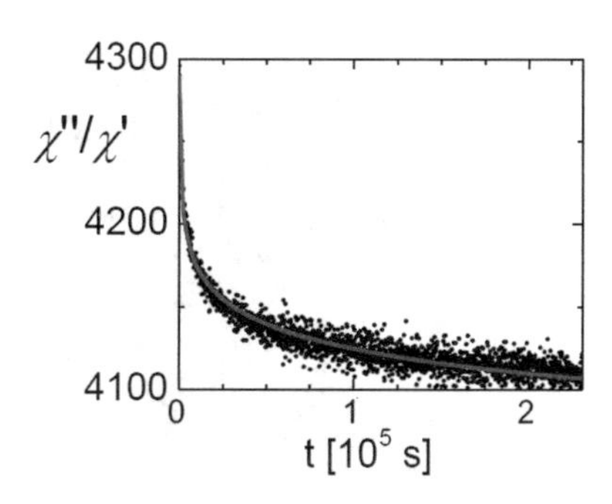

Fig. 5.6 Temporal relaxation of $\chi''/\chi'(f = 25$ Hz) of PMN at $T_a = 200$ K after zero-field cooling and best fitted to Eq. (5.3) (red line). Reproduced with permission from [31]

5.2.3 *Paraelectric PNR Precursor State*

Before going into more details of the cluster glassy state of PMN it appears useful to recall its genesis on the temperature scale. To this end we have drawn in Fig. 5.8 (see below) a schematic diagram of the stepwise evolution of the relaxor ground state upon cooling from the melting point ($T_m > 1300$ K [32]) to below the glass temperature $T_g = 238.8$ K [21]).

First of all, one has to bear in mind that PMN can be considered as a stoichiometric disordered solid solution of two hypothetic perovskite compounds, $(PbMgO_3)_{1/3}(PbNbO_3)_{2/3} \equiv PbMg_{1/3}Nb_{2/3}O_3$ (Fig. 5.7a), both of which cannot exist separately because of their internal ionic charge imbalance. The inherent charge disorder due to the random distribution of Mg^{2+} and Nb^{5+} ions at *B* sites of this ABO_3 perovskite crystal (Fig. 5.7a) offers largely uncorrelated and quenched electric fields (random fields, RFs) at the sites of the *ferroelectric-active* ions, Pb^{2+} and Nb^{5+}. As predicted within the *RF* theory of magnetic spin systems [34] a tremendous slowing-down of the order parameter dynamics is expected when approaching the critical temperature. Activated dynamic scaling controls, e.g., the transition of the *RF* Ising model (*RFIM*) [35], since the critical *order parameter fluctuations* (i.e., temperature-dependent correlated regions with a preferential uniform direction of the order parameter) become increasingly pinned to *spatial fluctuations* of the *RF*s (i.e., to temperature-independent regions revealing an excess of the corresponding field direction) as the correlation length grows in the vicinity of T_c. Unlike the widely studied RFs in magnets, which linearly couple to an order parameter of the Ising or Heisenberg type [36], the quenched electric RFs of perovskite-type relaxors couple to a cubic order parameter. As deduced recently within a microscopical model and a statistical mechanical solution of the combined effect of dipolar forces and quenched RFs [37] a ground state with no-long range FE order and anisotropic, long-ranged fluctuations of polarization was identified. It emerges for any amount of compositional disorder in the general case of PMN-PT solid solutions, including that of pure PMN.

However, this approach neglects novel insight into relaxor physics at the mesoscale. Charge-disordered relaxor ferroelectrics such as PMN experience primarily RF disorder, which favors the formation of PNRs by virtue of their spatial

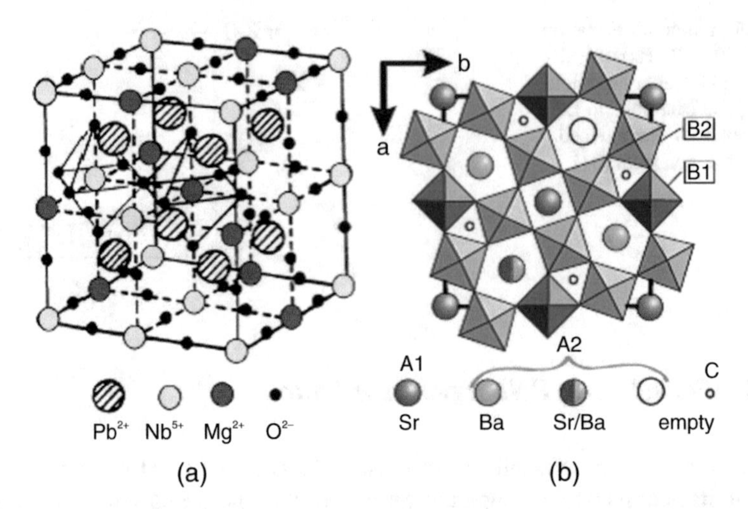

Fig. 5.7 Ionic distribution (**a**) in eight unit cells of PMN ($PbMg_{1/3}Nb_{2/3}O_3$: Pb^{2+} = hatched circles, O^{2-} = small solid circles, Nb^{5+} = yellow/light gray circles, Mg^{2+} = red/dark gray circles) and (**b**) in the (001) plane of SBN ($Sr_xBa_{1-x}Nb_2O_6$: Sr^{2+} = red/light gray spheres, Ba^{2+} = green/dark gray spheres, *A* site vacancies = large open circles, empty channels = small open circles). Reproduced with permission from [33]

fluctuations. It is common belief that the essential relaxor properties are initiated by RF-induced generation of metastable PNRs at the Burns temperature T_d, i.e., far above room temperature. Statistical fluctuations of arbitrarily weak quenched RFs are able to stabilize the PNRs above the Curie temperature via their local excess amplitudes. These mesoscopic fluctuations determine the further development on cooling. This concerns primarily the local stabilization of the PNR, their mutual frustrated dipolar interaction, and finally their freezing into a cluster glass as reported on PMN [18, 38]. Here, we adopt the idea of relaxors being "*ferroelectrics with multiple inhomogeneities*" [38] rather than that of breeding ferroic PNRs by virtue of Anderson localization treated within a pseudospin glass model [39]. Factually and asymptotically the *ac* susceptibility peak at T_m, approaches the glass temperature $T_g \approx 240$ K according to glassy dynamics, Eq. (5.1), and Fig. 5.4.

The appearance of PNR upon cooling down of a relaxor system toward T_g is characterized by two critical temperatures, T_d ("Burns temperature") and T^*. They are associated with the condensation and the stabilization of PNR, respectively. As noted in Fig. 5.8 for PMN the values $T_d \approx 630$ K and $T^* \approx 500$ K were obtained from characteristic acoustic emission signals [41]. Similar values of T_d have been attributed to the phenomenon of transverse ferroelectric *Anderson phonon localization* [42]. It is proposed that standing ferroelectric phonons develop with a coherence length equal to the PNR size. Alternatively, intrinsic (nonlinear) localized modes (ILM or "*discrete localized breathers*" [43]) have been proposed to originate PNRs. In our opinion both of these lattice vibrational mechanisms might well be vehicles for launching the PNR formation, but explicit stable pinning requires the

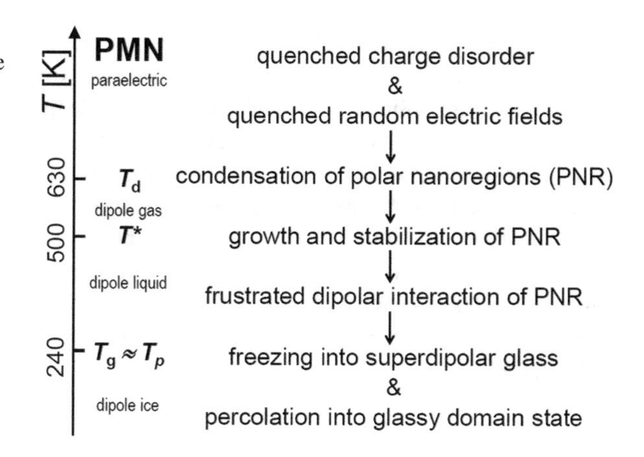

Fig. 5.8 Schematic diagram of the gradual evolution of the relaxor ground state of PMN upon cooling from the melting point ($T_m > 1300$ K) to below the glass temperature $T_g = 238.8$ K [21]. Specific temperatures T_d, T^*, T_g, and T_p separating different phases, sometimes denoted as paraelectric, dipole gas, dipole liquid, and dipole ice [40], are indicated

quenched forces of RF fluctuations. Hitherto no serious theoretical investigation of phonon localization under the influence of locally dominating Coulombic RFs has been performed, although scanning piezoforce microscopy has given clear hint at their spatial heterogeneity via the local reproducibility of PNR distribution in PMN-PT and SBN [44]. Although pinning of PNR at local fluctuations of field amplitudes has been proposed heuristically [8, 45], a complete theory is still missing. Dkhil et al. [41] referred to T^* as "*a local phase transition that gives rise to the appearance of static PNR.*" By contrast, Burton et al. [46] reject the term "*local phase transition,*" since phase transitions (strictly) only occur in infinite systems. However, their simulation results suggest a weakly first-order transition with a subtle stiffening of PNR-orientations below T^*.

At any rate, there is ample evidence of the intrinsic PNR distribution in relaxors. At the Burns temperature T_d, the soft TO phonon mode becomes overdamped near the zone center, and starts to condense into PNR [47]. These regions form with local polarizations along <111> directions and are shifted uniformly along their individual polarization direction. Number and size of PNR increase on cooling. At the phase transition, $T \approx T_C$, a large-scale overall "*freezing*" of the PNR occurs. Small PNR merge into larger ones and the total volume of PNR in the system keeps increasing. The related ferroelectric soft-mode lifetime increases below T_C, and the overdamping near the zone center disappears. A macroscopic ferroelectric polar phase without lattice distortion tends to become established.

Below T_C the size of the PNR can grow slowly with further cooling. However, if the coupling between the PNR and the surrounding lattice is not sufficiently strong, as is the case in pure PMN, then the energy barrier created by the uniform phase shift would prevent the PNR from merging further and forming macroscopic lattice distortions. The resulting phase will have a polar lattice of average cubic structure, but with embedded *rhombohedrally* polarized PNR. In addition to neutron and X-ray diffuse scattering measurements, Raman studies [48] and specific-heat measurements [49] have also provided useful information on PNR in PMN.

Probably the most exciting recent insight is the intimate coupling of the PNR to the ferroelectric polarization and anisotropy in relaxors [50], which raises hopes for tailoring new relaxor systems with unprecedented piezoelectric response. It has become clear that the observed giant electromechanical coupling of relaxor ferroelectrics is controlled by PNR vibrations. Actually the contribution of PNR to the room-temperature dielectric and piezoelectric properties of PMN is in the range of 50–80% [51]. A mesoscale mechanism is proposed to reveal the origin of the high piezoelectricity in relaxor ferroelectrics, where the PNR aligned in a ferroelectric matrix can facilitate polarization rotation. This mechanism emphasizes the critical role of local structure on the macroscopic properties of ferroelectrics.

5.2.4 Percolation Transition of PNR

The versatility of the PNR in PMN is crucial of their final structure in the cluster glass state below $T_g = 238.8$ K (Fig. 5.3). Obviously the relaxing elements in PMN undergo a fundamental reorganization in the intermediate temperature range. On one hand they are closely linked to the phase transition from *superparaelectric* disorder into the *superglass* state. On the other hand, and probably even more significant, a fundamental structural event takes place, viz. percolation of the continuously growing PNR under the control of electric RFs due to the still active cationic Mg^{+2}/Nb^{5+} charge disorder. Neutron scattering data (Fig. 5.9) have evidenced [55] that the volume fraction of PNR in PMN is overcoming the percolation threshold of 23% for the elliptical shape [53] at a conjectured ferroelectric phase transition at $T_c \approx 230$ K [54]. This opens the chance to form large coherent ferroelectric domains, whose dynamic phenomenology strongly reminds one of domain-wall (DW) dynamics in disordered ferroics represented by Cole-Cole (CC) diagrams in the complex permittivity plane [56, 57].

The resulting CC diagrams of PMN are shown in Fig. 5.10 [21]. Let us first consider the high-T region, $246 \le T \le 285$ K (Fig. 5.10, curves 1–11). A similar extreme "*blowing up*" of the superparaelectric CC curve from a "*dot*" (285 K) to an extremely broad distribution (246 K) was reported previously [58]. Axial ratios $\Delta\chi'/2\Delta\chi'' \approx 2.5$ and 6 at 275 and 250 K, respectively, are found and strongly exceed unity of monodispersive Debye relaxators. Our accessible spectral range does not suffice to correctly determine the width of the CC semicircles at $T < 246$ K, where recording of the very low-f branch would require submillihertz driving fields. Another important observation is the skewness of the CC plots and its temperature dependence. It starts with positive sign (i.e., peaking at the high-f side) within $285 \ge T \ge 262$ K (curves 1–6), passes through a "*crossover*" regime within $256 \ge T > 254$ K (curves 7–8), and ends up negatively (i.e., peaking at the low-f side) within $250 \ge T \ge 246$ K (curves 9–11). Positive skewness is well known

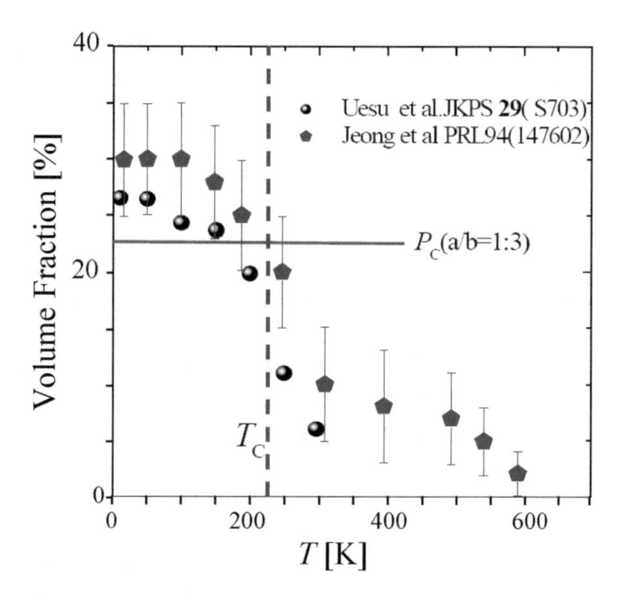

Fig. 5.9 Volume fraction of PNR in PMN estimated from neutron scattering data (circular [52] and pentagonal dots [23], respectively). Lines are marking the conjectured percolation transition for elliptical shape at 23% [53] and $T_C \approx 230$ K [47], respectively. Reproduced with permission from [54]

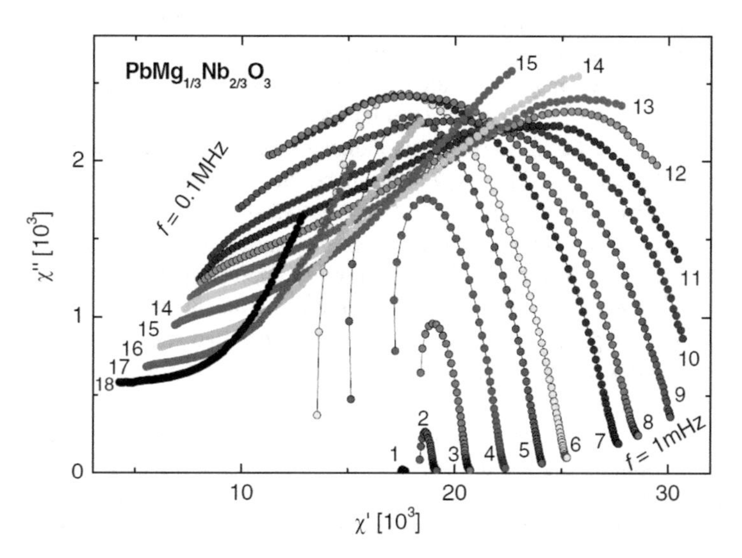

Fig. 5.10 Cole-Cole plots χ'' vs. χ' of PMN measured at $T = 285$ (1), 280 (2), 275 (3), 270 (4), 265 (5), 262 (6), 256 (7), 254 (8), 250 (9), 248 (10), 246 (11), 242 (12), 240 (13), 238 (14) 235 (15), 230 (16), 225 (17), and 220 K (18) within $10^{-3} \leq f \leq 10^5$ Hz. Reproduced with permission from [21]

from polydispersive polar cluster systems [59], as modeled in PMN-PT [60], and mathematically described with a Lacroix-Béné-type [61] distribution of relaxation times, $\tau = 1/2\pi f$,

$$G(\tau) = [\sin(\pi\beta)/\pi][\tau_0/(\tau - \tau_0)]^\beta \text{ for } \tau > \tau_0 \tag{5.4}$$

where $G(\tau) = 0$ for $\tau < \tau_0 =$ shortest relaxation time (ionic attempt time) and polydispersivity exponent $\beta \leq 1$. In the case of relaxor ferroelectrics this law was found to match with an exponentially decreasing volume distribution of PNR, $N(V) \propto \exp[(V_m - V)/V_0]$ with $V_m =$ minimum cluster size and $V_0 =$ width of the size distribution [60]. It determines the distribution of cluster relaxation times via Arrhenius-type activation, $\tau = \tau_0\exp(KV/k_BT)$, $\tau_0 =$ ionic attempt time, $K =$ anisotropy energy density, $k_B =$ Boltzmann constant. Within this theory the low-f tail of the CC "*semicircle*" is predicted to drop linearly under an angle $n\pi/2$, where the coefficient $n = k_BT/E_0$ decreases linearly with T. This subtle effect is confirmed for the asymptotic right-hand tails of curves 2–6 in Fig. 5.10, where n drops from 0.51 to 0.27 between $T = 280$ and 262 K. It implies both substantial broadening of the size distribution and growth of the PNR. Actually this result reinforces previous insight into PNR growth via quenched fluctuations of RFs in PMN-PT on cooling toward the glass temperature [62].

The crossover to negative skewness (Fig. 5.10: curves 7–11) signifies the upcoming relevance of intercluster interaction within $256 \geq T \geq 246$ K, which was neglected by Lu and Calvarin [60]. It is indispensable for the freezing process of the percolating glassy cluster as $T \rightarrow T_g$ and marks the preponderance of large clusters with steadily growing relaxation times. The polydispersive *Cole-Davidson* [63] or *Havriliak-Negami* [64] models, both showing negative skewness, might be chosen to approximately describe the CC semicircles enclosing the glass temperature, $242 \geq T_g \geq 238$ K (Fig. 5.10: curves 12–14). Within this interval the glass transition is more accurately pinpointed by the divergence of the characteristic relaxation time related to the peak temperatures $T_m(f)$ of χ' via the critical power law, Eq. (5.1), as discussed in Sect. 5.1.

While at $T > T_g$ the relaxor refers to fluctuating, hence, mobile clusters, these become immobile and stick domain-like together at $T < T_g$ as a result of glassy freezing (curves 14–18). All mobility under low external electric fields is now restricted to the interfaces between the clusters. These behave essentially like ferroic DWs under the constraint of pinning forces due to the still existing quenched electric RFs. In this situation a driving electric field will merely excite different modes of DW motion as observed in disordered ferroic materials such as periodically poled ferroelectric $KTiOPO_4$ [56]. They are described in terms of the "*universal DW dynamics of disordered ferroics*" [57] by CC diagrams in the complex permittivity plane:

1. The essentially flat polydispersive relaxation response of oscillating DW segments as defined by a high density of pinning centers yields the CC equation,

$$\chi\left(\omega\right) = \chi_s + \left(\chi_0 - \chi_s\right) / \left[1 + \left(i\omega\tau_c\right)^{1-\alpha}\right]. \tag{5.5}$$

2. Owing to the wide distribution of "*Larkin lengths*" between pinning centers the CC semicircles in PMN are essentially flat and result in a white noise-like spectrum as, e.g., also observed in the soft superferromagnetic discontinuous metal-insulator multilayer $[Co_{80}Fe_{20}(1.4\ nm)/Al_2O_3(3\ nm)]_{10}$ [65].
3. The creep regime describes the thermally activated net propagation of the DWs after overcoming the depinning threshold at very low frequencies, $\omega < \omega_p$,

$$\chi' - i\chi'' = \chi_\infty\left[1 + \left(i\omega t\right)^{-\delta}\right]\ 0 < \delta < 1, \tag{5.6}$$

which is readily transformed into the linear function $\chi''(\chi') = (\chi' - \chi_\infty)\tan(\delta\pi/2)$ as observed (Fig. 5.10). For example, at T = 220 K and $\chi' \approx 7500$ the imaginary component starts rising from $\chi''(f_p \approx 10^2$ Hz) = 630 to $\chi''(10^{-3}$ Hz) = 1640. The slope of this function results from scaling of the dynamical relaxation-to-creep transition of DWs [56] and yields $\delta = (2-x)/z$, where the *fractality exponent* $x \geq 1$ represents the roughness of the interfacial ("DW") contour line. From the experimental value $\delta(220\ K) \approx 0.45$ and $z = 1.56$ [65] the fractal dimension $x \approx 1.30$ is obtained (Fig. 5.10). For example, at T = 220 K and $\chi' \approx 7500$ the imaginary component starts rising from $\chi''(f_p \approx 10^2$ Hz) = 630 to $\chi''(10^{-3}$ Hz) = 1640. The slope of this function results from scaling of the dynamical relaxation-to-creep transition of DWs [57] and yields $\delta = (2-x)/z$, where the *fractality exponent* $x \geq 1$ represents the roughness of the interfacial ("DW") contour line. For example, from the experimental value $\delta(220\ K) \approx 0.45$ and $z = 1.56$ the fractal dimension $x \approx 1.30$ of interfacial contour lines in glassy PMN is obtained. Upon approaching T_g the slope decreases to $\delta(238\ K) \approx 0.10$. Thus $x \approx 1.84$ characterizes utmost wall roughness preceding the total loss of cluster connectivity at $T \geq T_g$. The transitions between the horizontal and inclined relaxation and creep lines, respectively, in Fig. 5.10 become more and more rounded as T increases. This phenomenon is well known, e.g., from periodically poled $KTiOPO_4$ [56], where the relaxation and the creep processes are controlled by distribution functions of local double well potentials and DW mobilities, respectively. Since these are not identical by nature, their transition frequencies f_p cannot be identical at all temperatures and the dynamical transition will become smeared.

4. Finally, at still lower frequencies the dynamical transitions "*creep-to-slide*" and "*slide-to-switching*" are expected [57], but remain to be shown at much higher *ac* voltages.

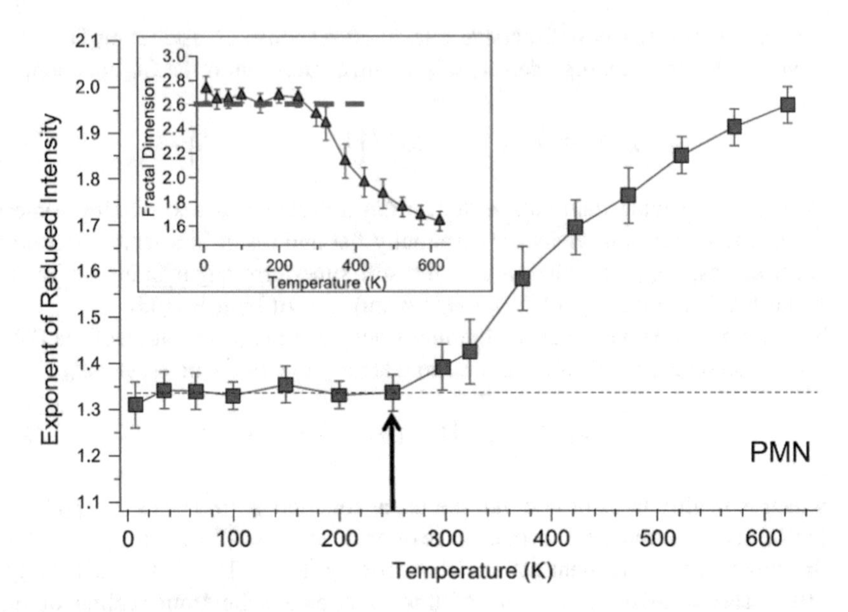

Fig. 5.11 Temperature dependence of the power exponent μ of the reduced light scattering intensity obtained in PMN, where the horizontal dotted line indicates the fitted asymptotic value $\mu \approx 1.33$ below $T_p \approx 240$ K (vertical arrow) and the horizontal broken line in the inset the estimated fractal dimension $d_p \approx 2.6$. Reproduced with permission from [66]

At the first glance the appearance of domain walls in the glassy state of PMN is surprising. However, one has, again, to take into account the action of quenched electric RFs in the charge-disordered host material. They will become active at the percolation transition of the PNR occurring in close vicinity of the glass transition (Fig. 5.9). Since RFs are known to favor local mesoscopic order (via their statistical fluctuations), they are also expected to take care of maximizing polar partial volumes, viz. condensing nano- into microdomains as observed on PMN in recent low-T transmission electron micrographs [54]. Such a process was also conjectured from diffuse neutron scattering experiments on PMN [47] and from quasi-elastic light scattering data, which clearly hint at percolation of the polar nanoregions into a fractal with dimension $d_p \approx 2.6$ at $T_p \approx 240$ K, Fig. 5.11 [66].

An important, if not decisive ingredient of domain growth is finally the occurrence of a global lattice instability due to the softening of the ferroelectric F_{1u} lattice mode in PMN as monitored by Raman spectroscopy [67] and neutron inelastic scattering [68] at $T_p \approx T_g$. These features and the observation of ferroelectric microdomains have been considered as signatures of a factual ferroelectric phase transition in PMN [54]. However, one still has to respect the undisputed existence of crucial cluster glass properties discussed above (Figs. 5.2 and 5.3) and to accept that the orientation of the microdomains and the topography of their walls remain controlled by RFs (Fig. 5.12 [54]). They eventually form what might be called a "*dipolar microdomain glass*" ground state with standard critical and non-ergodic properties.

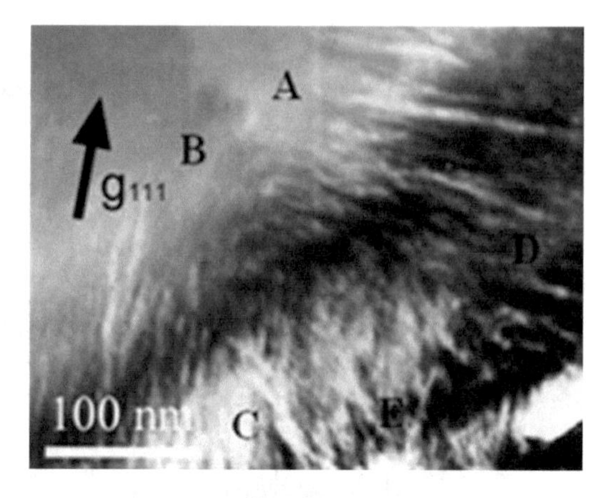

Fig. 5.12 TEM micrograph of nanoscale substructure in the *microdomain glass* phase of PMN at $T = 130$ K. Reproduced with permission from [54]

5.3 Superglass Transition in Uniaxial Relaxor Ferroelectric SBN

5.3.1 *Anisotropic PNR*

In contrast with the complete series of solid solutions from defect-free (PT) to extremely defective (PMN) in the binary system PMN-PT (Sect. 5.2.1), the end-members, $x = 0$ and 1, of the tetragonal tungsten bronze-type strontium barium niobate, $Sr_xBa_{1-x}Nb_2O_6$ (SBN), crystal family are unstable. The existence region for the structure is obviously delimited by $0.26 < x < 0.87$ [69]. Furthermore, unlike the system PMN-PT there are no "*pure*" and "*defect*" cations defined by their chemical structure. Nonetheless, also in this case the concentration ratio [Sr]/[Ba] $= x/(1-x)$ clearly measures the degree of disorder by virtue of the monotonically rising amount of mixing entropy due to the architecture of the underlying lattice geometry [70]. Recent attempts to determine the crossover concentration between relaxor and ferroelectric phases, $x_c \approx 0.4$, were, however, not conclusive [71].

Starting from the unfilled tungsten bronze structure of SBN (Fig. 5.7b), it is seen that one out of four A2- and two A1-sites per elementary lattice cell, respectively, must be empty [72]. This gives rise to missing charges with an effective disorder, which are the most intense sources of RFs. They cause shifts of Nb^{5+} cations from their central positions in the NbO_6 octahedra. On the other hand, occupation of A_2 sites by different cations, Sr^{2+} and Ba^{2+}, introduces disorder of the oxygen ion positions due to both different ionic sizes ($r \approx 113$ pm and 135 pm, respectively) and Ba-O and Sr-O bonding lengths. The presence of vacancies at both A_1 and A_2 sites enhances this disorder [72]. Note that A_1 sites are exclusively occupied by Sr^{2+} ions.

Accommodation misfits of the different oxygen octahedra give rise to local buckling and tilting deformation. The emerging localized electric multipole moments

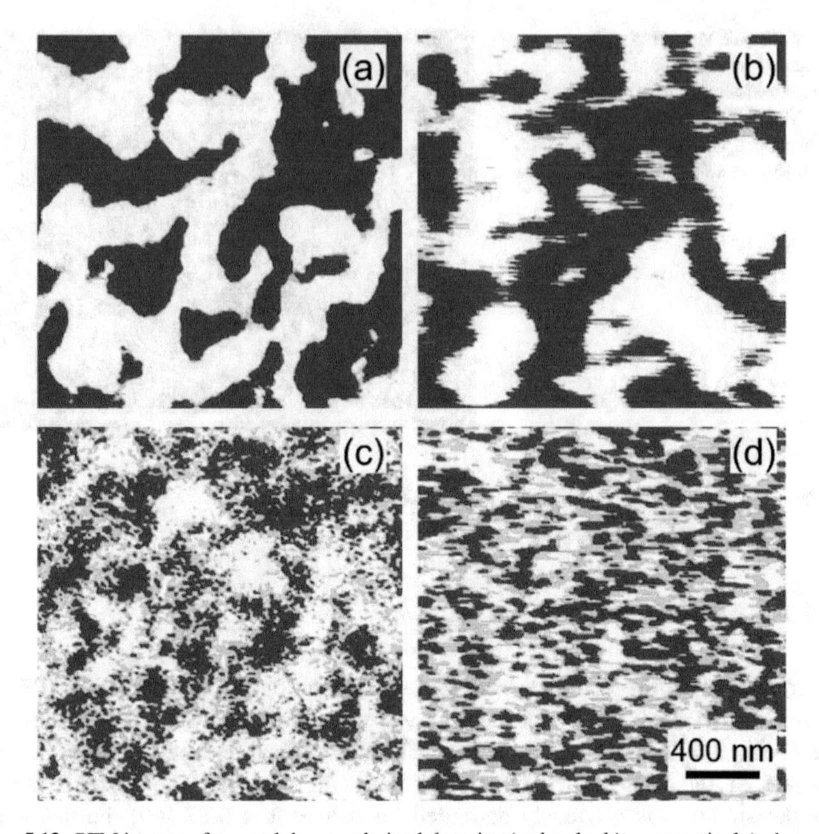

Fig. 5.13 PFM images of up- and down-polarized domains (*red* and *white*, respectively) observed at room temperature on c-cut single crystals of SBN40 (**a**), SBN50 (**b**), SBN61 (**c**), and SBN75 (**d**). *Yellow* contrast denotes vanishing piezoresponse. Reproduced with permission from [70]

are sources of electric *RF*s, which add to those of the bare vacancies. Thus we can assume that larger structural disorder will induce stronger *RF*s. By statistical consideration the most ordered structure is expected in (hypothetical) SBN20, where all A_2 sites are solely occupied by Ba^{2+} cations, while the Sr^{2+} ions and vacancies are randomly distributed over the A_1 sites. Upon increasing the Sr^{2+} content, the mixing entropy S first increases, then reaches a maximum level at about $x \approx 0.65$, and decreases gently at large x [33]. In accordance with the *R*andom *F*ield *I*sing *M*odel (*RFIM*) [34] this explains the increasing fine-graining of the low-T domain state as experienced by *PFM* images in the c-plane of single crystals SBN40, 50, 61, and 75 after being aged for 1 year at room temperature in Fig. 5.13 [70].

The decisive difference of SBN as compared to PMN, however, is the uniaxiality of the structure and, hence, the one-dimensionality of the polar order parameter. This changes the topology of the RF-induced domains, which appear maze-like within the c-plane (Fig. 5.13), but stripe-like along the c-axis (Fig. 5.14). Within the framework of an anisotropic Ising model [70] one finds the approximate relationship

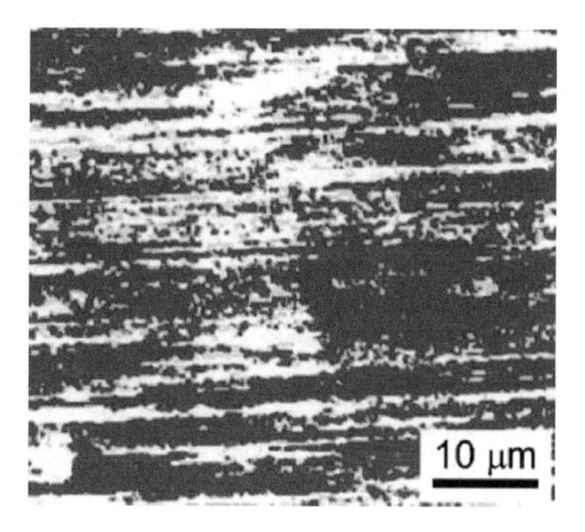

Fig. 5.14 PFM image of left- and right-polarized domains (*red* and *white*, respectively) observed on an *ab*-cut single crystal of SBN40 in the lateral mode. Reproduced with permission from [70]

$L_a/L_c = J_a/J_c$, where $L_a(L_c)$ are the lateral (longitudinal) domain dimensions and $J_a(J_c)$ the intra(inter)layer interaction energies. Since $L_a << L_c$ as observed experimentally, also $J_a << J_c$ is expected in accordance with the tetragonal lattice symmetry. This has consequences on the anisotropy of the relaxation spectra as discussed below.

5.3.2 Glass Transition of SBN80

Figure 5.15 shows the temperature dependence of the susceptibility components, χ'(a) and χ''(b) vs. T, for logarithmically equidistant frequencies $10^{-2} \leq f \leq 5 \times 10^5$ Hz, and temperatures $263 \leq T \leq 350$ K. The $\chi'(T)$ curves reveal extremely broad peak widths, FWHM $\approx$ 50 K, and strong shifts of their peak temperatures, T_m, with decreasing frequency very similar to those of PMN (Fig. 5.1). The cluster glass temperature of SBN80 is determined from the asymptotic low-f peak shift of $\chi'(T_m)$ as displayed in Fig. 5.4 (right curve). By use of the low-f data points within $10^{-2} \leq f \leq 2 \times 10^{-1}$ Hz (arrows) the dynamic scaling relation, Eq. (5.1), delivers $T_g = (299.0 \pm 0.2)$ K. We notice that this result seems to differ considerably from that reported by Dul'kin et al. [74] for the "*ferroelectric transition temperature*," $T_C \approx 322$ K, of the closely related compound SBN75. Actually, however, this latter value was obtained from χ'_{max} ($f = 100$ Hz) and thus considerably overestimates the real T_C (or T_g) in view of the large frequency dispersion of χ' (Fig. 5.15a).

Peculiarly, the loss curves, $\chi''(T)$ (Fig. 5.15b), reveal two distinct regimes of frequencies and temperatures by an obvious "clustering" into two groups of curves, labeled "*high f & T*" and "*low f & T*", respectively [73]. Their crossover takes place at $f \approx 10^2$–10^3 Hz and $T \approx 310$ K. This puzzling situation becomes transparent

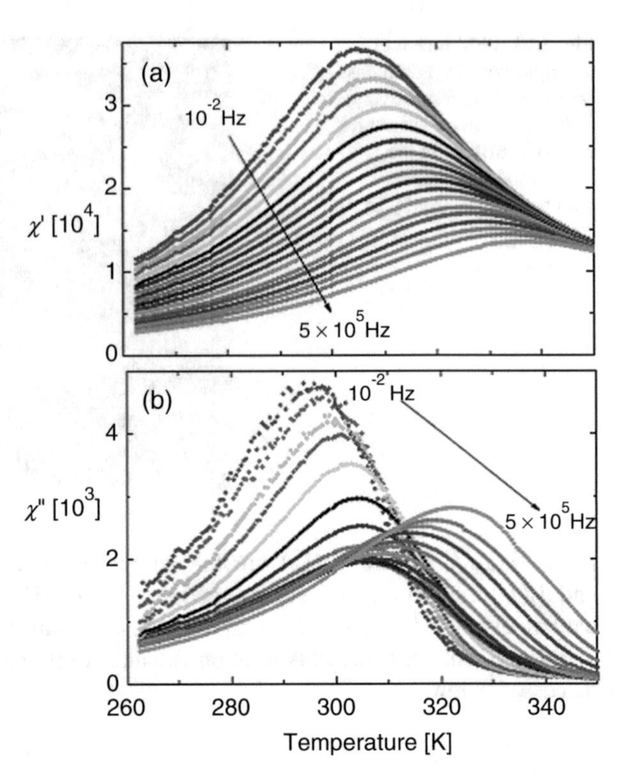

Fig. 5.15 Temperature dependence of the longitudinal susceptibility components χ' (**a**) and χ'' (**b**) of SBN80 measured within $263 \text{ K} \leq T \leq 350 \text{ K}$ at frequencies $10^{-2} \leq f \leq 5 \times 10^5$ Hz. Reproduced with permission from [73]

when examining the dielectric spectra in Fig. 5.16a $\chi'(f)$, and Fig. 5.16b, $\chi''(f)$, for temperatures within the range $260 \leq T \leq 380$ K. While all spectra are essentially flat for $T > 330$ K and monotonically descending for $T < 290$ K, distinct structures become visible within $290 < T < 330$ K. They consist of two smeared dispersion steps in $\chi'(f)$ and peaks in $\chi''(f)$ at $f \approx 10^0$ and 10^5 Hz.

We argue that the double-humped relaxation spectrum mirrors the spatial anisotropy of the PNRs, which was discovered by vertical and lateral mode PFM, respectively, on SBN40 (Figs. 5.13a and 5.14 [70]). The estimated needle-like width-to-length ratio $L_a/L_c \approx 10^{-3}$ results from the anisotropy of the dipolar intra- and interlayer interactions of O-Nb-O chains, $J_a << J_c$. This has an impact on the response of a weak driving field $E||c$, which will not flip the static PNR owing to the high activation energy of switching the dipole moment of the individual RF stabilized PNR. Instead, a breathing motion under RF pinning constraints occurs with pronounced longitudinal ("headways") and lateral ("sideways") contributions at different frequencies. They are controlled by Debye-type relaxation processes, which contribute to the dielectric susceptibility via

$$\chi(L) \propto 1/[1 + i\omega\tau(L)]. \tag{5.7}$$

The relaxation time is expected to obey dynamic scaling of wall motion under weak pinning, $\tau(L) \propto L^z$, with a dynamic exponent $z \approx 1.6$ [75]. Since peak

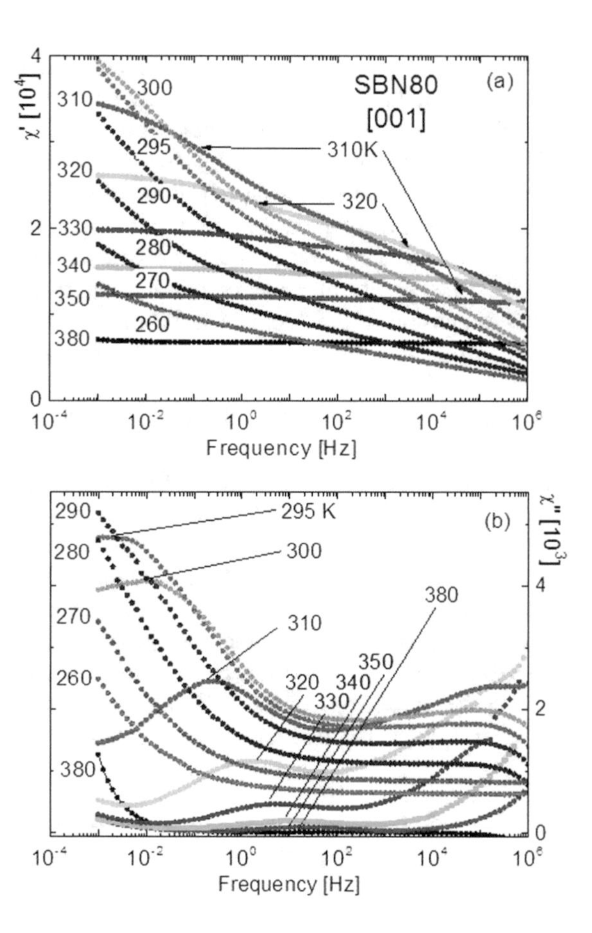

Fig. 5.16 Frequency dependence of the longitudinal susceptibility components χ' (**a**: double-logarithmic) and χ'' (**b**) of SBN80 measured within $10^{-3} \leq f \leq 10^6$ Hz at temperatures $260 \leq T \leq 380$ K. Reproduced with permission from [73]

contributions to χ'' satisfy the condition $\omega\tau = 1$, we can estimate the ratio of the relevant relaxation frequencies by $\omega_c/\omega_a = \tau_a/\tau_c = (L_a/L_c)^z \approx 10^{-5}$ in agreement with the experimental observation.

It is seen in Fig. 5.16b that the fast headways relaxation enters the spectral window ($f_c = \omega_c/2\pi \approx 10^5$ Hz) only at 320 K and fades out below 290 K, a signature of longitudinal coalescence of the PNRs, which form domains in the ferroelectric regime below $T_C \approx 289$ K. On the other hand, the slow sideways relaxation continuously shifts on cooling from 380 to 295 K from $f_a = \omega_a/2\pi \approx 10$ Hz to $\approx 10^{-2}$Hz, probably a signature of lateral PNR growth on cooling. Most spectacular, however, is the replacement of the sideways relaxation peak below 295 K by continuously decreasing hyperbolic curves, $\chi'' \propto f^{-\beta}$, which look very similar in the real part curves. In logarithmicscale (Fig. 5.16a) they correspond to parallel straight lines with negative slope as shown for $\chi'(290\ \text{K})$ by a best-fitted line with slope $\beta \approx 0.09$. This is a clear signature of domain wall creep, which obeys the dispersion law [76]

$$\chi'' \propto \chi' - \chi'_\infty \propto \omega^{-(2-x)/z}, \tag{5.8}$$

where $\chi'_\infty << \chi'\left(10^{-3}\text{Hz}\right)$ can be neglected in first approximation. Note that the creep formula, Eq. (5.8), is valid only under the condition $\omega < \omega_\text{p}$ (= Larkin or depinning frequency), which holds up to $f > 1$ Hz, hence, way above the sideways relaxation peaks at $T \geq T_\text{C}$. This confirms that a qualitatively new situation arises below T_C, where "*infinitely long*" domain walls (along the c direction) have replaced PNR boundaries with finite length. From the exponent in Eq. (5.8), $\beta = (2-x)/z \approx 0.09$, and $z \approx 1.6$ we calculate the exponent $x \approx 1.95$, which describes the density of states of pinning lengths, $g(L) \propto L^{-x}$, and likewise the fractal dimension of the "domain wall" in the (001) plane. Remarkably, x agrees within errors with the exact value of the fractal dimension of the 2D RFIM, $D = 1.96$ [75].

Additional evidence of DW dynamics is delivered by the CC plots χ'' vs. χ' in Fig. 5.17. The low-T curves (12)–(18) for $295 \geq f \geq 260$ K resemble pretty much the hockeystick-like ones found for PMN (Fig. 5.10), where the flat "blades" designate the flat high-f local relaxation spectra of pinned DW segments, while the oblique "handles" signify the creep regimes. A major modification, however, is encountered in the high-T curves (1)–(11) for $350 \geq T \geq 300$ K as a consequence of the double-humped loss curves in Fig. 5.16b. All of them show marked low and high frequency Debye semicircles peaking at f_c and f_a, respectively (e.g., for $T = 310$ K at 10^{-1} and 10^5 Hz, curve 9, vertical arrows). This phenomenon is due to the spatial anisotropy of the PNR and best observed within $315 \geq T \geq 295$ K (curves 8–12) for the frequency range available. The percolation transition into

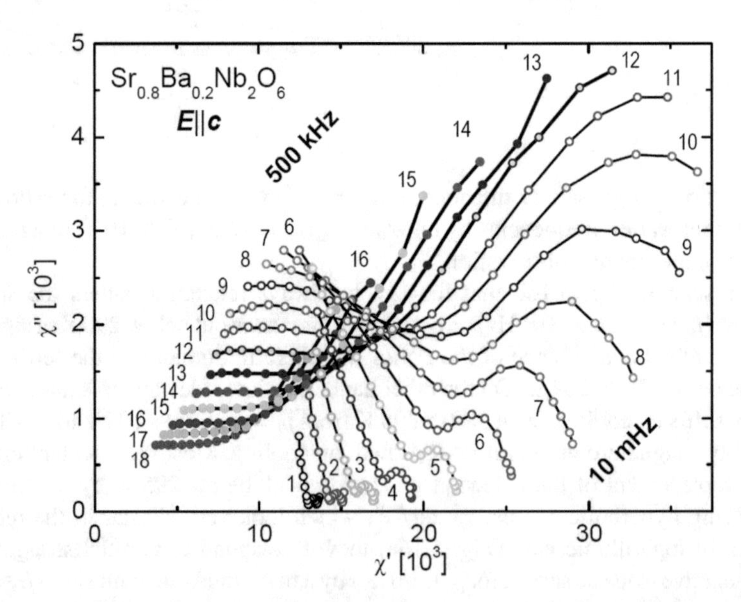

Fig. 5.17 Cole-Cole plots χ'' vs. χ' of SBN80 measured at $T = 350$ (1), 345 (2), 340 (3), 335 (4), 330 (5), 325 (6), 320 (7), 315 (8), 310 (9), 305 (10), 300 (11), 295 (12), 290 (13), 285 (14), 280 (15), 275 (16), 272 (17), and 260 K (18) within $10^{-2} \leq f \leq 5\times10^5$ Hz. Assembled from Fig. 5.16 [73]

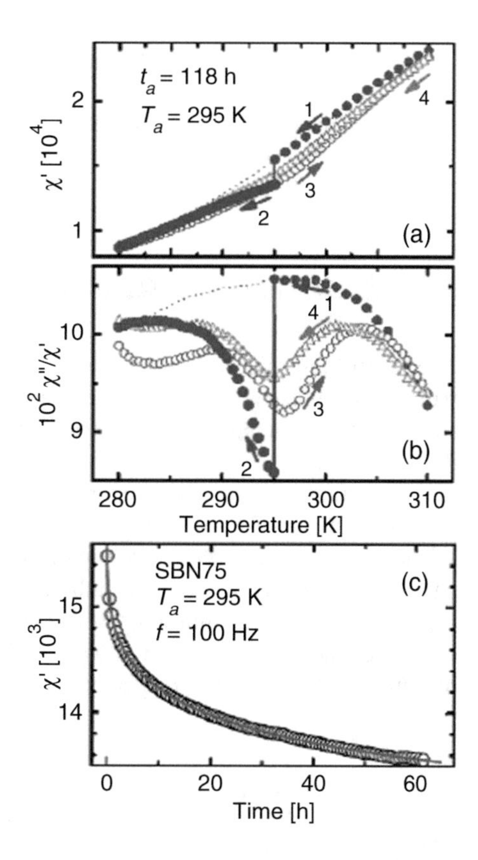

Fig. 5.18 (**a**, **b**) Aging of χ' and χ''/χ' vs. T of SBN75 measured after ZFC from $T = 450$ K at rates $dT/dt = \pm 0.2$ K min^{-1} and frequency $f = 100$ Hz on first cooling to 280 K with intermittent halt at 295 K (curves 1 and 2), continuous reheating to 310 K (curve 3), and subsequent cooling back to 280 K (curve 4). Dotted line = continuous cooling from 450 K with the same rate [77]. (**c**) $\chi'(f = 100$ Hz) vs. t of SBN75 after ZFC at $T = 100$ K and best fitted to Eq. (5.9) [77]. Reproduced with permission from [78]

the ferroelectric microdomain state with "*hockeystick-like*" DW dynamics is thus located at $T_C \approx 295$ K, below which the cluster glass regime with typical attributes of non-ergodicity is encountered similarly as in PMN (Figs. 5.3 and 5.4).

These phenomena are shown for the compound SBN75 by a classic stop-and-wait procedure as shown in Fig. 5.18a, b [77]. Aging of the susceptibility components χ' (a) and χ''/χ' (b) of SBN75 is recorded vs. T after zero-field cooling (ZFC) from $T = 450$ K at rates $dT/dt = \pm 0.2$ K min^{-1} and frequency $f = 100$ Hz, on first cooling to halt for 118 h at the wait temperature $T_w = 295$ K (curves 1 and 2, respectively), continuous reheating to 310 K (curve 3), and subsequent continuous cooling back to 280 K. The dotted line refers to continuous cooling from 450 K under identical cooling rate. Distinct drops of both χ' and χ''/χ' are observed at T_w. They remain clearly visible albeit slightly smeared on subsequent heating (curve 3) and recooling (curves 4). Obviously a distinct memory effect due to glassy domain growth (in the sense of "*locally optimized disorder*") is encountered, which confirms that the experiment has been carried out in the regime $T_w < T_g$.

The temporal relaxation of χ' during t_w as measured after ZFC at $T = 295$ K is shown in Fig. 5.18c. The observed decay is excellently fitted to a stretched exponential corresponding to Eq. (5.3),

$$\chi'(t) = A + B\exp\left(-[t/\tau]^{\beta}\right), \tag{5.9}$$

with parameters $A = 12{,}783(20)$, $B = 2803(26)$, $\tau = 30.9(9)$ s, and $\beta = 0.367(9)$. The small value of $\beta < 1$ reflects stretching due to polydispersive relaxation processes as in PMN (Fig. 5.6).

5.4 Strain Glass as a Random Field System

The controlling role of random electric fields in the formation of relaxor systems with cluster glass ground states has been demonstrated in Sects. 5.2 and 5.3. A similar mechanism applies to the structural "*strain glass*." As proposed by Ren [2] the formation of ferroelastic "*martensitic nanoregions*" (MNR) in $Ti_{50-x}Ni_{50+x}$ at the shear transformation is hampered by point defects, *viz.* random misfits due to non-stoichiometry for $x \neq 0$ or to doping with Fe or Cr [79]. Figure 5.19 shows the schematics of "*strain glass*" evolving from ferroelastic systems exhibiting random nanostress fields as in the nonstoichiometric martensitic compound $Ti_{50-x}Ni_{50+x}$ [3]. The proximity to spin glass is demonstrated by Fig. 5.20, where $Ti_{48.5}Ni_{51.5}$ undergoes the glass transition at $T_g = 168$ K under shear stress $\tau = 40$ MPa on heating after ZFC and FC, respectively.

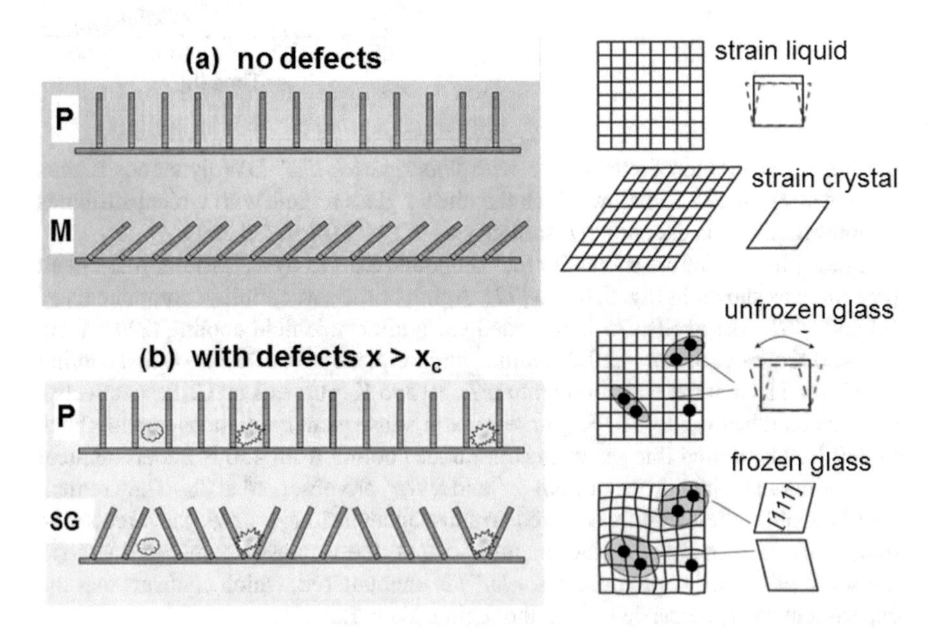

Fig. 5.19 Schematic illustration of martensitic long-range shear strain without defects (**a**) and stabilization of strained nanodomans due to point defects (**b**). Reproduced with permission from [80]

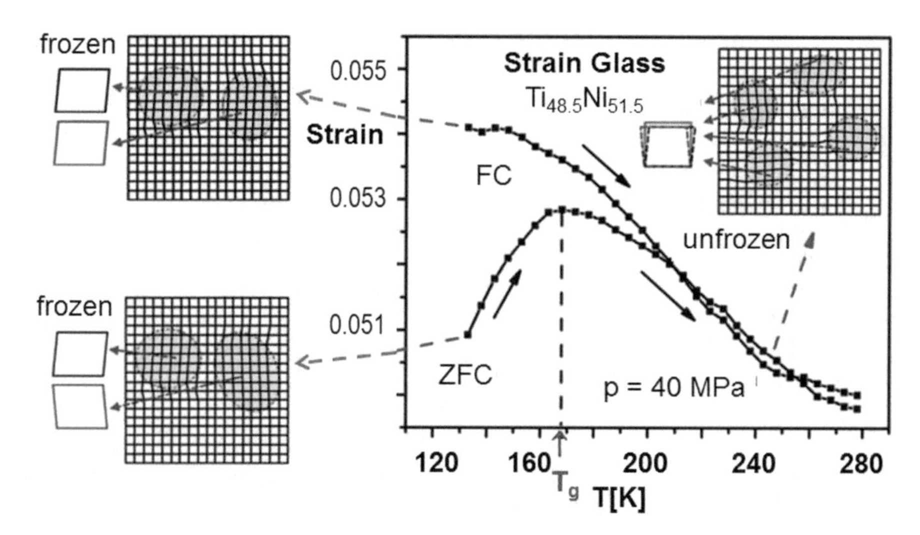

Fig. 5.20 ZFC/FC curves of average strain under 40 MPa shear stress vs. T of $Ti_{48.5}Ni_{51.5}$ strain glass showing a large deviation below $T_g = 168$ K. Inset sketches depict the unfrozen strain states above T_g and the two differently frozen states below T_g upon field heating after different thermal histories, respectively. Reproduced with permission from [80]

The different initial strain states contain disordered and ordered MNR, respectively. Direct visualization of differently sheared frozen MNR separated by antiphase boundaries was first demonstrated on $Ti_{50}Pd_{41}Cr_9$ by HRTEM at room temperature [81]. Figure 5.21 shows the *storage modulus* (= inverse mechanical susceptibility, $d\varepsilon'/d\sigma$) of $Ti_{50}Pd_{41}Cr_9$ together with the loss tangent, *tan* δ, as functions of the temperatures for various low frequencies. In the insert the *Vogel-Fulcher* relation, Eq. (5.2), is shown to apply satisfactorily and tempting to deliver a formal "*transition temperature*," $T_{VF} = T_g(f = 0) = 298$ K. However, as we have discussed in the case of uniaxial relaxors (Sect. 5.3), Eq. (5.2) is not expected to describe the asymptotic cluster glass criticality. To the best of our knowledge sufficiently precise asymptotic data are not yet available in order to extrapolate the true value of T_g by checking the asymptotic dynamic criticality via Eq. (5.1).

It has been stressed [82] that the random fields emerging from point defects changes normal long-range ordered, polytwinned domain structure ("*strain crystal*") into MNR of individual variants of martensite ("*strain glass*"), full of strain disorder at interfaces between martensite and retained austenite (Fig. 5.22). It alters the overall characteristics of the martensitic transformation from sharp first-order to continuous as shown for $Ti_{50}Ni_{50-x}Co_x$ with $0 \leq x \leq 25$ in Fig. 5.23 [83]. Such an apparently continuous martensitic transformation is accompanied by a gradual softening of the elastic modulus upon cooling that compensates the normal modulus hardening associated with anharmonic atomic vibration, leading to the so-called *Elinvar* anomaly. The first *Elinvar* alloy, $Fe_{52}Ni_{36}Cr_{12}$, which has an invariant elastic modulus over a wide temperature range, was discovered almost 100 years ago [84].

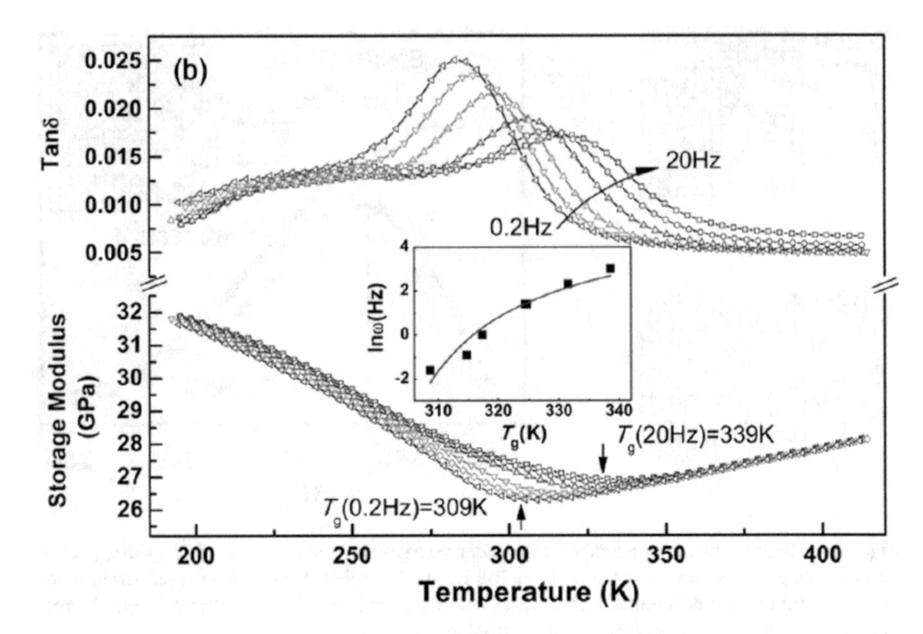

Fig. 5.21 Dynamic freezing of $Ti_{50}Pd_{41}Cr_9$ as shown by the T and f dependence of the real part and the loss tangent of the inverse mechanical susceptibility around the "*Vogel-Fulcher freezing*" temperature $T_{VF} = 298$ K. Reproduced with permission from [81]

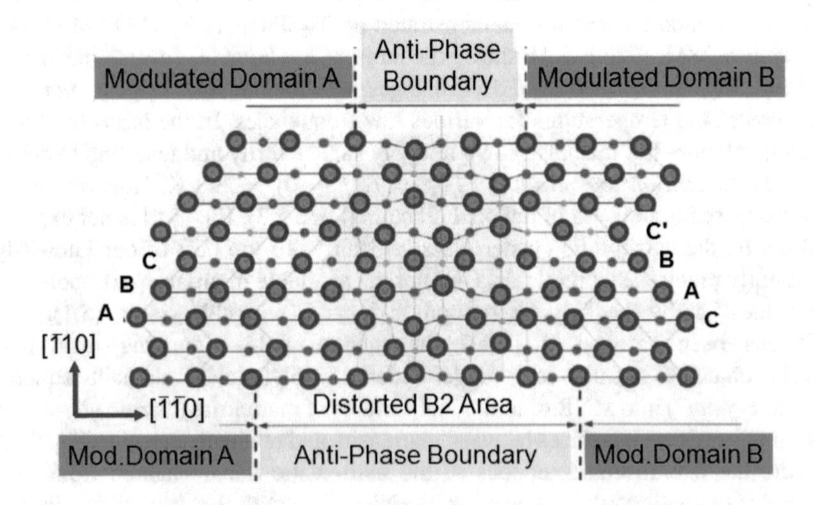

Fig. 5.22 Sketch of the strain glass structure of $Ti_{50}Pd_{41}Cr_9$ with martensitic twins (A and B) and their austenitic antiphase boundary. Cr atoms (smallest circles) exert compressive strain and thus control the MNR formation via local majority rules. Reproduced with permission from [81]

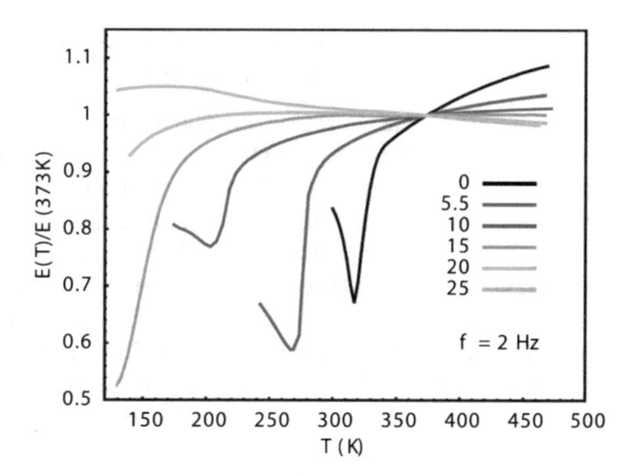

Fig. 5.23 Temperature dependence of the elastic modulus of $Ti_{50}Ni_{50-x}Co_x$ with $0 \leq x \leq 25$ normalized to $E(373\text{ K})$ Reproduced with permission from [83]

Similarly to the relaxor case (Sect. 5.3) a complete theory of strain glass has not yet been developed. In the case of the NiTi-system this would be a modification of the conventional theory of the martensitic transition from cubic B2 into monoclinic B19′ martensite under the nanostress action of random lattice defects in $Ni_{50-x}Ti_{50+x}$ [3]. However, it has been argued [1] that the defect-free system ($x = 0$) might be described by a Landau-Ginzburg free energy function including long-range elastic interaction in order to reproduce the ferroelastic transition with the observed domain morphology. Random defects ($x > 0$) are then expected to nucleate random MNR via a spatial distribution of M_s temperatures over the system. This appears similar to the range of local Curie temperatures within $T_d \leq T \leq T^*$ in relaxors (Fig. 5.8). The free energy density will then be dominated by the harmonic term,

$$F = a\,[T - T_c - \eta(r)]\,{e_2}^2 - \ldots, \tag{5.10}$$

where T_c is the lower stability limit of the high temperature phase in the absence of disorder and $\eta(\boldsymbol{r})$ is the disorder field, e.g., Gaussian distributed with zero mean. The disorder thus creates a distribution of transition temperatures, $T(\mathbf{r}) = T_0 + \eta(\boldsymbol{r})$, where T_0 represents the transition temperature in the clean limit. This comparatively simple model has been used in numerical simulations [85, 86] with remarkable success. In good agreement with experiments it was found that

1. Static premartensitic nanostructures are absent in defect-free ($\eta(\boldsymbol{r}) \equiv 0$) systems, but arise for $\eta(\boldsymbol{r}) \neq 0$. At a critical defect concentration, x_c, the crossover from the normal martensitic transition at T_c into a frozen "strain glass" transition at T_g is observed.
2. The glass transition at T_g is proven by non-ergodic behavior of the shear-induced strain between ZFC and FC procedures (Fig. 5.20).
3. External stress can transform a strain glass containing MNR into normal martensite with large domains or even a single domain.

4. Inherent elastic anisotropy A can crucially affect the critical defect concentration x_c, which usually increases as A increases.

The unmistakable success of the above model theory certainly calls for a more complete series of test experiments, e.g., the fulfillment of dynamic criticality according to Eq. (5.1), since the *Vogel-Fulcher* test using Eq. (5.2) is definitely unsuitable. On the other hand, "*less developed regions*" in the field of ferroic glasses might profit from the successful strain glass discussion [1]. This applies in particular to the relaxor problem (Sects. 5.2 and 5.3), where the distribution of local phase transition temperatures has hitherto not yet become subject of discussion, although local clustering of PNR and their mutual dipolar interaction are widely accepted [8]. On the other hand, the elastic quadrupolar interaction between the MNR and the random-field action of the defects is not yet explicitly manifest in the actual discussion of strain glass [1]. Deeper insight into the crossover from microscopic processes into mesoscopic glass formation seems indispensable in both fields.

5.5 Cluster Spinglass

Cluster spinglass systems have entered mesoscopic solid state physics in two versions—as matrix isolated nanoparticular "*superspin glass*" [87] and as point defect-activated intrinsic "*ferroic glass*" [1], respectively. Both undergo glassy dynamic criticality as $T \rightarrow T_g$ and non-ergodicity at $T < T_g$. While a "*superspin glass*" (SSG for short) largely resembles an atomic spin glass [88], the "*ferroic glass*" states are proposed to emerge from random field-controlled nanoclustering and subsequent dipolar glass formation as discussed for ferroelectric relaxors like $PbMg_{1/3}Nb_{2/3}O_3$ (Sects. 5.2 and 5.3) and ferroelastic martensites like $Ti_{48.5}Ni_{51.5}$ (Sect. 5.4). Although only the latter realization had been in mind in the discussion of "*ferroic glasses*" [1], the SSG concept should shortly be visited here in order to recognize its spirit and to specify its uniqueness as a matrix isolated nanoparticular system.

Indeed, neither ferroelectric nor ferroelastic nanoparticles have ever been observed to enter a "*neutral*" matrix environment like metallic magnetic nanoparticles in a nonmetallic insulating embedding material. Let us consider soft magnetic nanoparticles of $Co_{80}Fe_{20}$ embedded in a film of insulating alumina, α-Al_2O_3 [88], where they establish granular (discontinuous) metal–insulator multilayer (DMIM) systems $[Co_{80}Fe_{20}(t_n)/Al_2O_3(3\ \text{nm})]_m$ at nominal single-layer thickness t_n with $m = 1$–10 bilayer periods after growth using sequential Xe-ion beam sputtering on sapphire glass substrates [89]. Under non-wetting condition the CoFe alloy coagulates into quasi-spherical granules (*Volmer-Weber* growth mode), while their individual size and spatial density vary via t_n and thus tune the inter-particle interactions. Up to a nominal thickness $t_n = 1.8$ nm nonpercolating discontinuous distributions of immiscible CoFe clusters ("*superspins*") embedded within adjacent alumina layers have been grown. Figure 5.24a shows a transmission

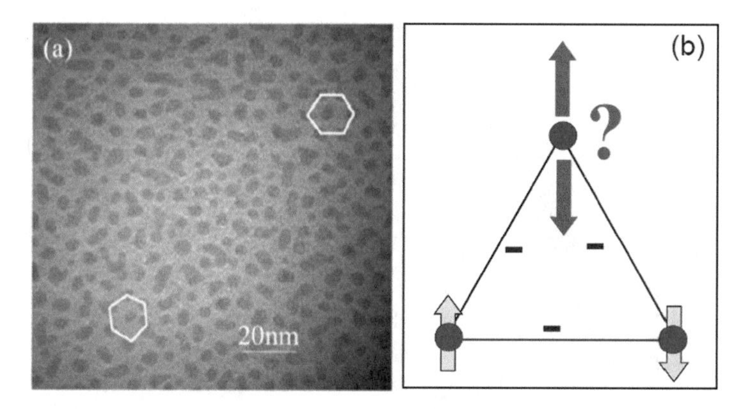

Fig. 5.24 (**a**) TEM top view micrograph of $Co_{80}Fe_{20}$(0.9 nm)/Al_2O_3(3 nm) bilayer. Reproduced with permission from [89] via *CCC Rights Link*. (**b**) Frustrated dipolar nearest-neighbor configuration on a triangular lattice. Reproduced with permission from [88]

electron micrograph (TEM) of the irregular cluster distribution (dark dots) with a near Gaussian distribution of diameters, $d \approx (2.8 \pm 1.0)$ nm and partial hexagonal short-range ordering (white contours indicated) in a bilayer ($m = 1$) with $t_n = 0.9$ nm [89]. Typically they exhibit frustrated nearest-neighbor configurations of magnetic dipoles on a triangular lattice (Fig. 5.24b).

Figure 5.25 shows the temperature dependence of the susceptibility components, χ' and χ'' vs. T, of $[Co_{80}Fe_{20}(0.9\ \text{nm})/Al_2O_3(3\ \text{nm})]_{10}$ for frequencies $10^{-2} \leq f \leq 10^0$ Hz and temperatures $30 \leq T \leq 90$ K [90]. The $\chi'(T)$ curves reveal a broad full-width-at-half-maximum, FWHM ≈ 40 K, a monotonic increase of their heights with decreasing f, and simultaneous low-T shifts of their peak positions, T_m. Similarly as observed on PMN (Fig. 5.3) the isothermal frequency dependence of χ'' changes sign at $T \approx T_g = (45.6 \pm 4.6)$ K. While this low value is due to the sparse spatial density of the MNP, the relatively large attempt time, $\tau_0 = (2.8 \pm 1.3) \times 10^{-7}$ s, as obtained from fitting to Eq. (5.1) reflects their mesoscopic dimension (Fig. 5.24a).

Non-ergodic behavior (aging, rejuvenation, and memory) of the FC and ZFC magnetization is observed below the glass temperature T_g in Fig. 5.26 (see caption for details) and resembles those of PMN (Fig. 5.4) and of SBN75 (Fig. 5.17). This confirms the glassy character of the SSG. Nevertheless, important differences arise in the glassy regimes, $T < T_g$. While relaxors like PMN and SBN80 enter domain states with domain wall susceptibility dynamics as documented by the spectra in Figs. 5.10 and 5.16, respectively, the SSG continues to map single particle (MNP) dynamics as evidenced by CC plots in Fig. 5.27. All of them are truncated semicircle-like and described by the CC equation [93],

$$\chi(\omega) = \chi_s + (\chi_0 - \chi_s) / \left[1 + (i\omega t_c)^{1-\alpha}\right], \tag{5.11}$$

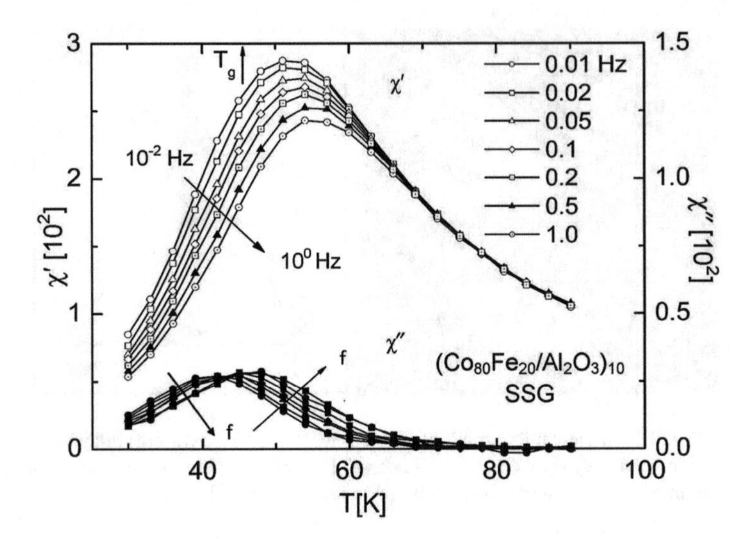

Fig. 5.25 Temperature dependences of the magnetic susceptibility components χ' and χ'' of the SSG [$Co_{80}Fe_{20}$(0.9 nm)/Al_2O_3(3 nm)]$_{10}$ at decade stepped frequencies within $10^{-2} \leq f \leq 10^0$ Hz. $T_g = 45.6$ K is marked by an arrow. Reproduced with permission from [90]

where the exponent, $0 < \alpha < 1$, describes the degree of polydispersivity as distinguished from the mono-disperse Debye process, $\alpha = 0$. As discussed previously [92], α increases from 0.75 to 0.87 as T decreases for 60 to 45 K, while it does not increase any more for $T \leq 40$ K (uppermost panel). This is a signature of the frozen glass state at $T < T_g \approx 46$ K. On decreasing T, the limited range of frequencies makes the data point groups continuously shift away from the rightmost intersection with the χ' axis, $f = 0$. Indeed, the fits reflect the tremendous slowing down on cooling to the glass transition with "characteristic" relaxation times $10^{-4} \leq \tau_c \leq 10^4$ s at the apex points, $2\pi f\tau_c = 1$. Thus the CC plots of the SSG clearly reflect the structural invariance of the cluster system (Fig. 5.24a) under cooling to below T_g.

The situation is, hence, much simpler as compared to the "ferroic glasses" PMN and $Ni_{48.5}Ti_{51.5}$, respectively, the glassiness of which is based on the disorder-controlled generation of PNR/MNR, their subsequent dipolar/quadrupolar interaction, glassy freezing, and eventual percolation. We are now left to unravel a similar scenario in "spin-cluster glasses" such as $La_{0.7}Ca_{0.3}Mn_{0.8}Cd_{0.2}O_3$ [6] as suggested by Ren et al. [1].

The Cd doped CMR material $La_{0.7}Ca_{0.3}Mn_{1-x}Cd_xO_3$ with $0 \leq x \leq 0.2$ shows both ferromagnetism and apparent cluster spin-glass behavior. A metal-insulator transition is exhibited by samples with $x < 0.1$, while samples with $x = 0.15$ and 0.2 are semiconductors. With increasing Cd content the system changes from para- to ferromagnetic at $x < 0.10$, as well as from paramagnetic to spin-glass-like for $x > 0.1$. Spin-glass-like behavior is indicated by the typical non-ergodicity of *ZFC* and *FC* magnetization data, while χ'_{ac} is found to follow dynamical critical slowing-down, Eq. (5.1). Obviously, by slight variation of the Cd dopant with $4d^{10}$ closed electron

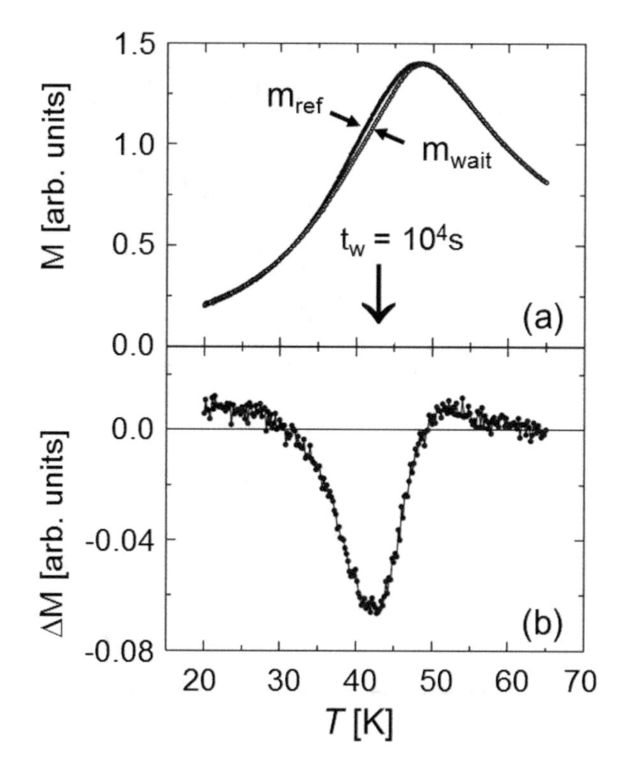

Fig. 5.26 Aging and rejuvenation of the magnetization $M(T)$ of the SSG $[Co_{80}Fe_{20}(0.9\ nm)/Al_2O_3(3\ nm)]_{10}$ measured in $\mu_0 H = 40\ \mu T$: (**a**) Temperature dependencies of continuously measured $M_{ref}\ (T)$ (solid circles) and of $M_{wait}(T)$ (open circles) measured with intermittent zero-field wait period of $\Delta t = 10^4$ s at $T = 42$ K. (**b**) Magnetic hole $\Delta M = M_{wait}(T) - M_{ref}\ (T)$ vs. T. Reproduced with permission from [91]

shell configuration, a spin glass seems to form at $x = 0.2$ with best-fit parameters $T_g = 42.1 \pm 0.5$ K, $z\nu = 10.8 \pm 1.6$, and $\tau_0 = 10^{-12}$ s. While T_g and $z\nu$ come up to expectation, the value of τ_0 corresponds to an atomic spin glass, $\tau_0 \approx 10^{-13}$ s, rather than to a spin cluster system, for which values of $\tau_0 \approx 10^{-6}$ s are expected as in our above SSG. The situation in the $La_{0.7}Ca_{0.3}Mn_{1-x}Cd_xO_3$ system is not quite clear, since nearby compositions, e.g., $x = 0.15$, were reported [6] to come closer to the spin cluster model, but dynamical critical scaling turned out to fail, supposedly because of phase impurity issues (Fig. 5.28).

Further research is needed for clarifying the spin-clustering conjecture, if any, in $La_{0.7}Ca_{0.3}Mn_{1-x}Cd_xO_3$. Another candidate is the related CMR material $La_{0.7}Ca_{0.3}Mn_{1-x}Co_xO_3$, which has shown the ZFC-FC anomaly of a spin cluster glass for $x = 0.3$ (Fig. 5.29) [94]. The drop of M_{ZFC} for $T < T_g \approx 100$ K has been interpreted as a signature of cluster glass behavior. The Co substitution breaks down the long-range ferromagnetic order seen in the pure manganates, replacing it with

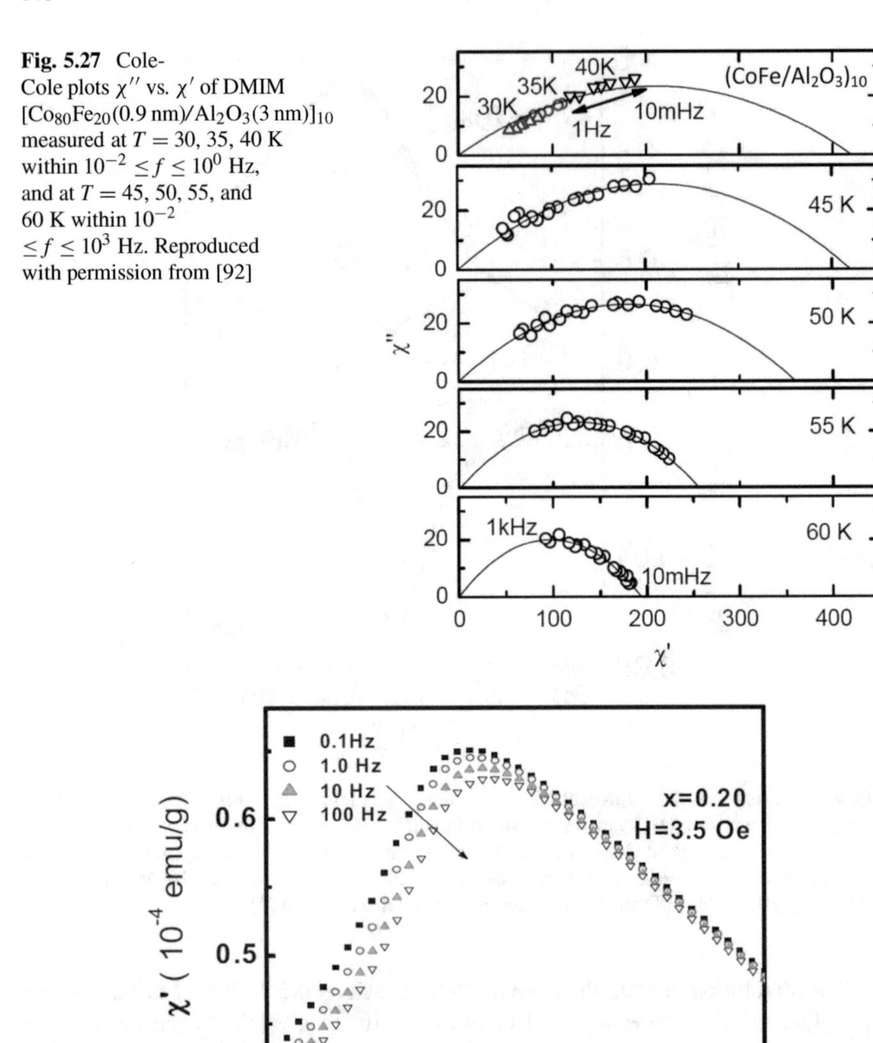

Fig. 5.27 Cole-Cole plots χ'' vs. χ' of DMIM $[Co_{80}Fe_{20}(0.9\ nm)/Al_2O_3(3\ nm)]_{10}$ measured at $T = 30, 35, 40$ K within $10^{-2} \le f \le 10^0$ Hz, and at $T = 45, 50, 55$, and 60 K within $10^{-2} \le f \le 10^3$ Hz. Reproduced with permission from [92]

Fig. 5.28 χ'_{ac} of $La_{0.7}Ca_{0.3}Mn_{0.8}Cd_{0.2}O_3$ within $28 \le T \le 68$ K for $10^{-1} \le f \le 10^2$ Hz and $H_{ac} = 3.5$ Oe. Reproduced with permission from [6]

cluster-type ferromagnetic order. This breakdown of ferromagnetic order can arise both from the mixed exchange and from the dilution by low spin diamagnetic Co^{3+} ions. A similar mechanism is assumed to work in the above $La_{0.7}Ca_{0.3}Mn_{0.8}Cd_{0.2}O_3$ system.

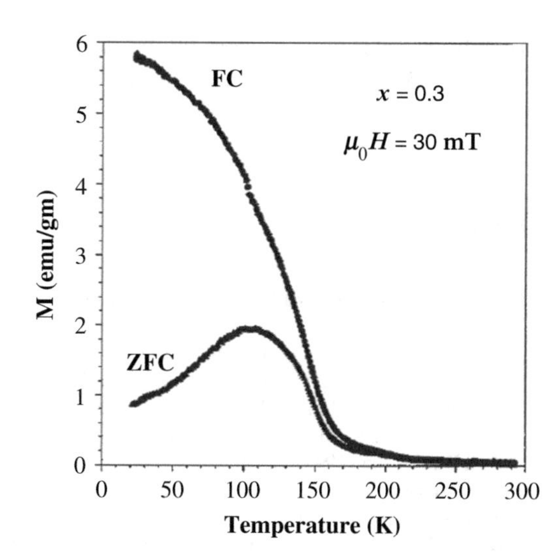

Fig. 5.29 FC and ZFC magnetization of $La_{0.7}Ca_{0.3}Mn_{0.7}Co_{0.3}O_3$ measured in $B = 30$ mT. Reproduced with permission from [94]

Although nanodomain structures have not yet been reported in the $La_{0.7}Ca_{0.3}Mn_{0.8}M_{0.2}O_3$ cluster glass and precursor phases (M = Cd and Co, respectively), it appears highly probable that they do exist and are preceded by magnetic tweed patterns above T_g similarly as in melt-spun magnetic Ni_2(Mn,Fe)Ga Heusler alloys [95]. These were recently predicted as structural and precursory tweed in shape memory Heusler alloys, $Ni_{50-x}Co_xMn_{39}Sn_{11}$, at the upper bound of the Co concentration, $0 \leq x \leq 10$ [96].

5.6 Conclusion

Dipolarly interacting magnetic nanoparticles in an insulating matrix paved the way toward the field of "*supermagnetism*." Its first manifestation was a system of nanosized particles of amorphous $Fe_{0.78}C_{0.22}$ under the name "*super spin-glass*" [97]. Later on the modified name "*superspin glass*" [98] became commonly accepted in order to pinpoint the main actor of the physics involved: "*superspin*" denoting the nanoparticular magnetic moment. The physics of "*supermagnetism*" has meanwhile become standard as manifested by recent reviews [88, 99].

This does not yet fully apply to "*ferroic glasses*" [1], which involve nontrivial interactions between point defects and matrix. They are mesoscopic cluster glasses, where "*point defects*" trigger nanoscale glassy freezing in martensitic "*strain glass*" [2] or charge disorder initiates "*polar nanoregions*" condensing into "*superdipolar glass*" in relaxors [8], while "*cluster spin glass*" emerges from complex spin and charge frustration in dilute magnetic materials [6]. The mechanisms of martensitic shear, electric random field polarity, or double-exchange and competing magnetic instabilities, respectively, need thorough experience for adequate understanding. These mechanisms are only partly deciphered, where most questions seem to remain open for the "*cluster spin glass*" in CMR materials.

References

1. X. Ren, Y. Wang, K. Otsuka, P. Lloveras, T. Castán, M. Porta, A. Planes, A. Saxena, MRS Bull. **34**, 838 (2009)
2. X. Ren, Phys. Status Solidi a **251**, 1982 (2014)
3. T. Sarkar, X. Ren, K. Otsuka, Phys. Rev. Lett. **95**, 205702 (2005)
4. Y. Murakami, D. Shindo, K. Oikawa, R. Kainuma, K. Ishida, Acta Mater. **53**, 2173 (2002)
5. N. Gayathri, A.K. Raychaudhuri, S.K. Tiwary, R. Gundakaram, A. Arulraj, C.N.R. Rao, Phys. Rev. B **56**, 1345 (1997)
6. S. Karmakar, S. Taran, B.K. Chaudhuri, H. Sakata, C.P. Sun, C.L. Huang, H.D. Yang, Phys. Rev. B **74**, 104407 (2006)
7. D. Viehland, J.F. Li, S.J. Jang, L.E. Cross, M. Wuttig, Phys. Rev. B **46**, 8013 (1992)
8. W. Kleemann, Phys. Status Solidi a **251**, 1993 (2014)
9. G.A. Smolenskii, V.A. Isupov, Dokl. Acad. Nauk SSSR **97**, 653 (1954)
10. Z.-G. Ye, B. Noheda, M. Dong, D. Cox, G. Shirane, Phys. Rev. B **64**, 184114 (2001)
11. J.C. Slater, Phys. Rev. **78**, 748 (1950)
12. W. Cochran, Phys. Rev. Lett. **3**, 412 (1959)
13. D. Viehland, J.F. Li, S.J. Jang, L.E. Cross, M. Wuttig, Phys. Rev. B **43**, 8316 (1991)
14. V. Westphal, W. Kleemann, M.D. Glinchuk, Phys. Rev. Lett. **68**, 847 (1992)
15. R. Blinc, J. Dolinsek, A. Gregorovic, B. Zalar, C. Filipic, Z. Kutnjak, A. Levstik, R. Pirc, Phys. Rev. Lett. **83**, 424 (1999)
16. W. Kleemann, J. Dec, R. Blinc, B. Zalar, R. Pankrath, Ferroelectrics **267**, 157 (2002)
17. A.A. Bokov, Z.-G. Ye, J. Mater. Sci. **41**, 31 (2006)
18. N. Novak, R. Pirc, M. Wencka, Z. Kutnjak, Phys. Rev. Lett. **109**, 037601 (2012)
19. C.W. Ahn, C.-H. Hong, B.Y. Choi, H.P. Kim, H.-S. Han, Y. Hwang, W. Jo, K. Wang, J.-F. Li, J.-S. Lee, I.W. Kim, J. Korean Phys. Soc. **68**, 1481 (2016)
20. L.E. Cross, Ferroelectrics **76**, 241 (1987)
21. W. Kleemann, J. Dec, Phys. Rev. B **94**, 174203 (2016)
22. A.T. Ogielski, Phys. Rev. B **32**, 7384 (1985)
23. I.K. Jeong, T.W. Darling, J.K. Lee, T. Proffen, R.H. Heffner, J.S. Park, K.S. Hong, W. Dmowski, T. Egami, Phys. Rev. Lett. **94**, 147602 (2005)
24. J.A. Quilliam, S. Meng, C.G.A. Mugford, J.B. Kycia, Phys. Rev. Lett. **101**, 187204 (2008)
25. A.K. Tagantsev, Phys. Rev. Lett. **72**, 1100 (1994)
26. D. Viehland, S.L. Jang, L.E. Cross, M. Wuttig, J. Appl. Phys. **68**, 2916 (1990)
27. Z.G. Lu, G. Calvarin, Phys. Rev. B **51**, 2694 (1995)
28. K. Binder, J.D. Reger, Adv. Phys. **41**, 547 (1992)
29. K. Binder, A.P. Young, Rev. Mod. Phys. **58**, 801 (1996)
30. A.J. Bray, M.A. Moore, Phys. Rev. Lett. **58**, 57 (1987)
31. S. Miga, J. Dec (unpublished)
32. H. Luo, G. Xu, H. Xu, P. Wang, Z. Yin, Jpn. J. Appl. Phys. **39**, 5581 (2000)
33. W. Kleemann, J. Advan. Dielectr. **2**, 124001 (2012)
34. I. Imry, S.K. Ma, Phys. Rev. Lett. **35**, 1399 (1975)
35. D.S. Fisher, Phys. Rev. Lett. **56**, 416 (1986)
36. R.A. Cowley, S.N. Gvasaliya, S.G. Lushnikov, B. Roessli, G.M. Rotaru, Adv. Phys. **60**, 229 (2011)
37. J.R. Arce-Gamboa, G.G. Guzmán-Verri, NPJ Quantum Mater. **2**, 28 (2017)

38. D.S. Fu, H. Taniguchi, M. Itoh, S.-Y. Koshihara, N. Yamamoto, S. Mori, Phys. Rev. Lett. **103**, 207601 (2009)
39. D. Sherrington, Phys. Rev. B **88**, 064105 (2014)
40. B. Hehlen, M. Al-Sabbagh, A. Al-Sein, J. Hlinka, Phys. Rev. Lett. **117**, 155501 (2016)
41. B. Dkhil, P. Gemeiner, A. Al-Barakaty, L. Bellaiche, E. Dul'kin, E. Mojaev, M. Roth, Phys. Rev. B **80**, 064103 (2009)
42. M.E. Manley, J.W. Lynn, D.L. Abernathy, E.D. Specht, O. Delaire, A.R. Bishop, R. Sahul, J.D. Budai, Nat. Commun. **5**, 3783 (2014)
43. A. Bussmann-Holder, A.R. Bishop, T. Egami, Europhys. Lett. **71**, 249 (2005)
44. V.V. Shvartsman, W. Kleemann, D. Kiselev, I. K. Bdikin, and A. L. Kholkin, in: *Scanning Probe Microscopy of Functional Materials: Nanoscale Imaging and Spectroscopy*, eds. By A. Gruverman, S.V. Kalinin, Chap. 12 (Springer, Berlin, 2011), p. 345
45. W. Kleemann, J. Mater. Sci. **41**, 129 (2006)
46. B.P. Burton, D. B. Gopman, G. Dogan, E. Cockayne, S. Hood, arXiv 1612.00398 (2016)
47. G. Xu, G. Shirane, J.R.D. Copley, P.M. Gehring, Phys. Rev. B **69**, 064112 (2004)
48. O. Svitelskiy, J. Toulouse, Z.-G. Ye, Phys. Rev. B **68**, 104107 (2003)
49. Y. Moriya, H. Kawaji, T. Tojo, T. Atake, Phys. Rev. Lett. **90**, 205901 (2003)
50. M.E. Manley, D.L. Abernathy, R. Sahul, D.E. Parshall, J.W. Lynn, A.D. Christianson, P.J. Stonaha, E.D. Specht, J.D. Budai, Sci. Advan. **2**, e1501814 (2016)
51. F. Li, S. Zhang, T. Yang, Z. Xu, N. Zhang, G. Liu, J. Wang, J. Wang, Z. Cheng, Z.-G. Ye, J. Luo, T.R. Shrout, L.-Q. Chen, Nat. Commun. **7**, 13807 (2016)
52. Y. Uesu, H. Tazawa, K. Fujishiro, Y. Yamada, J. Korean Phys. Soc. **29**, S703 (1996)
53. E.J. Garboczi, K.A. Snyder, J.F. Douglas, M.F. Thorpe, Phys. Rev. E **52**, 819 (1995)
54. D. Fu, H. Taniguchi, M. Itoh, S. Mori, in: *Advan. Ferroelectrics*, Ch. 3 (InTech, 2012), p. 51
55. Y. Uesu, H. Tazawa, K. Fujishiro, Y. Yamada, J. Korean Phys. Soc. **29**, S703 (1996)
56. T. Braun, W. Kleemann, J. Dec, P.A. Thomas, Phys. Rev. Lett. **94**, 117601 (2005)
57. W. Kleemann, Annu. Rev. Mater. Res. **37**, 415 (2007)
58. A. Levstik, Z. Kutnjak, C. Filipic, R. Pirc, Phys. Rev. B **57**, 11204 (1998)
59. U. Bianchi, J. Dec, W. Kleemann, J.G. Bednorz, Phys. Rev. B **51**, 8737 (1995)
60. Z.G. Lu, G. Calvarin, Phys. Rev. B **51**, 2694 (1995)
61. R.P. Lacroix, G.J. Béné, Arch. Sci. **4**, 430 (1931)
62. V.V. Shvartsman, J. Dec, T. Łukasiewicz, A.L. Kholkin, W. Kleemann, Ferroelectrics **373**, 77 (2008)
63. D.W. Davidson, R.H. Cole, J. Chem. Phys. **18**, 1417 (1950)
64. S. Havriliak, S. Negami, Polymer **8**, 161 (1967)
65. T. Nattermann, S. Stepanow, L.H. Tang, H. Leschhorn, J. Phys. II **2**, 483 (1992)
66. A. Koreeda, H. Taniguchi, S. Saikan, M. Itoh, Phys. Rev. Lett. **109**, 197601 (2012)
67. H. Taniguchi, M. Itoh, D. Fu, J. Raman Spectrosc. **42**, 706 (2011)
68. S. Wakimoto, C. Stock, R.J. Birgeneau, Z.-G. Ye, W. Chen, W.J.L. Buyers, P. Gehring, G. Shirane, Phys. Rev. B **65**, 172105 (2002)
69. M. Ulex, R. Pankrath, K. Betzler, J. Cryst. Growth **271**, 128 (2004)
70. V.V. Shvartsman, W. Kleemann, T. Lukasiewicz, J. Dec, Phys. Rev. B **77**, 054105 (2008)
71. M. Aftabuzzaman, M. A. Helal, R. Paszkowski, J. Dec, W. Kleemann, and S. Kojima, Sci. Rept. **7**, 11615 (2017)
72. P.B. Jamieson, S.C. Abrahams, J.L. Bernstein, J. Chem. Phys. **48**, 5048 (1968)
73. J. Dec, W. Kleemann, V.V. Shvartsman, D.C. Lupascu, T. Łukasiewicz, Appl. Phys. Lett. **100**, 052903 (2012)
74. E. Dul'kin, S. Kojima, M. Roth, J. Appl. Phys. **110**, 044106 (2011)
75. J. Esser, U. Nowak, K.D. Usadel, Phys. Rev. B **55**, 5866 (1997)
76. W. Kleemann, J. Dec, S.A. Prosandeev, T. Braun, P.A. Thomas, Ferroelectrics **334**, 3 (2006)
77. J. Dec, W. Kleemann, S. Miga, V.V. Shvartsman, T. Łukasiewicz, M. Swirkowicz, Phase Transit. **80**, 131 (2007)
78. W. Kleemann, In: Mesoscopic Phenomena in Multifunctional Materials, A. Saxena, A. Planes (eds.). Chap. 10, Springer Ser. Mater. Sci. **198**, 249 (2014)

79. X. Ren, Y. Wang, Y. Zhou, Z. Zhang, D. Wang, G. Fan, K. Otsuka, T. Suzuki, Y. Ji, J. Zhang, Philos. Mag. **90**, 141 (2010)
80. Y. Wang, X. Ren, K. Otsuka, A. Saxena, Phys. Rev. B **76**, 132201 (2007)
81. Y. Zhou, D. Xue, Y. Tian, X. Ding, S. Guo, K. Otsuka, J. Sun, X. Ren, Phys. Rev. Lett. **112**, 025701 (2014)
82. L. Zhang, D. Wang, X. Ren, Y. Wang, Sci. Rep. **5**, 11477 (2015)
83. J. Cui, X. Ren, Appl. Phys. Lett. **105**, 061904 (2014)
84. C.E. Guillaume, Proc. Phys. Soc. Lond. **32**, 374 (1919). https://www.nobelprize.org/nobel_prizes/physics/laureates/1920/ guillaume-lecture.html
85. P. Lloveras, T. Castán, M. Porta, A. Planes, A. Saxena, Phys. Rev. Lett. **100**, 165707 (2008)
86. P. Lloveras, T. Castán, M. Porta, A. Planes, A. Saxena, Phys. Rev. B **80**, 054107 (2009)
87. C. Djurberg, P. Svedlindh, P. Nordblad, M.F. Hansen, F. Bødker, S. Mørup, Phys. Rev. Lett. **79**, 5154 (1997)
88. S. Bedanta, W. Kleemann, J. Phys. D. Appl. Phys. **42**, 013001 (2009)
89. S. Sahoo, O. Petracic, W. Kleemann, S. Stappert, G. Dumpich, P. Nordblad, S. Cardoso, P.P. Freitas, Appl. Phys. Lett. **82**, 4116 (2003)
90. S. Sahoo, Ph.D. thesis, University Duisburg-Essen (2003)
91. S. Sahoo, O. Petracic, W. Kleemann, P. Nordblad, S. Cardoso, and P. P. Freitas, J. Magn. Magn. Mater. **272–276**, 1316 (2004)
92. O. Petracic, S. Sahoo, C. Binek, W. Kleemann, J.B. Sousa, S. Cardoso, P.P. Freitas, Phase Transit. **76**, 367 (2003)
93. A.K. Jonscher, *Dielectric Relaxation in Solids* (Chelsea Dielectrics, London, 1983)
94. N. Gayathri, A.K. Raychaudhuri, S.K. Tiwary, R. Gundakaram, A. Arulraj, C.N.R. Rao, Phys. Rev. B **56**, 1345 (1997)
95. R.V.S. Prasad, M. Manivel Raja, G. Phanikumar, Adv. Mater. Res. **74**, 215 (2009)
96. P. Entel, M.E. Gruner, D. Comtesse, V.V. Sokolovskiy, V.D. Buchelnikov, Physica Status Solidi B **251**, 2135 (2014)
97. C. Djurberg, P. Svedlindh, P. Nordblad, M.F. Hansen, F. Bødker, S. Mørup, Phys. Rev. Lett. **79**, 5154 (1997)
98. W. Kleemann, O. Petracic, C. Binek, G.N. Kakazei, Y.G. Pogorelov, J.B. Sousa, S. Cardoso, P.P. Freitas, Phys. Rev. B **63**, 132423 (2001)
99. S. Bedanta, O. Petracic, W. Kleemann, in *Handbook of Magnetic Materials*, vol. 23, Chap. 10, ed. By K.F.H. Buschow (Elsevier, Amsterdam2014), pp. 1–83

Chapter 6
Probing Glassiness in Heuslers via Density Functional Theory Calculations

P. Entel, M. E. Gruner, M. Acet, A. Hucht, A. Çakır, R. Arróyave, I. Karaman, T. C. Duong, A. Talapatra, N. M. Bruno, D. Salas, S. Mankovsky, L. Sandratskii, T. Gottschall, O. Gutfleisch, S. Sahoo, S. Fähler, P. Lázpita, V. A. Chernenko, J. M. Barandiaran, V. D. Buchelnikov, V. V. Sokolovskiy, T. Lookman, and X. Ren

Abstract Heusler compounds and alloys form a unique class of intermetallic systems with functional properties interfering with basic questions of fundamental aspects of materials science. Among the functional properties, the magnetic shape memory behavior (Planes et al., J Phys: Condens Matter 21:233201 (29 pp), 2009) and the ferrocaloric effects like the inverse magnetocaloric effect which is

---

P. Entel · M. E. Gruner (✉) · M. Acet · A. Hucht
Faculty of Physics and CENIDE, University of Duisburg-Essen, Duisburg, Germany
e-mail: entel@thp.uni-due.de; markus.gruner@uni-due.de; mehmet.acet@uni-due.de; fred@thp.uni-due.de

A. Çakır
Muğla Üniversitesi, Metalurji ve Malzeme Mühendisliği Bölümü, Muğla, Turkey
e-mail: cakir@mu.edu.tr

R. Arróyave · I. Karaman · T. C. Duong · A. Talapatra · N. M. Bruno · D. Salas
Department of Materials Science & Engineering, A&M University, College Station, TX, USA
e-mail: raymundo@fastmail.fm; ikaraman@tamu.edu; terryduong84@tamu.edu; anjanatalapatra@tamu.edu; dsm073@tamu.edu

S. Mankovsky
Department Chemie, Ludwig-Maximilian-University Munich, Munich, Germany
e-mail: Sergiy.Mankovskyy@cup.uni-muenchen.de

L. Sandratskii
Max-Planck-Institut für Mikrostrukturphysik, Halle, Germany
e-mail: leonid.sandratskii@mpi-halle.mpg.de

T. Gottschall · O. Gutfleisch
Technical University Darmstadt, Institute of Materials Science, Darmstadt, Germany
e-mail: gottschall@fm.tu-darmstadt.de; gutfleisch@fm.tu-darmstadt.de

S. Sahoo
Institute of Materials Science, University of Connecticut, Storrs, CT, USA
e-mail: sanjubala.sahoo@uconn.edu

T. Lookman, X. Ren (eds.), *Frustrated Materials and Ferroic Glasses*, Springer Series in Materials Science 275, https://doi.org/10.1007/978-3-319-96914-5_6

associated with the first order magnetostructural transformation with a jump-like change of the magnetization with lowering of temperature (Acet et al., Magnetic-field-induced effects in martensitic Heusler-based magnetic shape memory alloys. In: Bushow KHJ (ed) Handbook of magnetic materials, vol 19. North-Holland, Amsterdam, pp 231–289, 2011) have been intensively investigated in various reviews. Important references can be found in Acet et al. (Magnetic-field-induced effects in martensitc Heusler-based magnetic shape memory alloys. In: Bushow KHJ (ed) Handbook of magnetic materials, vol 19. North-Holland, Amsterdam, pp 231–289, 2011). Besides magnetocaloric effects, other ferroic cooling mechanisms of Heuslers (electrocaloric, barocaloric, and elastocaloric ones) have recently been discussed by Xavier Moya et al. (Nat Mater 13:439–450, 2014). A discussion of caloric effects in ferroic materials including a brief discussion of the importance of correlating time and length scales can be found in Fähler et al. (Adv Eng Mater 14:10–19, 2012). In the present article, we emphasize this item further by showing that, in particular, the physics at different time scales leads to markedly different properties of the Heusler materials. "Rapidly quenched" alloys behave differently from "less rapidly quenched" alloys. In the latter case, the so-called magnetostructural transformation may vanish altogether because of segregation of the alloys into the stoichiometric $L2_1$ Heusler phase and $L1_0$ Ni-Mn occurs. We argue that this tendency for segregation is at the origin of glassiness in Heuslers.

S. Fähler
IFW Dresden, Dresden, Germany
e-mail: s.faehler@ifw-dresden.de

P. Lázpita · V. A. Chernenko · J. M. Barandiaran
BCMaterials and Department of Electricity and Electronics, University of Basque Country (UPV/EHU), Bilbao, Spain
e-mail: patricia.lazpita@ehu.eus; volodymyr.chernenk@ehu.eus; manu@bcmaterials.net

V. D. Buchelnikov · V. V. Sokolovskiy
Condensed Matter Physics Department, Chelyabinsk State University, Chelyabinsk, Russia
e-mail: buche@csu.ru

T. Lookman
Theoretical Division, Los Alamos National Laboratory, Los Alamos, NM, USA
e-mail: txl@lanl.gov

X. Ren
Frontier Institute of Science and Technology and State Key Laboratory for Mechanical Behaviour of Materials, Xi'an Jiaotong University, Xi'an, China

Center for Functional Materials, National Institute for Materials Science, Tsukuba, Ibaraki, Japan
e-mail: ren.xiaobing@nims.go.jp

6.1 Introduction

Glassiness near martensitic transformations has been observed experimentally in intermetallic alloys where precursor effects above the martensitic transformation determine the ferroelastic properties. The disordered nature of the alloys or disorder arising from impurities may lead to a disordered array of strain fields (as for the systems studied in [1, 2]). For an overview on strain-glass effects in ferroelastic systems, where premartensitic tweed competes with strain glass, see Ren et al. [3]. The exciting point is that the ferroelastic systems may show magnetic order like the ternary intermetallic Heusler alloys and that the resulting strain glass exhibits the same competing ferromagnetic–antiferromagnetic interactions as bulk austenite and martensite.

This article describes the physical properties of rapidly quenched magnetic Heusler alloys with frozen compositional disorder and competing magnetic interactions. These emerge as a consequence of the excess Mn atoms interacting ferromagnetically because of the occupation of atomic sites in the original Mn sublattice and interacting antiferromagnetically because of the occupation of the excess Mn of atomic sites of the Z element in Ni-Mn-Z with Z = Al, Ga, In, Sn, Sb. In contrast to the rapidly quenched alloys we describe also the physics of the less rapidly quenched alloys where the temper-annealed samples of $Ni_{50}Mn_{45}$(Al, Ga In, Sn)$_5$ are not stable but decompose into ferromagnetic $L2_1$ Heusler Ni_2Mn(Al, Ga, In, Sn) and antiferromagnetic $L1_0$ NiMn. This type of segregation leads to new functional properties as the quenching under magnetic fields generates ferromagnetic Heusler-type precipitates with a paramagnetic core. These precipitates affect the magnetic response in a peculiar way by shifting the magnetization loops vertically [4]. Shell-ferromagnetism, noncollinear magnetism, and skyrmions may lead to new exciting functional properties of Heusler alloys [4–7].

The prototype magnetic Heusler system Ni_2MnGa consists of four interpenetrating fcc lattices with a phase transition to a tetragonal structure upon cooling below 202 K [8] and a magnetic transition to ferromagnetic order (where experimental and calculated Curie temperatures, using ab initio exchange coupling constants, are close to each other, $T_C \approx 376$ K [8, 9]). Phonon softening in Heuslers underlines the importance of Fermi surface nesting to explain the origin of martensitic instabilities [10]. With respect to shape memory properties, the reorientation of the tetragonal unit cell can be induced either by a magnetic field or by mechanical stress [11–15]. In this context it is important to note that the coexistence of austenite, 14M phase and tetragonal martensite in Ni-Mn-Ga and other Heusler alloys as well as the presence of adaptive martensite in magnetic shape memory alloys which follows from elastic energy minimization are beneficial for the shape memory properties [16, 17] and ferroic cooling [35, 36, 86, 87]. Furthermore, the modulated structure is a nanoscale microstructure of non-modulated martensite [89].

For the quenched magnetic Heusler alloys Ni-Mn-Ga and Ni-Mn-Sn with Mn excess Fig. 6.1 shows the typical behavior of the critical temperatures (T_C and M_s) versus electron concentration with austenite in the $L2_1$ structure and martensite

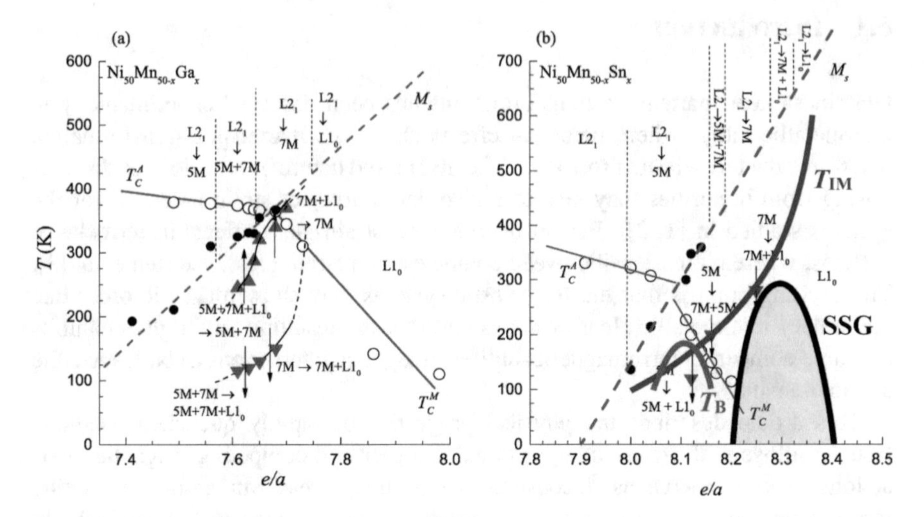

Fig. 6.1 Phase diagrams of (**a**) $Ni_{50}Mn_{50-x}Ga_x$ and (**b**) $Ni_{50}Mn_{50-x}Sn_x$ with austenite-martensite transitions (filled circles) and intermartensitic transitions (triangles up and down). Open circles mark the Curie temperatures of austenite and martensite. T_{IM}, SSG, and T_B mark the intermartensitic transition line, the super-spin-glass region, and the blocking temperature, respectively. The experimental phase diagram has been adapted from Çakır et al., copyright Elsevier (2015) [1]

in the $L1_0$ structure and intermartensitic modulated structures. Besides spin-glass phases we also find strain-glass phases in most of the Mn excess region, where the strain-glass emerges because of local disorder beyond a certain threshold.

An example for a strain-glass phase in Ni-Mn-Ga alloys with Co as impurity is shown in Fig. 6.2. We find a ferromagnetic strain-glass phase because the underlying intermetallic phase is ferromagnetic (FM austenite, FM martensite, and FM strain glass). The strain-glass phase arises because of the disorder induced frozen strain of the intermetallic alloy and the additional local strains arising from the Co impurities, which destroys the long-range strain features of martensite. This phenomenon leads to a ferromagnetic strain glass with coexisting short range strain ordering and long-range ordering of the magnetic moments, where Co essentially suppresses the long-range strain ordering of martensite and enhances the ferromagnetic exchange [2].

We will see below that the segregation into an antiferromagnetic background matrix of NiMn and precipitates of ferromagnetic $L2_1$ Heuslers is also a non-equilibrium phenomenon which leads to a mixture of complex nanophase material. This, however, is not beneficial for the magnetocaloric effect, since the jump of the magnetization at the magnetostructural phase transition can be considerably reduced.

We first investigate the trend of the intermetallic alloys and compounds of the rapidly quenched Heusler alloys for noncollinear magnetic features and tendencies to form skyrmions like magnetic excitations.

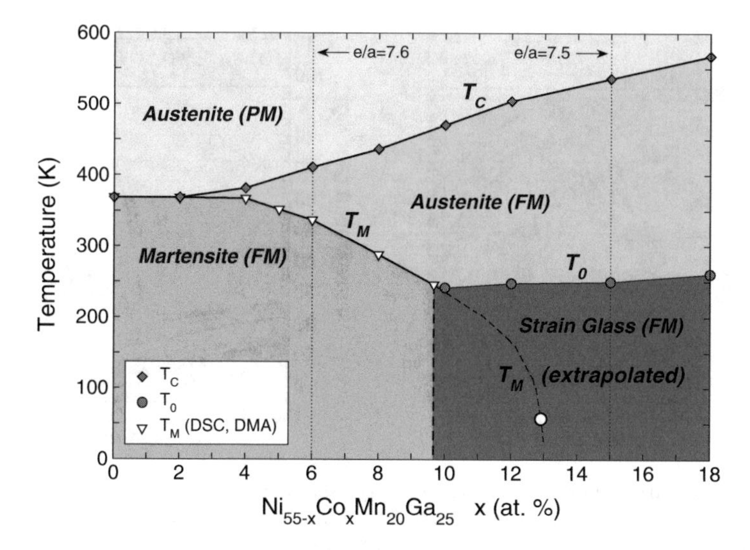

Fig. 6.2 Phase diagram of $Ni_{55-x}Co_xMn_{20}Ga_{25}$ showing the paramagnetic and ferromagnetic austenitic phases, ferromagnetic martensite as well as the ferromagnetic strain-glass phase (calculation of the extrapolated T_M is also shown). Ferromagnetic martensite exists to rather high concentrations of Co impurities and the strain-glass phase exists between 10% Co and more than 18% Co. Phase diagram adapted from Wang et al. [2]

6.2 Magnetostructural Phase Transition of Rapidly Quenched Heusler Alloys

Figure 6.3 shows the structural transformation between the high-temperature ferromagnetic austenite and the low-temperature weak magnetic, antiferromagnetic or paramagnetic martensite as well as the shift of the martensitic transformation, with applied magnetic field in $Ni_{50}Mn_{34.5}In_{15.5}$ (experimental data provided by P. Lazpita [18]).

Note that the isofield magnetization curves of Fig. 6.3 are not the result of a very fast quenching of the alloy, since some partial order prevails although the degree of order retained has never been specified in detail [18].

The occurrence of the isothermally magnetic-field induced transformation at different temperatures is evident and allows to discuss magnetization and entropy changes, ΔM and ΔS, as a function of the applied field as predicted by the Clausius-Clapeyron equation $dT/dH = \mu_0 \, \Delta M/\Delta S$. The direct martensitic transformation is accompanied by a drastic drop in magnetization, favoring the magnetic field-induced reverse transformation, demonstrating the metamagnetic behavior.

In disordered $Ni_{50}Mn_{34.5}In_{15.5}$ alloy (rapid quenching) the jump-like curve of ΔM across the martensitic transition is steeper and increases with field while it remains nearly constant for the more ordered sample (slower cooling) [19].

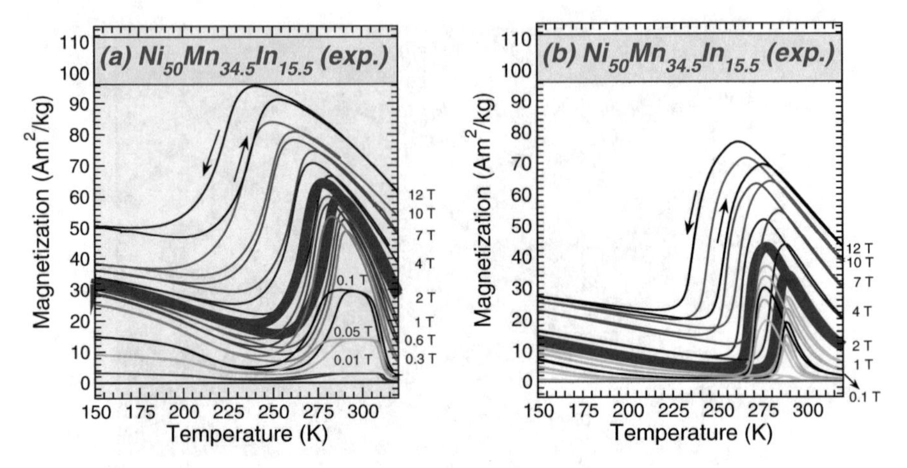

Fig. 6.3 Magnetization curves of $Ni_{50}Mn_{34.5}In_{15.5}$ showing first-order magnetostructural transformation from weak magnetic martensite to ferromagnetic austenite. The $M(T)$ data from more ordered alloys (**a**) are very different from disordered alloys shown in (**b**). We think that this could be related to the order and its influence on the Curie temperature. For both cases, the shift of the martensitic transformation with applied field is clearly visible. Data provided by P. Lazpita Arizmendiarreta. (**a**) Copyright AIP 2013 [18], (**b**) copyright IOP 2013 [19]. With increasing magnetic field the jump is reduced in both cases and vanishes for very large fields (saturation magnetization) or the transformation can become kinetically arrested, when the driving force $\Delta G \approx \Delta S \times \Delta T$ becomes smaller and smaller

Hence, the change of slope of $dT_M/\mu_0 dH = \Delta M/\Delta S$ may be related to the decrease of ΔS with applied field. However, for both, more ordered and more disordered alloys, the entropy change amounts to $\Delta S \approx 5$ J/(kg K), which originates mostly from the lattice contribution being larger than the magnetic entropy change across the magnetostructural transformation [20].

With respect to the theoretical modeling of magnetostructural transition, we have adopted a combined effort consisting of ab initio modeling of the complex scenario of magnetic exchange interactions in Ni-Co-Mn-In combined with Monte Carlo simulations of an effective spin model (Potts model for the multi-spin interactions in Heusler alloys), where the exchange integrals serve as input.

The Hamiltonian (1) consists of the Potts model in Eq. (6.2), the coupling to the structural (elastic) component which allows martensitic transformation with the help of Eq. (6.3), and the magnetoelastic interaction (in Eq. (6.4)):

$$\mathscr{H} = \mathscr{H}_{Potts} + \mathscr{H}_{elastic} + \mathscr{H}_{magneto-elastic} \tag{6.1}$$

with

$$\mathscr{H}_{Potts} = \sum_{\langle ij \rangle} J_{ij} \delta_{S_i,S_j} - g\mu_B H_{ext} \sum_i \mu_i \delta_{S_i,S_g} - K_{ani} \sum_i \mu_i^2 \delta_{S_i,S_k}, \tag{6.2}$$

$$\mathscr{H}_{elastic} = -\left(J + K_1 g \mu_B H_{ext} \sum_i \mu_i \delta_{S_i,S_g}\right) \sum_{\langle ij \rangle} \sigma_i \sigma_j - K \sum_{\langle ij \rangle} (1-\sigma_i^2)(1-\sigma_j^2) - k_B T \ln(p) \sum_i (1-\sigma_i^2), \quad (6.3)$$

$$\mathscr{H}_{magneto-elastic} = 2 \sum_{\langle ij \rangle} U_{ij} \mu_i \mu_j \delta_{S_i,S_j} \{(\tfrac{1}{2} - \sigma_i^2)(\tfrac{1}{2} - \sigma_j^2) - \tfrac{1}{4}\}. \quad (6.4)$$

The J_{ij} are the magnetic exchange parameters,

$$J_{ij} = \frac{1}{4\pi} \int_{-\infty}^{\epsilon_F} dE \, \mathrm{Im\,Tr}[\Delta_i \tau_\uparrow^{ij} \tau_\downarrow^{ji}], \quad (6.5)$$

calculated with the Munich SPR-KKR package [21]. We make use of them in the Monte Carlo simulations of the Potts model [22–24]. In order to describe the magnetostructural transformation, we use the Blume-Emery-Griffiths (BEG) model following the work of Castán [25]. Since mapping of magnetic ab initio energies is onto J_{ij} and unit length of spins, field terms must include explicitly μ_i and μ_i^2, where μ_i is the ab initio value of the magnetization of the atom at site i taken to be dimensionless. J and K are the elastic and K_1 (U_{ij}) the magnetoelastic interaction parameters. The Kronecker symbol restricts the spin–spin interactions to those between the same Potts-q states. The spin moment of Mn is $S = \frac{5}{2}$ and we identify the $2S + 1$ spin projections with $q_{Mn} = 1 \ldots 6$. Likewise, we assume $S = 1$ for Ni and $S = \frac{3}{2}$ for Co. The BEG model defines $\sigma_i = 0, \pm 1$ for austenite and two martensite variants, respectively [22–25]. The model allows first-order martensitic phase transformation with thermal hysteresis. Because of the magnetoelastic coupling term, the jump $\Delta\mu(T_m)$ is coupled to the martensitic transformation and exhibits hysteresis as well. For further technical details, see [22–24]. A recent summary of novel experimental achievements of the magnetocaloric effect can be found in [26]. We find that both conventional and inverse magnetocaloric effects are well reproduced by this model and even allows a quantitative description of the effect.

Note that spin-glass effects and strain-glass-effects can also be described by using the same model Hamiltonian and ab initio description of magnetic interactions for the atomically disordered Heusler alloys. This is in progress, but, the simulations require large supercells to treat disorder adequately. Nevertheless spin-glass effects and zero-field cooling and field-cooling protocols of Ni-Mn-In alloys have been successfully simulated using a standard Heisenberg model [27].

Figure 6.4 shows typical results of Monte Carlo simulations obtained for the magnetostructural transition for a series of Ni-Mn-In based alloys using the model defined by Eq. (6.1) with large magnetocaloric effects (MCE) [28].

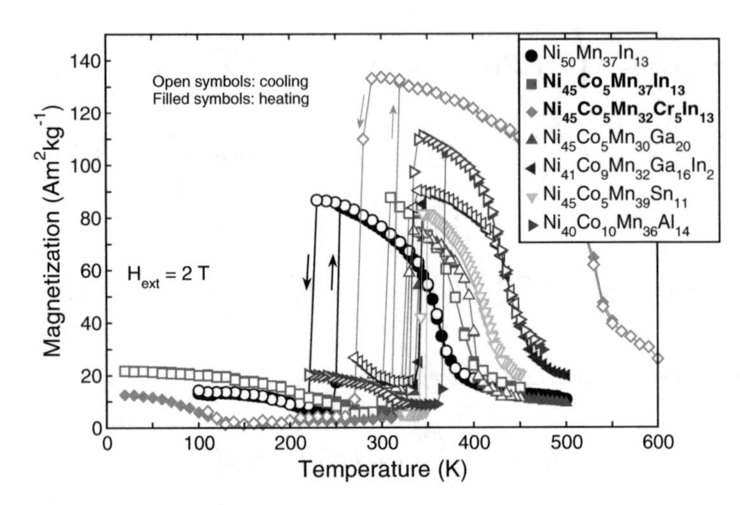

Fig. 6.4 Results of Monte Carlo simulations for a series of Ni-Mn-In alloys. The magnetostructural transition shows hysteresis where the hysteresis width is in most cases of the same order although the jump $\Delta M(T)$ is different. The largest jump is shown when Co and Cr are added to Ni-Mn-In. This alloy also exhibits the largest MCE which is of the order of $\Delta T_{ad} \approx 10\,\mathrm{K}$ [28]. Reprinted from Ref. [28]. © (2015) by the American Physical Society

Note that the magnetostructural transitions shown in Fig. 6.4 correspond to those of very rapidly quenched samples, since the standard atomic relaxations during the ab initio calculations do not consider diffusion or similar relaxation processes.

6.2.1 Order–Disorder Transitions and Classification of Phase Transitions

Since Heusler alloys undergo an ordering process between 1100 and 900 K when cooling from high temperatures, rapidly or less rapidly quenched alloys show different degrees of order and show time dependence effects of the nucleation of the microstructure depending on the details of the cooling process. This has never been discussed in deeper details for magnetic Heusler alloys by trying to relate, for example, the degree of order and atomic density function theory to the evolution of microstructure at the atomic scale as discussed by Khachaturyan [29] in relation to martensite transformation [30].

Also we do not want to go into details, most classification schemes of phase transitions use a simple thermodynamic concept for systems in thermal equilibrium, where the transition is a compromise of the energy, which tends to order, and the entropy associated to temperature which tends to break the order. Crystal chemistry then provides a basis for the classification of phase transitions at a critical temperature T_c where the system absorbs thermal energy leading to a higher internal energy of the transformed phase, the bonding between neighboring atoms is weaker in the low-temperature phase. This results in a change in the nature of mainly the first and second-nearest neighbor bonds. We distinguish three phase transitions. Displacive (martensitic) transitions with small distortion of the bonds. Characteristic is the group-subgroup relationship between the phases. Reconstructive transitions through the breaking of primary or secondary bonds and order–disorder transitions through substitution between atoms which is usually followed by small displacements commonly found in metals. Some systems keep a group-subgroup relationship, others are reconstructive. Now, the Heusler alloys considered here are mostly rapidly quenched materials showing features of martensitic transformations as Ni-Mn-Z with Z = (Al, Ga, In, Sn, Sb) with, for example, simple tetragonal transformation $L2_1 \rightarrow L1_0$, and disorder-order transformation from the completely disordered high-temperature A2 phase to partially ordered B2 to ordered $L2_1$: A2 → B2 → $L2_1$.

Since we will discuss below the case of segregation of Heusler alloys in case of slower cooling, where some partial order is kept, we have to envisage systems where structural and order–disorder phase transitions may overlap in some temperature or composition range. This requires a kind of unified model description of order–disorder and displacive structural phase transitions. Indeed such an approach has been formulated and applied to ferroelectric systems which show mixed type of phase transitions [31].

But to our knowledge, this concept unified model description has never been applied to metallic Heusler systems with order–disorder transition using some pseudo-spin formulation which coexists with a displacive transformation originating from some cooperative phonon mechanism which would yield a mixed type of phase transition, where the individual order parameters can be varied continuously depending on the coupling strength. Note, however, that for a one-dimensional model system with strong anharmonic displacement fields new solutions in the form of domain walls have been found [32], which is also some kind of order–disorder transition, where domain-like excitations induce the formation of microdomains which act as precursor-clusters of the ordered phase.

We think that this is an important point because the influence of atomic disorder on the martensitic transformation may also cause the appearance of strain-glass and spin-glass effects. We just argue further below that the nature of glassy effects is related to the existence of some kind of atomic disorder in the alloy and so, the alloys which show tendencies for segregation should also show tendencies for strain-glass behavior. And due to the existence of competing ferromagnetic and antiferromagnetic interaction spin-glass may appear quite naturally.

6.2.2 *Influence of Annealing Process on the Isofield Magnetization Curves*

The magnetostructural transformation and the form of the magnetization curve over the structural change at the martensitic transformation depend to a large degree on the annealing process. For high annealing temperature A_Q, for example, $A_Q \approx$ 1173 K and rapid quenching of $Ni_{50.2}Mn_{33.4}In_{16.4}$, the form of the magnetization curve is rectangular like and very steep, but with decreasing annealing temperature the rage of forward martensite transformation temperatures T_{FMT} and reverse martensite transformation temperatures T_{RMT} increases and thus the magnetization curves are smoothed [33]. But for high annealing temperature and rapid quenching, the form of the magnetization curve is nearly rectangular (in the case of a very low external field of 10 mT).

This becomes obvious if one compares further experimental results of the influence of the atomic order in Ni-Mn-In metamagnetic shape memory alloys on the martensitic transformation and magnetic transition by Recarte et al. and similar results observed by other authors. Recarte et al. considered $Ni_{50.2}Mn_{33.4}In_{16.4}$ polycrystalline alloy, which was homogenized at 1173 K. In order to retain states with different degrees of LRO (long-range order), the alloys were subjected to a 30 min annealing treatment at three different temperatures (samples labeled AQ 1173 K, AQ 823 K, and AQ 723 K in the Fig. 6.5), followed by quenching into ice water. Another piece of the same alloy was slowly cooled from 1173 K (labeled AQ 300 K) for comparison with the quenched samples. The temperature dependence of the low-field magnetization ($H = 0.1$ T) curves is displayed in Fig. 6.5.

A "quadratic-block" like form of magnetization curves was obtained recently for $Ni_{50}Mn_{35}In_{15}$ in a higher field of 0.5 T together with a splitting of ZFC, FC, and FH curves [34]. Note that such block like form of the magnetization curves is very common for the magnetic Heusler alloys after rapid quenching, references can be found in [35, 36]. For recent work on Ni-Mn-Sn alloy systems, see [37, 38].

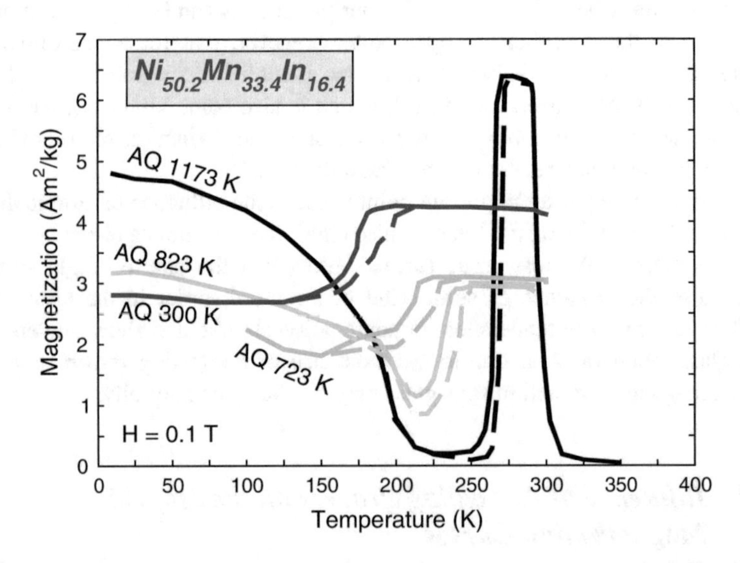

Fig. 6.5 Temperature dependence of the magnetization in a field of 0.1 T for the $Ni_{50.2}Mn_{33.4}In_{16.4}$ polycrystalline alloy subject to four different heat treatments labeled AQ 1173 K, AQ 823 K, AQ 723 K and AQ 300 K, which yield different degrees of order. The slowly cooled sample labeled AQ 300 K exhibits the largest degree of order (without further specifying the order here). Figure adapted from [33]. Copyright Elsevier (2012)

6.2.3 Effect of Cobalt on the Isofield Magnetization Curves

If Co is added to the rapidly quenched Ni-Mn-In alloys, the magnetostructural transformation does not vanish, but the transformation becomes steeper because of the enhanced ferromagnetic component due to Co, compare Fig. 6.6.

The magnetic nature has been recently studied in other Ni-Mn based metamagnetic systems, primarily for Co-added Ni-(Co)-Mn-Z with Z = In, Sn and Sb and various explanations of the magnetic behavior of martensite have been proposed: paramagnetism, antiferromagnetism, superparamagnetism, re-entrant spin-glass, super-spin glasses, etc. [35, 39–41]. Note that all spin-glass discussions automatically involve the possibility of strain-glass formation due to the coupling of the spins to the local strain fields arising from atomic disorder.

Superparamagnetic domains in a paramagnetic matrix have been shown to exist in the martensitic phase of $Ni_{45}Mn_{36.5}In_{13.5}Co_5$ evolving to a superspin glass on cooling below a critical temperature [42]. Superparamagnetic and superspin-glass behavior have also been observed in $Ni_{50-x}Co_xMn_{39}Sn_{11}$ ($0 \leq x \leq 10$), where the superparamagnetic state is formed by magnetic clusters distributed in a weak magnetic matrix, which has directly been confirmed by small-angle neutron scattering [43].

For completeness, we give reference to the work of Kainuma's and Chaddah's group which performed experiments like those shown in Fig. 6.6 some years before for Ni-Co-Mn-In alloys, see Fig. 6.7 [44–49]. Figs. 6.6 and 6.7 show results of rapidly quenched samples.

Note that the shift of the isofield magnetization curves over a temperature interval covering the magnetostructural transition in Figs. 6.6 and 6.7 has already been postulated by a 30-year-old model calculation of structural and magnetic

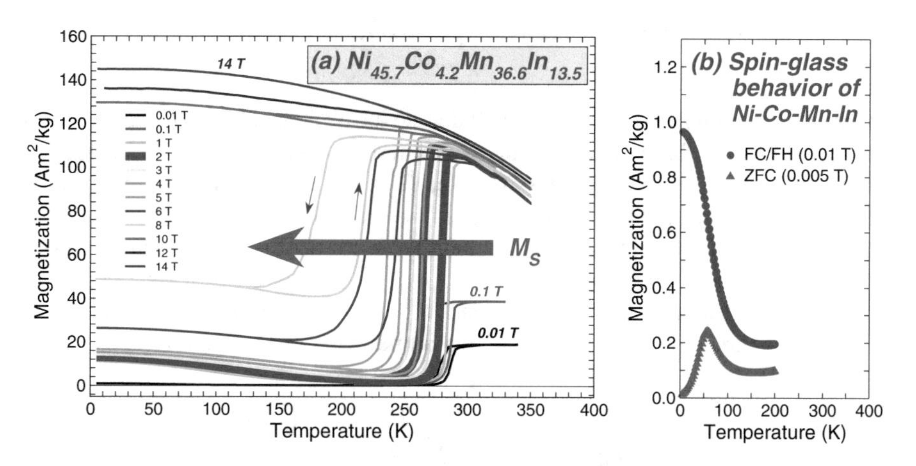

Fig. 6.6 (**a**) Shift of isofield $M(T)$ curves of Ni-Co-Mn-In with external magnetic field. For 14 T the magnetization curve is saturated. Figure (**b**) shows that spin-glass behavior emerges if the measuring protocol of the magnetization is done according to field cooling and field heating. Reprinted from Ref. [88]. © (2018) by Wiley-VCH

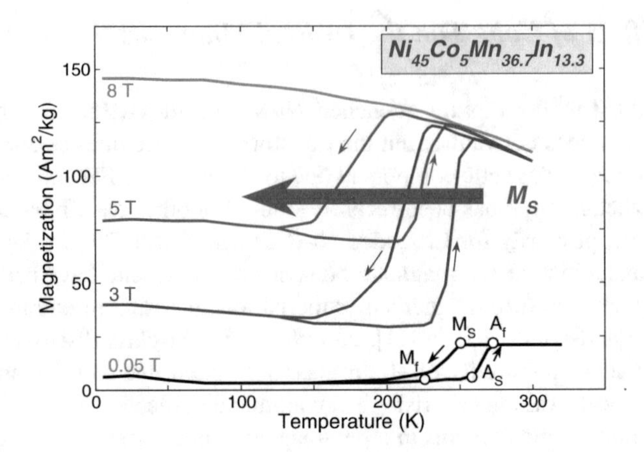

Fig. 6.7 Isofield magnetization curves of $Ni_{45}Co_5Mn_{36.7}In_{13.3}$ across the first-order magnetostructural transformation in fields of 0.05, 3, 5, and 8 T showing the shift of martensitic transformation with applied field to lower temperatures. From the fields H_{Af}, H_M, and $H_0 = (H_{Af}+H_M)/2$ the magnetic field-temperature phase diagram scan be evaluated showing the kinetic arrest phenomena. Figure adapted from Ito et al., copyright AIP (2008) [44]

interactions in a twofold degenerate band model of e_g symmetry using the model Hamiltonian [50, 51]:

$$\begin{aligned}\mathscr{H} = &\sum_{ij\sigma} t_{ij}(c^{\dagger}_{i1\sigma}c_{j1\sigma} + c^{\dagger}_{i2\sigma}c_{j2\sigma}) + U\sum_{i}(n_{i1\uparrow}n_{i1\downarrow} + n_{i2\uparrow}n_{i2\downarrow}) + U'\sum_{i\sigma\sigma'} n_{i1\sigma}n_{i2\sigma'} \\ &- J\sum_{i\sigma} n_{i1\sigma}n_{i2\sigma} + Ge\sum_{i\sigma}(n_{i1\sigma} - n_{i2\sigma}) + \tfrac{3}{4}N(C_{11} - C_{12})e^2 \\ &- \mu_B H \sum_{i}[(n_{i1\uparrow} - n_{i1\downarrow}) + (n_{i2\uparrow} - n_{i2\downarrow})],\end{aligned} \tag{6.6}$$

where n is the number of atoms, 1 and 2 denote the orbitals, U and U' are the intra-orbital and inter-orbital Coulomb interaction terms, J is the exchange interaction between the two orbitals at the same site, G is the interaction between e_g-type electrons and strain modes, $(C_{11} - C_{12})$ is the tetragonal-type elastic constant for the lattice, and the last term is the Zeeman term, for details and the Hartree-Fock approximation, see [50, 51].

This degenerate e_g-band model yields for the shift of the martensitic transformation temperature in an external field with a Stoner-like enhancement factor,

$$\frac{\Delta T_M}{T_M^0} \propto \left(\frac{\mu_B H}{k_B T_M^0(1 - \rho_{\epsilon_F} U_{eff})}\right)^2, \tag{6.7}$$

where $U_{eff} = U + J$.

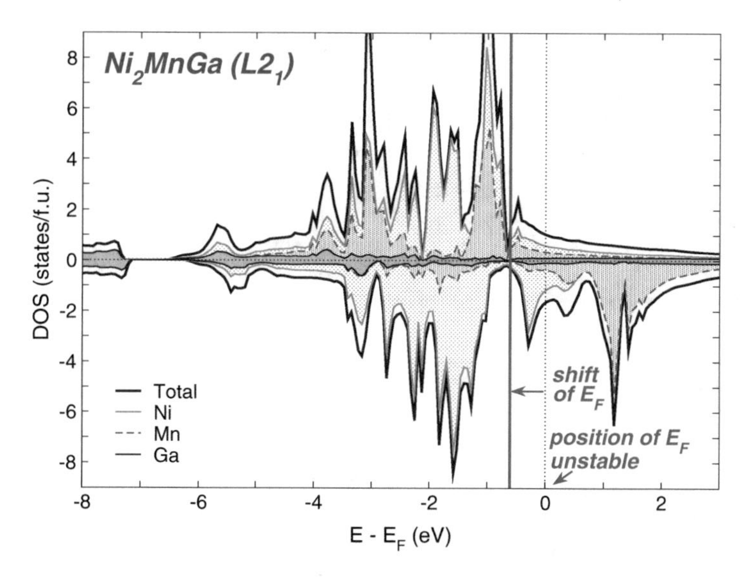

Fig. 6.8 Ab initio electronic density of states of prototype Ni_2MnGa showing the pseudogap of austenite and the Ni e_g peak in the spin-down channel. The corresponding shift of the Fermi level with tetragonal distortion in the martensitic state places E_F in the pseudogap region and the Ni e_g peak to the unoccupied region above E_F. This stabilizes the martensitic phase

This result shows that the shift of the transformation temperature varies approximately with the square of the magnetic field. Although ΔT_M follows from a simple e_g band model, it gives at least an explanation why the shift is non-linear with the external field. However, we believe that a more realistic band structure involving all d-bands and the s, p bands are necessary for an adequate description, since all d-bands contribute to the martensitic transformation as follows from the density of states and its orbital decomposed e_g and t_{2g} contributions as illustrated in Figs. 6.8 and 6.9.

A further interesting phenomena in magnetic Heusler alloys is the so-called kinetic arrest and de-arrest, for example, in $Mn_{50}Ni_{36}Sn_9Co_5$, which has recently been summarized in [52]. Usually, the Curie temperature of austenite is higher than the magnetostructural transformation temperature and 100% austenite is transformed to 100% martensite. But the first-order magnetostructural transformation can be significantly influenced by the magnetic field, by pressure and by the alloy composition. In $Mn_{50}Ni_{36}Sn_9Co_5$ the magnetostructural transformation becomes kinetically arrested below 35 K, when the cooling field exceeds 1 T; the amount of frozen austenite depends on the cooling field. Kinetic arrest has been observed in Ni-Mn-In [53, 54], Ni-Co-Mn-In [44], Ni-Co-Mn-Ga [45], Ni-Co-Mn-Sn [46, 47], and Ni-Mn-Co-Al [48, 49].

Figure 6.10 shows experimental results of the kinetic arrest phenomenon for the Heusler alloy $Ni_{45}Co_5Mn_{36.6}In_{13.4}$.

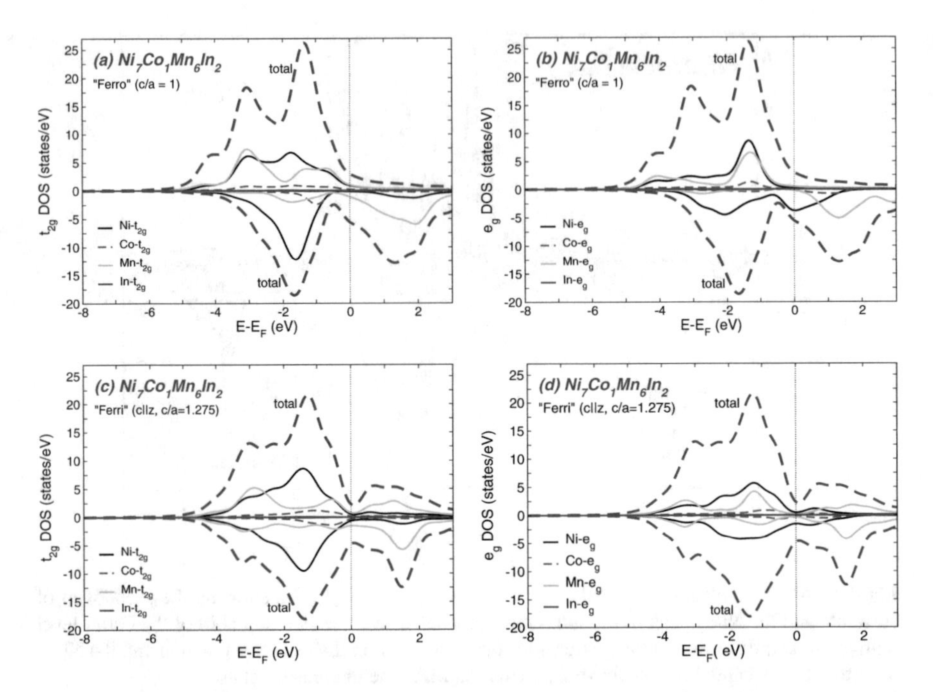

Fig. 6.9 Orbital-projected electronic density of states of t_{2g} and e_g orbitals for austenite in (**a**) and (**c**) and for martensite in (**b**) and (**d**), respectively. In spite of the atomic disorder in the supercell with 16 atoms, it is evident that martensite with a tetragonal distortion of $(c/a = 1.275)$ is stabilized by the dips in the density of states at the Fermi level. The electronic contributions of s, p electrons is omitted from the plots

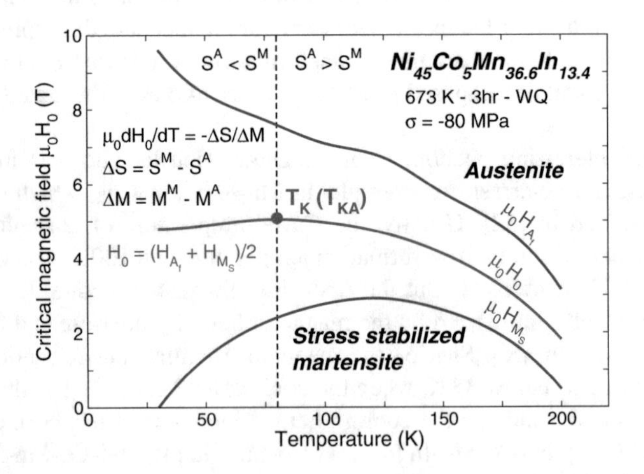

Fig. 6.10 Martensite start (H_{M_s}), austenite finish (H_{A_f}) and equilibrium (H_0) transformation magnetic fields as a function of temperature under −80 MPa stress showing a change of sign of entropy at 80 K [55]. This is the kinetic arrest phenomenon (T_{KA} = 80 K). Alternatively, this has been interpreted as a Kauzmann point T_K, where entropies of high and low temperatures become identical [56]. Figure adapted form [55] (copyright Acta Mater. (2015))

Although in [57], the martensitic and magnetic properties of $Ni_{45}Co_5Mn_{30}Ga_{20}$ were investigated in the full range of aging times (from as quenched conditions up to 2×10^4 min at 470 K in a small magnetic field of 80 mT) trends for decomposition have not been investigated, nor in the other recent papers [37, 38, 42]. Therefore, the report on decomposition tendencies of Ni-Mn-(Al, Ga, In, Sn, Sb) in this paper is rather new information as well as its impact on segregation and formation of structural glasses.

In the context of magnetostructural transition, the magnetization curves of austenite is usually a high-spin state (HS) while martensite is paramagnetic or a low-spin state (LS). Indeed, on the basis of the e_g-band model it was shown that a high-spin austenite and a high-spin martensite are mutually exclusive [50, 51]. This reminds of Invar [58], where LS and HS states can be connected by a bunch of non-collinear solutions.

From the discussion of the magnetostructural transition leading to the inverse magnetocaloric effect, we deduce that beside both spin-glass effects accompanied by glassy effects involving the strains in the alloys may exist, although we have not yet simulated the dynamic mechanical properties of $Ni_{55-x}Co_xMn_{20}Ga_{25}$ as reported in [2]. Ab initio calculations for $Ni_{55-x}Co_xMn_{20}Ga_{25}$ are so far restricted to the evaluation of the martensitic transformation temperature. Strain-glass effects still require attention.

To summarize this part, we may say that in addition to fast cooling (rapidly quenched alloys) and magnetic frustration arising from ferromagnetic and anti-ferromagnetic exchange interactions, a third independent cause for glassiness is the presence of quenched-in disorder, which is particularly relevant in strain-glasses since it leads to anisotropic forces as is evident from Fig. 6.11 showing the differences in tetragonal and orthorhombic distortions. We notice here that these three triggering factors are independent, although magnetic frustration and quenched-in disorder, mostly appear together. This non-equilibrium situation in Fig. 6.11 may well accompany segregation tendencies discussed further below. Also the frozen-in disorder will help the strain-glass phase to form. From this non-equilibrium situation we expect the most important driving force for segregation to arise. A deeper discussion concerning ferroic systems and their corresponding cluster glasses is presented in the contribution by P. Lloveras et al. in this volume [59].

Concerning cluster-spin glasses, we just like to comment that the cause for magnetic spin cluster glasses has indeed been observed in the ab initio simulations (fixed spin moment calculations) when simulating the effect of reversing clusters of Mn spins in a supercell and considering their energetics, which is discussed in [60].

In the next section we check whether indeed noncollinear spin states in Heusler alloys may exist.

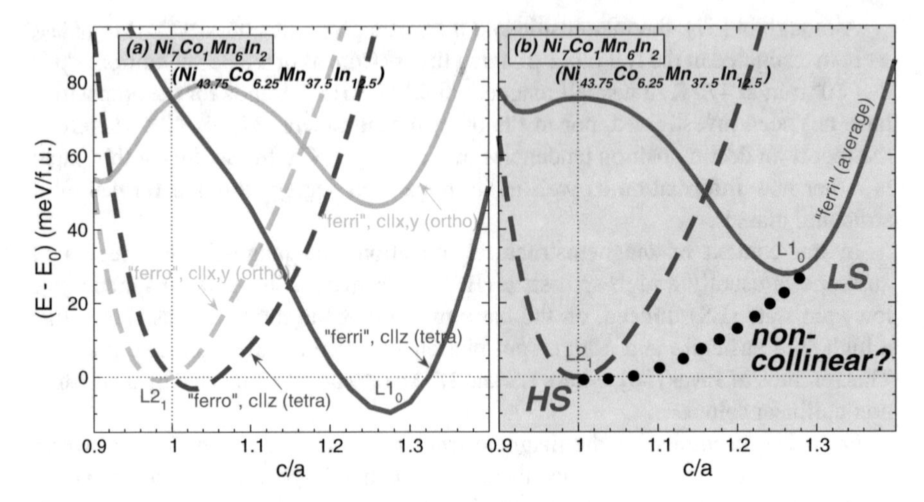

Fig. 6.11 Energy landscape of austenite and martensite formation for $Ni_7Co_1Mn_6In_2$ (16-atom supercell, which corresponds to $Ni_{43.75}Co_{6.25}Mn_{37.5}In_{12.5}$). Because of non-stoichiometry, the symmetry is not cubic, hence the tetragonal distortion in x and z direction yields different results in (**a**). Plotted in (**b**) are average values $E_{average} = 1/3E_{c||z} + 2/3E_{c||x}$ for ferromagnetic and ferrimagnetic (spins of Mn on In sites reversed) solutions, respectively. We assume that high spin (HS) and low spin (LS) states can be joined smoothly as in Invar [58] as indicated by the black dots

6.3 Noncollinear Magnetism

We have searched for tendencies that the Heusler alloys form noncollinear spin configurations or spin spirals which could be related to spin-glass features brought about the atomic disorder of the alloys. All calculations are based on the evaluation of magnetic exchange constants using ab initio calculations. Since we highlight in this article particular the properties of disordered Ni-Mn-In-Z as a representative case for Ni-Mn-(Al, Ga, In, Sn)-Z, with Z = Co, where Co can be used to tune the strength of ferromagnetic interactions with respect to the strength of antiferromagnetic interactions, we show in Fig. 6.12 the rich scenario of the exchange constants of $Ni_{50}Mn_{34}In_{16}$ and $Ni_{45}Co_5Mn_{34}In_{16}$. Ferromagnetic and antiferromagnetic exchange interactions, which arise from the excess Mn, which contribute to the metamagnetic behavior of the Heusler alloys leading to a first-order magnetostructural transition have been discussed in detail in [24]. We would like to stress that the antiferromagnetic interactions lead to low-spin-behavior of martensite below the magnetostructural transition. When decomposing the exchange interaction into their orbital contributions and resulting mixed terms as in magnetic-exchange it becomes obvious that the antiferromagnetic contributions result more from the localized e_g orbitals. We would like to point out that these competing magnetic exchange interactions may well explain the tendencies to form a spin glass in martensite and may explain the absence of a high-spin ferromagnetic state in martensite. When adding Co to the Heusler, ferromagnetic trend increases as

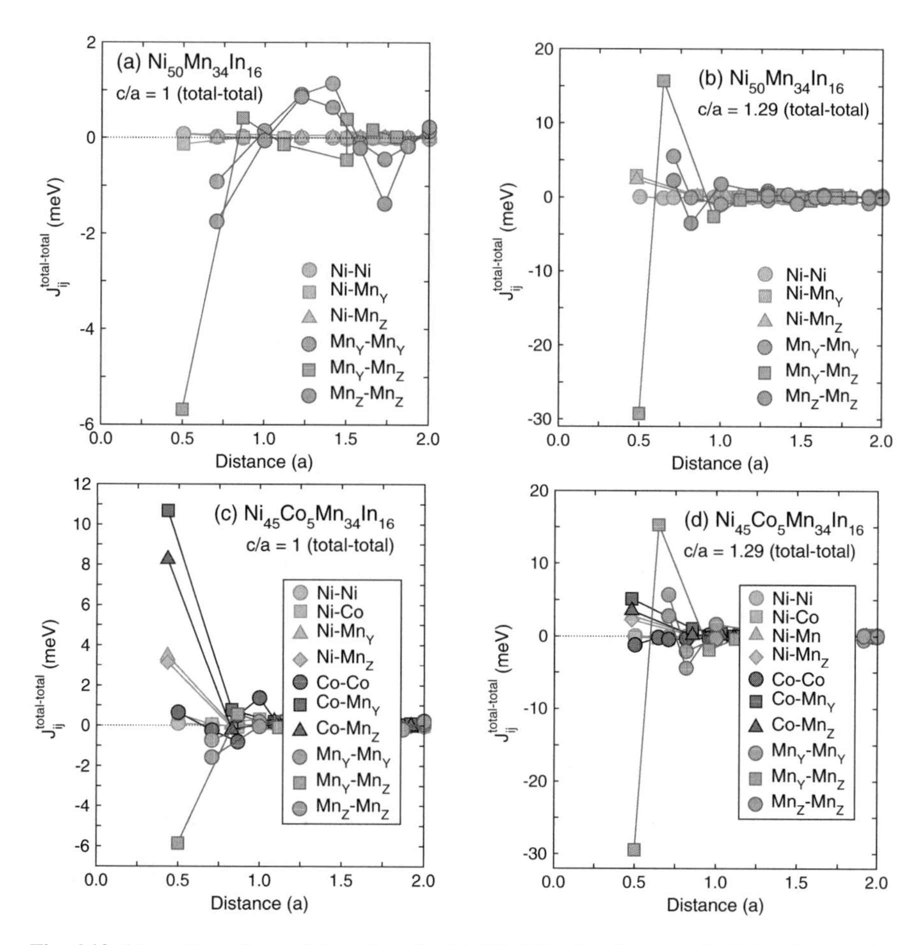

Fig. 6.12 Magnetic exchange interaction of cubic $Ni_{50}Mn_{34}In_{16}$ for austenite (**a**) and martensite with a tetragonal distortion of $c/a = 1.29$ (**b**) and for the corresponding alloy containing 5% Co in (**c**) and (**d**). Plotted are the total of the interaction constants consisting of the $s - p$ and d electron contributions as well as mixed terms from the different orbital contributions. For details, see the corresponding decomposition of Mn-rich $Ni_{50}Mn_{30}Ga_{20}$ in [24]. The notation used is obvious: The first atom is always at the origin and the distance between the atom sites is n units of the lattice constant. Positive exchange interactions are ferromagnetic and negative exchange interactions are antiferromagnetic ones. Reprinted from Ref. [88]. © (2018) by Wiley-VCH

is obvious from Fig. 6.12. We would like to emphasize that the disordered array of magnetic atoms will also favor the formation of a strain-glass phase as in $Ni_{55-x}Co_xMn_{20}Ga_{25}$ [2].

In order to see whether the competing magnetic interactions may enhance the tendency for noncollinear spin configurations, we have performed a few model calculations by allowing for different spin orientations of the neighboring Mn spins. Figure 6.13 shows typical resulting configurations which are indeed lower in energy by a few meV than the collinear one (calculations by S. Mankovsky [61] and L. Sandratskii [62]).

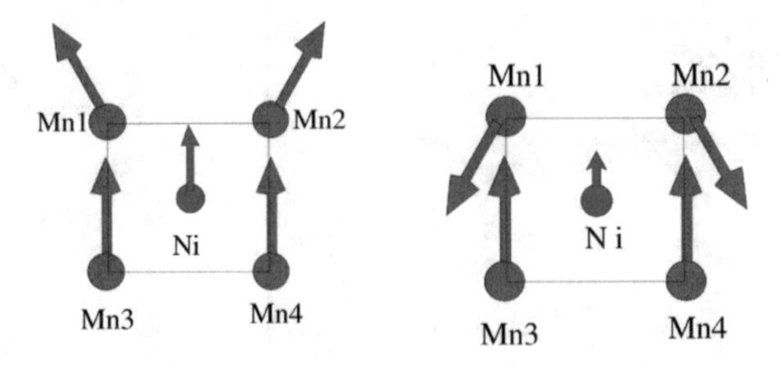

Fig. 6.13 Noncollinear magnetic moments between neighboring Mn atoms in a 16-atom supercell for $Ni_7Co_1Mn_6In_2$ Heusler alloy. Left: Spin configuration obtained with the SPR KKR method [61] which is slightly lower in energy than the configuration in the right panel [62]. Reprinted from Ref. [88]. © (2018) by Wiley-VCH

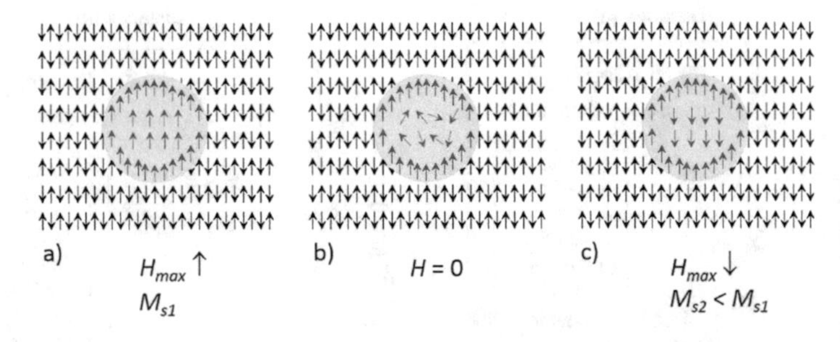

Fig. 6.14 A model description of the resulting magnetic configuration after the sample decomposition in a magnetic field adapted from [4]. The yellow area represents the precipitated Heusler in the surrounding antiferromagnetically aligned Ni-Mn. The red arrows depict the spins in the applied field direction during segregation which are strongly pinned by the interlayer exchange. The blue spins are those which align only in an external magnetic field. In (**a**) all spins are aligned in the external field. In (**b**) the external field is zero. In (**c**) the core spins align opposite the reverse field direction

A new functionality and on the edge of noncollinearity have indeed been observed in antiferromagnetic martensitic Heusler $Ni_{50}Mn_{45}In_5$, where shell ferromagnetism of nano-Heuslers is generated by segregation under magnetic field [4], which is shown in Fig. 6.14. Figure 6.15 shows recent observation of nanoscale skyrmions in a nonchiral multiferroic Ni_2MnGa Heusler material [7].

6.4 Decomposition in Less Rapidly Quenched Heusler Alloys

In a series of publications a systematic adjustment of composition for magnetic ordering and giant magnetocaloric effect has been discussed [63–65]. It appears that for the Ni-Mn-Sn alloy system, this compositional tuning has dramatic effects on the microstructural development influencing structural as well as magnetic phase transitions in that it can mask the magnetostructural behavior in $Ni_{50}Mn_{37}Sn_{13}$.

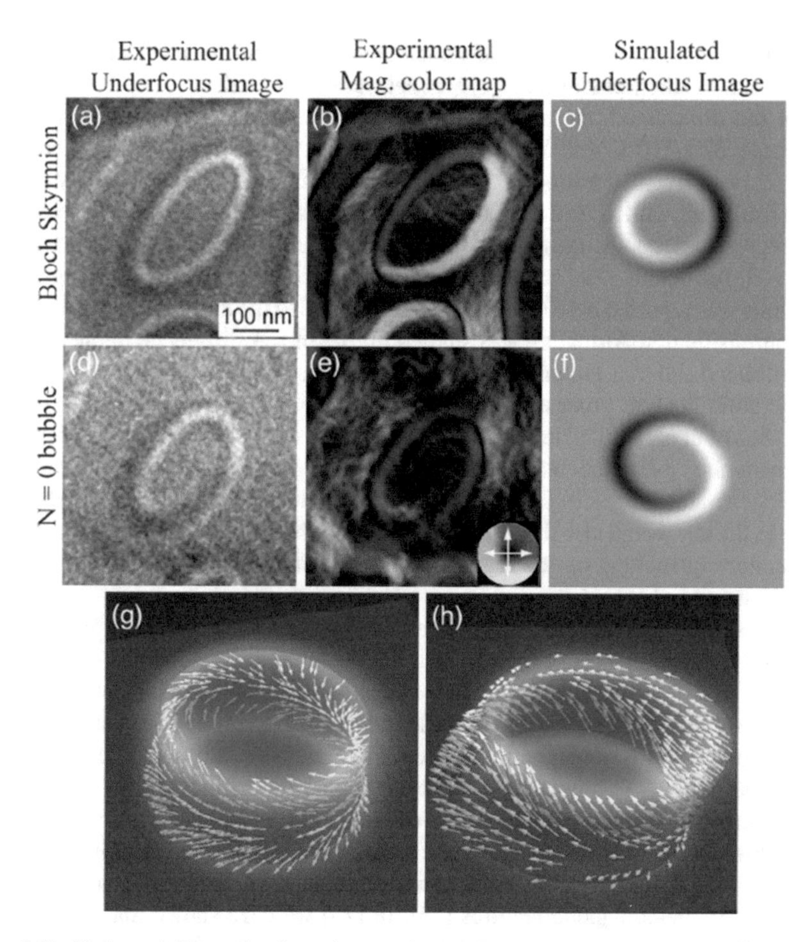

Fig. 6.15 High-resolution under-focus image of a bubble domain in (**a**) 0 mT and (**d**) 180 mT field with corresponding color maps in panels (**b**) and (**e**). (**c**) and (**f**) show the corresponding under-focus LTEM Images and (**g**) and (**h**) show the 3D representation of the simulated magnetic spin textures. Image adapted from [7]. Reprinted with permission from Phatak et al. [7]. Copyright (2016) American Chemical Society

Consistent with the magnetization data, transition electron microscopy examination confirms that $Ni_{50}Mn_{50-x}Sn_x$ is decomposed into two phases with $x = 20$ and $x = 1$. Hence, one may conclude that the martensitic transformation occurs only in those compositions where the single phase $L2_1$ has been retained in a metastable state on cooling [64]. In a subsequent experiment it was checked that 150 K below the disorder-order transition ($B2 \rightarrow L2_1$) on a 1223 K homogenized sample the magnetization measurements show drastic changes of the annealed samples. Transmission electron microscopy (TEM) and X-ray diffraction (XRD) analysis confirmed the decomposition of the single-phase Heusler alloys into two phases with compositions close to $Ni_{54}Mn_{45}Sn_1$ and $Ni_{50}Mn_{30}Sn_{20}$. The observed phase decomposition indicates that the off-stoichiometric Ni-Mn-Sn Heusler alloys, which feature martensitic transformations, are metastable at 773 K.

On the other hand, thermal stability of high-temperature Ni-Mn-Ga alloys in the range 620–770 K was investigated by Cesari et al. showing that increasing the e/a ratio by substitution of Ga with Mn results in very stable alloys under aging [66]. This work aimed to characterize the martensitic transformation evolution and microstructural changes produced by aging. Relatively large precipitates were observed, for details see [66]. 'Self-patterning' of epitaxial Ni-Mn-Ga/MgO(001) thin films was reported in [67]. Quite interestingly, also the influence of vacancies on the martensitic transformation was studied in [68, 69]. A difference in the behavior of modulated and non-modulated structures was observed.

The solidification process of metamagnetic Ni-Co-Mn-Sn was investigated by Perez-Sierra et al. [70]. The as-cast microstructures are composed of four phases: $L2_1$ austenite, 6-layered martensite, DO_3 phase, and γ phase (disordered fcc). Subsequent annealing at high temperatures at 1173 K completely dissolves the DO_3 phase in the alloys and also the γ phase in alloys with low Co content. Effect of high-temperature quenching on the martensitic transformation and magnetic properties was discussed in [71].

This is not a complete list of publications but shows that disorder-order transitions and related physical properties of shape memory and magnetocaloric Heusler materials is a field of active research. Other experimental work shows trends or new features like the exchange bias effect, which may arise from the competing ferromagnetic–antiferromagnetic interactions in Heusler alloys with excess of Ni or Mn atoms [72]. In any case, it has been shown by using ab initio calculations that anomalous change of the nature of magnetic exchange interactions above and below the magnetostructural phase transition from dominating ferromagnetic interactions in austenite to antiferromagnetic interaction in martensite can very often be related to metamagnetic features [73]. In [73] we have shown that for Mn-rich Ni-Mn-Ga the competing magnetic interactions in austenite transform to dominating antiferromagnetic interactions in martensite. The as-quenched austenitic alloys may easily micro-segregate to form nanophase strain-glasses. Hence, we have another driving force for forming strain-glasses in form of micro-segregation of the Heusler alloys.

Indeed, ab initio calculations show that all Ni-Mn-based Heusler alloys with excess of either Mn or Ni show tendencies for micro-segregation on the nano-scale in form of, taking Ni-Mn-In as example,

$$Ni_{50}Mn_{45}In_5 \rightarrow Ni_{10}Mn_5In_5 + Ni_{40}Mn_{40},$$

i.e. Ni-Mn-In with intrinsically competing ferromagnetic and antiferromagnetic interaction for the rapidly quenched alloys segregates into ferromagnetic $L2_1$ structure and an $L1_0$ antiferromagnetic Ni-Mn matrix [4]. Meanwhile, this micro-segregation has been confirmed for a series of Heusler alloys, namely $Ni_{49.6}Mn_{45.5}In_{4.9}$[4]. $Ni_{49.8}Mn_{45.1}Sn_{5.1}$[74], $Ni_{48.7}Mn_{46.2}Ga_{5.1}$ [75], $Ni_{51.87}Mn_{43.73}Al_{4.40}$ [76].

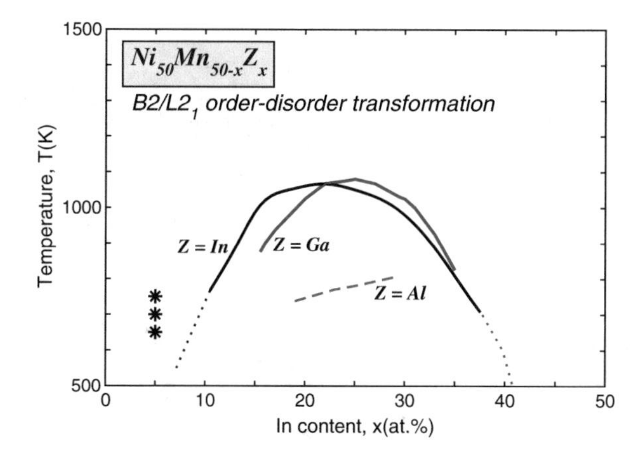

Fig. 6.16 Critical temperature of the $B2/L2_1$ order–disorder transformation determined by DSC measurements for the $Ni_{50}Mn_{50-x}In_x$ alloy system compared with the critical temperature for Ni-Mn-Al and Ni-Mn-Ga. The NiMnIn alloys which segregate after temper-annealing at 650, 700, and 750 K, are marked by the stars in the figure. They are outside the order–disorder transition region. Figure adapted from [78]. Copyright Elsevier (2010)

The tendencies for decomposition of the Ni-Mn-(Al, Ga, In, Sn) have to be compared with the corresponding phase stability and magnetic properties of Heusler type alloys, namely the critical temperature of the $B2/L2_1$ order–disorder transformation determined by the DSC measurements for $Ni_{50}Mn_{50-x}In_x$ [77, 78] compared with those for Ni-Mn-Al [79] and Ni-Mn-Ga [80], where it was shown that the ordering sequence $A2 \rightarrow B2' \rightarrow L2_1$ may also be modeled by using the Bragg-Williams approximation [81].

The order–disorder transformation temperature plays an important role when discussing segregation in Heusler alloys. The closer in composition the alloy is, for example, to the critical temperature of $B2/L2_1$ order–disorder transformation, the easier one can relate segregation phenomena to the ordering process. However, this is not always the case, as in $Ni_{50}Mn_{50-x}Z_x$ alloys, the temper annealing experiments to observe decomposition are done for alloy composition which are well outside the order–disorder transformation area as shown in Fig. 6.16.

Taking Ni-Mn-In as an example for decomposition when temper-annealing the alloy, we show in Fig. 6.17 that the $B2/L2_1$ ordering and the decomposition which we observe in experiment occur outside the disorder-order phase transition.

6.5 Calculation of Mixing Energies in Heusler Alloys

A convenient way to look for the stability of compounds and alloys is to calculate the mixing energy of the corresponding material system.

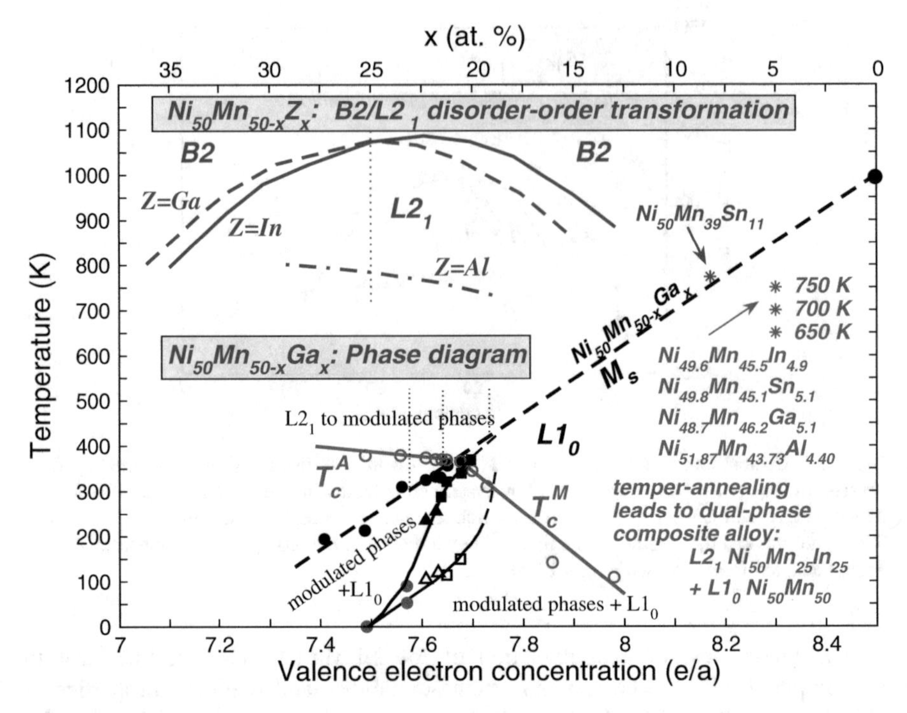

Fig. 6.17 Critical temperature of the B2/L2$_1$ disorder-order transformation of $Ni_{50}Mn_{50-x}Z_x$ as a function of x (upper axis) and e/a (lower axis) for $Z = Al, Ga, In$ [77, 78]. The disorder-order transformation temperature for $Z = Sn$ has not been measured. The phase diagram of $Ni_{50}Mn_{50-x}Ga_x$ [1] has been added to the plot to show the separation (in temperature) between the B2/L2$_1$ disorder-order transformation and $Ni_{50}Mn_{39}Sn_{11}$ [64] and Ni-Mn-In [4], Ni-Mn-Sn [74], Ni-MnGa [75], Ni-Mn-Al [76] alloy systems (blue and red stars) for which the segregation process of the alloys has been investigated. Figure based on [77, 78]. Reprinted from Ref. [88]. © (2018) by Wiley-VCH

Figure 6.18a, b shows the calculated mixing energies of the two Heusler alloy systems $Ni_2Mn_{1+x}In_{1-x}$ and $Ni_2Mn_{1+x}Sn_{1-x}$ while the antiferromagnetic ordering (columnar and staggered) is shown in Figs. 6.19 and 6.20. Spin ordering in binary NiMn has been discussed very early by Kaspar and Kouvel [82] and Krén et al. [84].

In order to explain the physical behavior of the influence of annealing and phase decomposition on the magnetostructural transitions in $Ni_{50}Mn_{39}Sn_{11}$ (blue star in Fig. 6.17) we refer to the work of [64] regarding the heat treatment of the alloys. To study the role of chemical ordering in fine-tuning their magnetostructural properties, the alloys were first annealed for 4 weeks at 1223 K to achieve structural and compositional homogeneity, and were then further annealed for 1 week (∼150 K below the reported B2 to L2$_1$ transition at 773 K to increase the degree of chemical ordering). For 11 at.% Sn, this anneal resulted in a dramatic change in the magnetic ordering temperature. Following the 1223 K anneal, the sample exhibited ferromagnetic ordering at 140 K. After the 773 K anneal, the ferromagnetic transition is at 350 K, a characteristic of the ferromagnetic austenite phase for alloys with

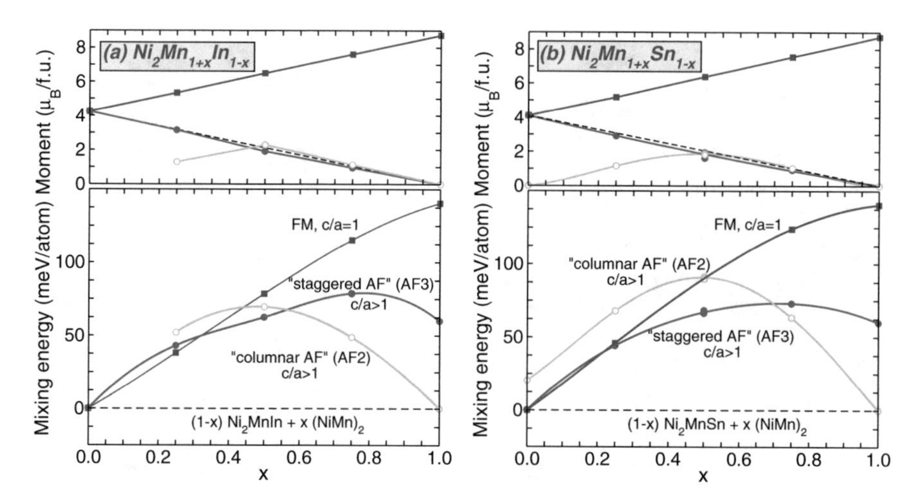

Fig. 6.18 Mixing energies and magnetic moments of (**a**) $Ni_2Mn_{1+x}In_{1-x}$ and (**b**) $Ni_2Mn_{1+x}Sn_{1-x}$ for two different antiferromagnetic spin orderings AF2 and AF3, which is illustrated in Figs. 6.19 and 6.20. The mixing energy is calculated by assuming that over the whole concentration range the decomposition of the alloy will lead to a dual-phase composite alloy like $L2_1$ $Ni_{50}Mn_{25}In_{25}$ and $L1_0$ $Ni_{50}Mn_{50}$. The Heusler precipitates are ferromagnetic while the antiferromagnetic matrix is assumed to have AF2 ordering. The structures of antiferromagnetic of Ni_8Mn_8 show considerable tetragonal deformations: "staggered AF" ($a = 5.076$, $c/a = 1.4209$), "columnar AF" ($a = 5.1066$, $c/a = 1.4019$), see also Figs. 6.19 and 6.20. Reprinted from Ref. [88]. © (2018) by Wiley-VCH

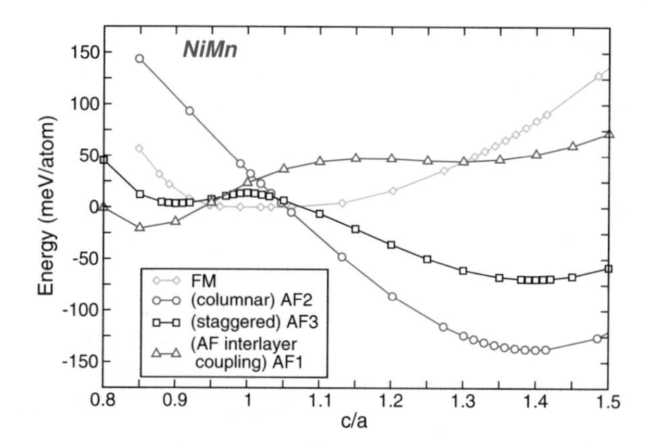

Fig. 6.19 Total energy as a function of the tetragonal distortion c/a for different spin configurations of binary NiMn. The orange curve denoted as FM corresponds to a ferromagnetic alignment of the Mn spins whereas the other curves are associated with different antiferromagnetic alignments of the Mn spins. The red curve denoted as AF2 shows the ground state, which has been reported by Kasper and Kouvel [82] (calculations by Mario Siewert [83])

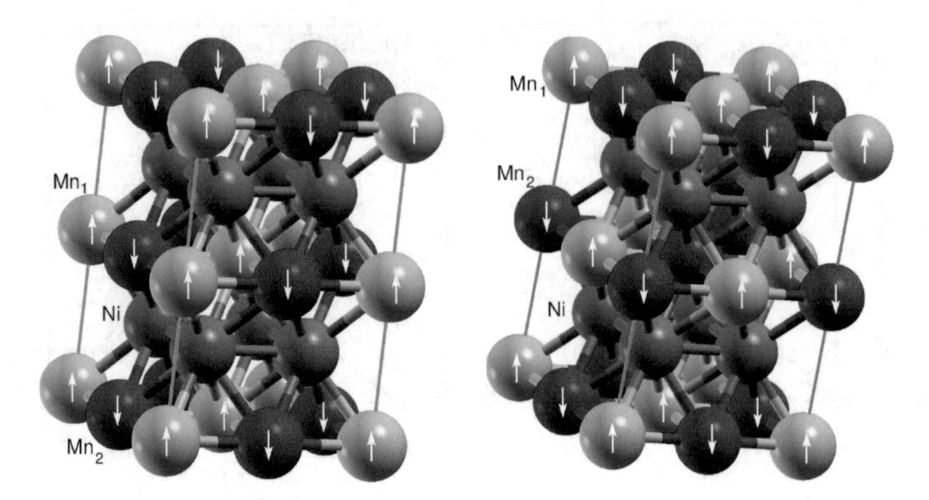

Fig. 6.20 Antiferromagnetic spin structures, for which the mixing energies have been evaluated. Left: "columnar AF" (AF2) and right: "staggered AF" (AF3) ordering. Reprinted from Ref. [88]. © (2018) by Wiley-VCH

$15 < x < 25$ content of Sn. The authors further find from transmission electron microscopy examination that the alloy decomposed into two phases with $x = 20$ and 1. From this result one can conclude that the martensitic transformation occurs only in those compositions where the single phase $L2_1$ has been retained in a metastable state on cooling.

Thus the driving force for the decomposition seems not to be directly related to the disordering-ordering $B2/L2_1$ phase transition, which is at higher temperatures, it occurs for compositions near the martensitic instability. Nonetheless, in the spirit of a unified theory of disorder-order and martensitic transformations we expect the disorder-order and martensitic phase transition lines to intersect, which would place the systems which segregate right in the interesting region of structural and disorder-order transformation and segregation. Experimentally it seems to be difficult to get information of disorder-order transformation at lower temperatures.

Our ab initio calculations of phonons for the disordered alloys confirm the existence of the martensitic instability near this critical concentration. Earlier calculations of Fermi surface nesting confirmed that nesting behavior is still present for the non-stoichiometric, disordered systems. The number of valence electrons which is larger in Ni-Mn-Sn ($e/a \approx 8.2$) compared to stoichiometric Ni_2MnGa ($e/a = 7.5$) just blows the Fermi surface up but does not lead to complete vanishing of nesting behavior.

There is a series of recent temper-annealing experiments on Ni-Mn-In [4], Ni-Mn-Sn [74], Ni-Mn-Ga [75], Ni-Mn-Al [76] alloy systems all with high e/a ratios and very well separated in temperature from the $B2/L2_1$ disordering-ordering transition but also close to the martensitic transformation, which are also marked in Fig. 6.17, with post-annealing temperatures from 650 to 750 K, which show similar effects of segregation than just explained for the Sn-sample. Indeed in the experimental setup one lets the samples undergo the martensitic transformation to the $L1_0$ structure and subsequently observes the decomposition or segregation into a dual-phase composite alloy, where the two phases are identified to be cubic $L2_1$ $Ni_{50}Mn_{25}In_{25}$ and $L1_0$ $Ni_{50}Mn_{50}$ [4].

The results of decomposition for the Ni-Mn-In and Ni-Mn-Sn alloys have been calculated by assuming that the segregation follows the prescription in Fig. 6.18,

$$E_{mixing} = E_{\mathrm{Ni_2Mn_{1+x}In_{1-x}}} - (1-x)\, E_{\mathrm{Ni_2MnIn}} - x\, E_{\mathrm{(NiMn)_2}}, \tag{6.8}$$

where the reference energy is the energy of the alloy decomposed into stoichiometric Ni_2MnIn and binary NiMn. We assume different antiferromagnetic configurations, columnar (AF2) and staggered (AF3). The energies of segregation are quite large and approximately correspond to the temperature scale in Fig. 6.17 of the decomposed alloys Ni-Mn-(Al, Ga, In, Sn).

It seems to be obvious that the decomposition into stoichiometric precipitates in a NiMn antiferromagnetic background of all alloy systems listed in Fig. 6.17 is energetically favored. The decomposed Heusler alloys all have high e/a rations near 8.3 which puts them in the martensitic region or very close to it. But segregation also occurs 150 K below the disorder-order transformation. So, it maybe debated whether coexistence of disorder-order and martensitic tendencies may play a role in these alloys and have to be considered within the frame of a unified description for order–disorder and structural instability.

6.6 Conclusions

There is clear tendency that the decomposition of the Heusler alloys with large (e/a) ratios leads to some nanophase materials which all bear glassy features ranging from cluster-spin glasses to strain-glasses. The very nature of the decomposition itself favors the formation of a nanocomposite alloy with glassy behavior. To be specific the dual phases of the composite alloy, $L2_1$ $(Co, Ni)_2MnIn$ + $L1_0$ $Ni_{50}Mn_{50}$, are separated by a three-dimensional array of nanophase or nanocomposite strain glass. All basic ingredients for a strain and spin glass are there: Intrinsic disorder and frustration as well as elastic anisotropy arising from symmetry breaking long-range displacive fields leading to non-ergodicity in C_{44} and C' on zero-stress-cooling and stress-cooling (ZSC/SC) cycles. In spite of this segregation phenomenon, reversible martensite transformation under low magnetic fields has been observed in $Ni_{45}Co_5Mn_{36.6}In_{13.4}$ without segregation [85].

Acknowledgements This work was supported by the DFG priority programme SPP 1599.

References

1. A. Çakır, L. Righi, F. Albertini, M. Acet, M. Farle, Intermartensitic transitions and phase stability in $Ni_{50}Mn_{50-x}Sn_x$ Heusler alloys. Acta Mater. **99**, 140 (2015)
2. Y. Wang, C. Huang, J. Gso, S. Yang, X. Ding, X. Song, X. Ren, Evidence for ferromagnetic strain glass in Ni-Co-Mn-Ga Heusler alloy system. Appl. Phys. Lett. **101**, 101913 (2012)
3. X. Ren, Y. Wang, Y. Zhou, Z. Zhang, D. Wang, G. Fan, K. Otsula, T. Suzuki, Y. Ji, J. Zhang, Y. Tian, S. Hou, X. Ding, Strain glass in ferroeleastic systems: premartensitic tweed versus strain glass. Philos. Mag. **90**, 141 (2010)
4. A. Çakır, M. Acet, M. Farle, Shell-ferromagnetism of nano-Heuslers generated by segregation under magnetic field. Sci. Rep. **6**, 28931 (2016)
5. O. Mesheriakova, S. Chadow, A.K. Nayak, U.K. Rößler, J. Kübler, G. André, A.A. Tsirlin, J. Kiss, S. Hausdorf, A. Kalache, W. Schnelle, M. Nicklas, C. Felser, Large noncollineraity and spin reorientation in the novel Mn_2RhSn Heusler magnet. Phys. Rev. Lett. **113**, 0897203 (2014)
6. S. Singh, S.W. D'Souza, J. Nayak, E. Suard, L. Chapon, A. Senyshyn, V. Petricek, Y. Skourskim, M. Nicklas, C. Felser, S. Chadov, Room-temperature tetragonal non-collinear Heusler antiferromagnet Pt_2MnGa. Nat. Commun. **7**, 12671 (2016)
7. C. Phatak, O. Heinonen, M. De Graef, A. Petford-Long, Nanoscale Skyrmions in a nonchiral metallic multiferroic: Ni_2MnGa. NanoLett. **16**, 4141 (2016)
8. P.J. Webster, K.R.A. Ziebeck, S.L. Town, M.S. Peak, Magnetic order and phase transformation in Ni_2MnGa. Philos. Mag. **49**, 295 (1984)
9. P. Entel, A. Dannenberg, M. Siewert, H.C. Herper, M.E. Gruner, V.D. Buchelnikov, V.A. Chernenko, Composition-dependent basics of smart Heusler materials from first-principles caculations. Mater. Sci. Forum **684**, 1 (2011)
10. P. Entel, V.D. Buchelnkov, V.V. Khovailo, A.T. Zayak, W.A. Adeagbo, M.E. Gruner, H.C. Herper, E.F. Wassermann, Modelling the phase diagram of magnetic shape memory alloys. J. Phys. D: Appl. Phys. **39**, 865 (2006)
11. K. Ullakko, J.K. Huang, C. Kantner, R.C. O'Handley, V.V. Kokorin, Large magnetic-field-induced strains in Ni_2MnGa, Appl. Phys. Lett. **69**, 1966 (1996)
12. S.J. Murray, M. Marioni, S.M. Allen, R.C. O'Handley, 6% magnetic-field-induced strain by twin-boundary motion in ferromagnetic Ni-Mn-Ga. Appl. Phys. Lett. **77**, 886 (2000)
13. O. Söderberg, Y. Ge, A. Sozinov, S.-P. Hannula, V.K. Lindroos, Recent breakthrough development of the magnetic shape memory effect in Ni-Mn-Ga alloys. Smart Mater. Struct. **14**, S223 (2005)
14. B. Kiefer, D.C. Lagoudas, Magnetic field-induced martensitic variant reorientation in magnetic shape memory alloys. Philos. Mag. **85**, 4289 (2005)
15. I. Karaman, B. Basaran, H.E. Karaca, A.I. Karsilayan, Y.I. Chumlyakov, Energy harvesting using martensite variant reorientation mechanism in NiMnGa magnetic shape memory alloy. Appl. Phys. Lett. **90**, 172505 (2007)
16. S. Kaufmann, U.K. Rößler, O. Heczko, M. Wuttig, J. Buschbeck, L. Schultz, S. Fähler, Adaptive modulations of martensites. Phys. Rev. Lett. **104**, 145702 (2010)
17. R. Niemann, U.K. Rößler, M.E. Gruner, O. Heczko, L. Schultz, S. Fähler. The role of adaptive martensite in magnetic shape memory alloys. Adv. Eng. Mater. **14**, 562 (2012)
18. J.M. Barandiaran, V.A. Chernenko, E. Cesari, D. Salas, P. Lazpitza, J. Gutierrez, I. Orue, Magnetic influence on the martensitic transformation entropy in Ni-Mn-In metamagnetic alloy. Appl. Phys. Lett. **102**, 071904 (2013)
19. J.M. Barandiaran, V.A. Chernenko, E. Cesari, D. Salas, J. Gutierrez, P. Lazpita, Magnetic field and atomic order effect on the martensitic transformation of a metamagnetic alloy. J. Phys.: Condens. Matter. **25**, 484005 (2013)
20. P.J. Stonaha, M.E. Manley, N.M. Bruno, I. Karaman, R. Arróyave, N. Singh, D.L. Abernathy, S. Chi, Lattice vibrations boost demagnetization entropy in a shape-memory alloy. Phys. Rev. B **92**, 140406(R) (2015)

21. H. Ebert et al., The Munich SPR-KKR package, version 6.3. http://ebert.cup.uni-muenchen.de/SPRKKR. H. Ebert, D. Ködderitzsch, J. Minár, Rep. Prog. Phys. **74**, 096501 (2011)
22. V.D. Buchelnikov, P. Entel, S.V. Taskaev, V.V. Sokolovskiy, A. Hucht, M. Ogura, H. Akai, M.E. Gruner, S.K. Nayak, Monte Carlo study of the influence of antiferromagnetic exchange interactions on the phase transitions of ferromagnetic Ni-Mn-X alloys (X = In, Sn , Sb). Phys. Rev. B **78**, 184427 (2008)
23. V.D. Buchelnikov, V.V. Sokolovskiy, S.V. Taskaev, V.V. Khovaylo, A.A. Aliev, L.N. Khanov, A.B. Batdalov, P. Entel, H. Miki, T. Takagi, Monte Carlo simulations of the magnetocaloric effect in magnetic Ni-Mn-X (X = Ga, In) Heusler alloys. J. Phys. D: Appl. Phys. **44**, 064012 (2011)
24. D. Comtesse, M.E. Gruner, M. Ogura, V.V. Sokolovskiy, V.D. Buchelnikov, A. Grünebohm, R. Arróyave, N. Singh, T. Gottsschall, O. Gutfleisch, V.A. Chernenko, F. Albertini, S. Fähler, P. Entel, First-pinciples calculation of the instability leading to giant inverse magnetocaloric effects. Phys. Rev. B **89**, 184403 (2014)
25. T. Castán, E. Vives, P. Lindgård, Modeling premartensitic effects in Ni_2MnGa: a mean-field and Monte Carlo simulation study. Phys. Rev. B **60**, 7071 (1999)
26. V.V. Sokolovskiy, M.A. Zagrebin, V.D. Buchelnikov, Novel achievements in the research field of multifunctional shape memory Ni-Mn-In and Ni-Mn-In-Z Heusler alloys. Mater. Sci. Found. **81/82**, 38 (2015)
27. N. Singh, R. Arróyave, P. Entel, Monte Carlo simulations of Spin-glass effects in Ni-Mn-In Heusler alloys (to be published)
28. V.V. Sokolovskiy, P. Entel, V.D. Buchelnikov, M.E. Gruner, Achieving large magnetocaloric effects in Co- and Cr-substituted Heusler alloys: predictions from first-principles and Monte Carlo studies. Phys. Rev. B **91**, 220409(R) (2015)
29. Y.M. Jin, A.G. Khachaturyan, Atomic density function theory and modeling of microstructure evolution at the atomic scale. J. Appl. Phys. **100**, 013519 (2006)
30. A.G. Khachaturyan, *Theory of Structural Transformation in Solids* (Dober Publications, New York, 1983)
31. S. Stamenković, The unified model description of order disorder and displacive structural phase transitions. Condens. Matter. Phys. **1**, 257 (1998)
32. J.A. Krumhansl, J.R. Schrieffer, Dynamics and statistical mechanics of a one-dimensional model Hamiltonian for structural phase transition. Phys. Rev. B **11**, 3535 (1975)
33. V. Recarte, J.I. Pérez-Landazábal, V. Sánchez-Alarcos, Dependence of the relative stability between autenite and martensite phases on the atomic order in a Ni-Mn-In metamagnetic shape memory alloy. J. Alloys Compd. **536s**, S5308 (2012)
34. C. Salazar Mejıa, A.K. Nayak, J.A. Schiemer, C. Felser, M. Nicklas, M.A. Carpenter, Strain behavior and lattice dynamics in $Ni_{50}Mn_{35}In_{15}$. J. Phys.: Condens. Matter. **27**, 415402 (2015)
35. A. Planes, L. Mañosa, M. Acet, Magnetovolume effect and its relation to shape-memory properties in ferromagnetic Heusler alloys. J. Phys.: Condens. Matter. **21**, 233201 (2009)
36. M. Acet, L. Mañosa, A. Planes, Magnetic-field-induced effects in martensitc Heusler-based magnetic shape memory alloys, in *Handbook of Magnetic Materials*, vol. 19, ed. by K.H.J. Bushow (North-Holland, Amsterdam, 2011), pp. 231–289
37. D.Y. Cong, S. Roth, L. Schultz, Magnetic properties and structural transfroations in Ni-Co-Mn-Sn multifunctional alloys. Acta Mater. **60**, 5335 (2012)
38. D.Y. Cong, S. Roth, Y.D. Wang, Superparamagnetism and superspin glass behaviors in multiferroic NiMn-based magnetic shape memory alloys. Phys. Status Solidi **251**, 2126 (2014)
39. R.Y. Umetsu, R. Kainuma, Y. Amako, Y, Taniguchi, T. Kanomata, K. Fukushima, A. Fujita, A. Oikawa, K. Ishida, Mössbauer study on martensitic phase in $Ni_{50}Mn^{57}_{36.5}Fe_{0.5}Sn_{13}$ metamagnetic shape memory alloy. Appl. Phys. Lett. **93**, 042509 (2008)
40. V.V. Khovaylo, T. Kanomata, T. Kanata, M. Nakashima, Y. Amako, R. Kainuma, R.Y. Umetsu, H. Morito, H. Miki, Magnetic properties of $Ni_{50}Mn_{34.8}In_{15.2}$ probed by Mössbauer spectroscopy. Phys. Rev. B **80**, 144409 (2009)
41. S. Chatterjee, S. Giri, S.K. De, S. Majumdar, Reentrant spin-glass state in $Ni_2Mn_{1.36}Sn_{0.64}$ shape-memory alloy. Phys. Rev. B **79**, 092410 (2009)

42. J.I. Pérez-Landazabal, V. Recarte, V. Sanchez-Alarcos, C. Gómez-Polo, E. Cesari, Magnetic properties of the martensitic phase in Ni-Mn-In-Co metamagnetic shape memory alloys. Appl. Phys. Lett. **102**, 101908 (2013)
43. S. Yuan, P.L. Kuhns, A.P. Reyes, J.S. Brooks, M.J.R. Hoch, V. Shrivastava, R.D. James, S. El-Khatib, C. Leighton, Magnetically nanostructured state in a Ni-Mn-Sn shape-memory alloy. Phys. Rev. B **91**, 214421 (2015)
44. W. Ito, K. Ito, R.Y. Umetsu, R. Kainuma, K. Koyama, K. Watanabe, A. Fujita, K. Oikawa, K. Ishida, T. Kanomata, Kinetic arrest of martensitic transformation in the NiCoMnIn metamagnetic shape memory alloy. Appl. Phys. Lett. **92**, 021908 (2008)
45. X. Xu, W. Ito, R.Y. Umetsu, K. Koyama, R. Kainuma, K. Ishida, Kinetic arrest of martenstic transformation in $Ni_{33.0}Co_{13.4}Mn_{39.7}Ga_{13.9}$ metamagnetic shape memory alloy. Mater. Trans. JIM **51**, 469 (2010)
46. A. Lakhani, S. Dash, A. Banerjee, P. Chaddah, X. Chen, R.V. Ramanjuan, Tuning the austenite and martensite phase fraction in ferromagnetic shape memory alloy ribbons of $Ni_{45}Co_5Mn_{38}Sn_{12}$. Appl. Phys. Lett. **99**, 242503 (2011)
47. R.Y. Umetsu, K. Ito, W. Ito, K. Koyama, T. Kanomata, K. Ishida, R. Kainuma, Kinetic arrest behavior in martensitic transformation of NiCoMnSn metamagentic shape memory alloy. J. Alloys Comp. **509**, 1389 (2011)
48. X. Xu, W. Ito, M. Tokunaga, R.Y. Umetsu, R. Kainuma, K. Ishida, Kinetic arrest of martensitic transformation in NiCoMnAl metamagnetic shape memory alloy. Mater. Trans. JIM **51**, 1357 (2010)
49. X. Xu, W. Ito, M. Tokunaga, T. Kihara, K. Oka, R.Y. Umetsu, T. Kanomata, R. Kainuma, The thermal transformation arrest phenomenon in NiCoMnAl Heusler alloys. Metals **3**, 298 (2013)
50. S.K. Ghatak, D.K. Ray, Structural and magnetic instabilities in a twofold-degenerate band. Phys. Rev. B **31**, 3064 (1985)
51. D.K. Ray, J.P. Jordan, Elastic and magnetic interactions in a narrow twofold-degenerate band. Phys. Rev. B **33**, 5021 (1986)
52. J.L. Shen, D.W. Zhao, G.K. Li, L. Ma, L.Y. Jia, C.M. Zhen, D.L. Hou, Kinetic arrest and de-arrest in $Mn_{50}Ni_{36}Sn_9Co_5$ ferromagnetic shape memory alloy. Phys. Status Solidi B **253**, 1923 (2016)
53. V.K. Sharma, M.K. Chattopadhyay, S.K. Roy, Kinetic arrest of the first-order austenite to martensite phase transition in $Ni_{50}Mn_{34}In_{16}$: DC magnetization studies. Phys. Rev. B **76**, 140401(R) (2007)
54. J.L. Sánchez Llamazares, B. Hernando, J.J. Suñol, C. Facia, C.A. Ross, Kinetic arrest of direct and reverse martensitic transfromation and exchange bias effect in $Mn_{49.5}Ni_{40.4}In_{10.1}$ melt spun ribbons. J. Appl. Phys. **107**, 09A956 (2010)
55. J.A. Monroe, J.E. Raymond, X. Xu, N. Nagasako, R. Kainuma, Y.I. Chumlyakov, R. Arróyave, I. Karaman, Multiple ferroic glasses via ordering. Acta Mater. **101**, 107 (2015)
56. F.H. Stillinger, P.G. Debenedetti, T.M. Truskett, The Kauzmann paradox revisited. J. Phys. Chem. B **105**, 11809 (2001)
57. C. Segui, E. Cesari, P. Lázpita, Magneitc properties of martensite in metamagnetic Ni-Co-Mn-Ga alloys. J. Phys. D: Appl. Phys. **49**, 165007 (2016)
58. M. van Schilfgaarde, I.A. Abrikosov, B. Johansson, Origin of the Invar effect in iron-nickel alloys. Nature **400**, 46 (1999)
59. P. Lloveras, T. Castán, M. Porta, A. Saxena, A. Planes, Mesoscopic modelling of strain glass, in *Frustrated Materials and Ferroic Glasses*, Springer Series in Materials Science 275 (Springer, Cham, 2018)
60. A. Grünebohm, H.C. Herper, P. Entel, On the rich magnetic phase diagram of (Ni, Co)-Mn-Sn Heusler alloys. J. Phys. D: Appl. Phys. **49**, 395001 (2016)
61. S. Mankovsky, unpublished data
62. L. Sandratskii, unpublished data
63. D.L. Schlagel, R.W. McCallum, T.A. Lograsso, Influence of solidification microstructure on the magnetic proerties of Ni-Mn-Sn Heusler alloys. J. Alloys Comp. **463**, 38 (2008)

64. W.M. Yuhasz, D.L. Schlagel, Q. Xing, K.W. Dennis, R.W. McCallum, T.A. Lograsso, Influence of annealing and phase decomposition on the magnetostructural transitions in $Ni_{50}Mn_{39}Sn_{11}$. J. Appl. Phys. **105**, 07A921 (2009)
65. W.M. Yuhasz, D.L. Schlagel, Q. Xing, R.W. Callum, T.A. Lograsso, Metastability of ferromagnetc Ni-Mn-Sn alloys. J. Alloys Comp. **492**, 681 (2010)
66. E. Cesari, J. Font, J. Muntashell, P. Ochin, J. Pons, R. Santamarta, Thermal stability of high-temperature Ni-Mn-Ga alloys. Scr. Mater. **58**, 259 (2008)
67. J.R. Aseguinoaza, V. Golub, O.Y. Salyuk, B. Muntifering, W.B. Knowlton, P. Müllner, J.M. Barandiaran, V.A. Chernenko, Self-patterning of epitaxial Ni-Mn-Ga/MgO(001) thin films. Acta Mater. **111**, 163 (2016)
68. D. Merida, J.A. Garcia, E. Apiãniz, F. Plazaola, V. Sanchez-Alarcos, J. Pérez Landazábal, V. Recarte, Positron annihilation spectroscopy study of Ni-Mn-Ga ferromagnetic shape memory alloys. Phys. Proc. **35**, 57 (2012)
69. D. Merida, J.A. Garćia, V. Sánchez-Alarcos, J.I. Perez-Landazábal, V. Recarte, F. Plazaola, Characterization and Modelling of vacancy dynamics in Ni-Mn-Ga ferromagnetic shape memory alloys. J. Alloys Comp. **639**, 180 (2015)
70. A.M. Pérez-Sierra, J. Pons, R. Santamarta, P. Vernaut, P. Ochin, Solidification process and effect of thermal treatments on Ni-Co-Mn-Sn metamagentic shape memory alloys. Acta Mater. **93**, 164 (2015)
71. V. Sánchez-Alarcos, J.I. Pérez-Landazábal, V. Recarte, I. Lucia, J. Vélez, J.A. Rodríguez-Velamazán, Effect of high temperatue quenching on the magnetostructural transformations and long-range atomic order of Ni-Mn-Sn adn Ni-Mn-Sb metamagetic shape memory alloys. Acta Mater. **61**, 4676 (2013)
72. M.K. Ray, K.Bagani, S. Banerjee, Effect of excess Ni on martensitic transition, exchange bias and inverse magnetocaloric effect in $Ni_{2+x}Mn_{1.4-x}Sn_{0.6}$ alloy. J. Alloys Comp. **600**, 55 (2014)
73. P. Entel, V.V. Sokolovskiy, V.D. Buchelnikov, M. Ogura, M.E. Gruner, A. Grünebohm, D. Comtessse, H, Akai, The metamagnetic behavior and giant inverse magnetocaloric effect in Ni-Co-Mn-(Ga, In, Sn) Heusler alloys. J. Magn. Magn. Mater. **385**, 193 (2015)
74. A. Çakır, M. Acet, Non-volatile high-temperature shell-magnetic pinning of Ni-Mn-Sn Heusler precipitates obtained by decomposition under magnetic field. J. Magn. Magn. Mater. **448**, 13 (2018)
75. T. Krenke, A. Çakır, F. Scheibel, M. Acet, M. Farle, Magnetic proximity effect and shell-ferromagnetism in metastable $Ni_{45}Mn_{45}Ga_5$. J. Appl. Phys. **120**, 243904 (2016)
76. A. Çakır, M. Acet, Shell ferromagnetism in Ni-Mn based Heuslers in view of ductile Ni-Mn-Al. AIP Adv. **7**, 056424 (2017)
77. T. Miyamoto, W. Ito, R.Y. Umetsu, R. Kainuma, T. Kanomata, K. Ishida, Phase stability and magnetic properties of $Ni_{50}Mn_{50-x}In_x$ Heusler type alloys. Scr. Mater. **62**, 151 (2010)
78. T. Miyamoto, W. Ito, R.Y. Umetsu, T. Kanomata, K. Ishida, R. Kainuma, Influence of annealing conditions on magnetic properties of $Ni_{50}Mn_{50-x}In_x$ Heusler type alloys. Mater. Trans. JIM **52**, 1836 (2011)
79. R. Kainuma, F. Gejima, Y, Sutou, I. Ohnuma, K. Aoki, K. Ishida, Ordering, martensitic and ferromagnetic transformations in Ni-Al-Mn Heusler shape memory alloys. Mater. Trans. JIM **41**, 943 (2000)
80. R.W. Overholser, M. Wuttig, D.A. Neumann, Chemical ordering in Ni-Mn-Ga Heusler alloys. Scr. Mater. **40**, 1095 (1999)
81. W.L. Bragg, E.J. Williams, The effect of thermal agitation on atomic arrangement in alloys. Proc. R. Soc. A **145**, 699 (1934)
82. J.S. Kaspar, J.S. Kouvel, The antiferromagnetic structure of NiMn. J. Phys. Chem. Solids **11**, 231 (1959)
83. M. Siewert, Electronic, magnetic and thermodynamic properties of magnetic shape memory alloys from first principles. PhD thesis, University of Duisburg-Essen, 2012

84. E. Krén, E. Nagy, I. Nagy, L. Pál, P. Szabó, Structures and phase transformations in the Ni-Mn system near equiatomic concentration. J. Phys. Chem. Solids **29**, 101 (1968)
85. N.M. Bruno, S. Wang, I. Karaman, Y.I. Chumlyakov, Reversible martensitic transformation under low magnetic fields in magnetic shape memory alloys. Sci. Rep. **7**, 40434 (2017)
86. X. Moya, S. Kar-Narayan, N.D. Mathur, Caloric materials near ferroic phase transitions. Nat. Mater. **13**, 439–450 (2014)
87. S. Fähler, U. Rößler, O. Kastner, J. Eckert, G. Eggeler, H. Emmerich, P. Entel, S. Müller, E. Quandt, K. Albe, Caloric effects in ferroic materials: new concepts for cooling. Adv. Eng. Mater. **14**, 10–19 (2012)
88. P. Entel, M.E. Gruner, S. Fähler, M. Acet, A. Çahır, R. Arróyave, S. Sahoo, T.C. Duong, A. Talapatra, L. Sandratskii, S. Mankowsky, T. Gottschall, O. Gutfleisch, P. Lázpita, V.A. Chernenko, J.M. Barandiaran, V.V. Sokolovskiy, V.D. Buchelnikov, Probing Structural and Magnetic Instabilities and Hysteresis in Heuslers by Density Functional Theory Calculations. Phys. Status Solidi B **255**, 1700296 (2018). https://doi.org/10.1002/pssb.201700296
89. M.E. Gruner, R. Niemann, P. Entel, R. Pentcheva, U.K. Rössler, K. Nielsch, S. Fähler, Modulations in martensitic Heusler alloys originate from nanotwin ordering. Sci. Rep. **8**, 8489 (2018)

Chapter 7
Strain Glasses

Yuanchao Ji, Shuai Ren, Dong Wang, Yu Wang, and Xiaobing Ren

Abstract Since its discovery in 2005 in a Ni-rich Ti-Ni alloy, strain glass has drawn much attention in martensite/ferroelastic community and has been reported to be a rather general phenomenon in martensitic/ferroelastic systems. In this chapter we present a review of strain glass, including a brief history, its physical origin, its generic phase diagram, followed by presenting strain glasses in various martensitic/ferroelastic systems induced by 0D defects (point defects), 1D defects (dislocations), and 3D defects (nano-precipitates), respectively. The material systems include Ti-Ni-based systems, Ti-Pd-based systems, Ti-based alloys, ferromagnetic shape memory alloys and ceramics. We further show that strain glass can result in technologically important properties such as the "Gum metal" properties, high damping, and giant magnetostriction at small field, thus making strain glass a promising candidate for novel structural/functional materials. Finally, a "ferroic glass" concept (the glass form of ferroic materials) is introduced by combining three physically parallel glasses: strain glass, relaxor ferroelectrics, and cluster spin glass. It is expected that ferroic glasses and generic ferroic glasses may yield unique mechanical, electrical, magnetic properties, and thus may lead to a new class of structural/functional materials.

7.1 A Brief History of Strain Glass

Since 1960s martensite community has been plagued by a puzzling phenomenon known as "premartensitic tweed," which refers to a cross-hatched or mottled

---

Y. Ji · S. Ren · D. Wang · Y. Wang
Frontier Institute of Science and Technology and State Key Laboratory for Mechanical Behaviour of Materials, Xi'an Jiaotong University, Xi'an, China

X. Ren (✉)
Frontier Institute of Science and Technology and State Key Laboratory for Mechanical Behaviour of Materials, Xi'an Jiaotong University, Xi'an, China

Center for Functional Materials, National Institute for Materials Science, Tsukuba, Ibaraki, Japan
e-mail: ren.xiaobing@nims.go.jp

T. Lookman, X. Ren (eds.), *Frustrated Materials and Ferroic Glasses*, Springer Series in Materials Science 275, https://doi.org/10.1007/978-3-319-96914-5_7

nanoscale microstructure appearing well above the onset of martensitic transformation in many systems [1], and it has been subject to intensive debate [2]. The term "premartensitic" signifies that the phenomenon occurs prior to a martensitic transformation [3, 4] and also implies the linkage to the forthcoming martensitic transformation.

Experiments indicate the tweed is an early stage of the low temperature phase, the martensite, and thus it can be viewed as a "baby martensite" [3]. However, this poses a big challenge to the classical theory of martensitic transformation, because it is hard to understand why static (not dynamic) baby martensite can stay stable over 100K above the martensitic transformation temperature. According to standard theory of martensitic transformation [5], if stable martensite nuclei (i.e., the baby martensite domains) are formed, martensitic transformation should have happened, as nucleation is the bottleneck of a martensitic transformation.

A number of models [6–8] have been developed to account for the mysterious stability of the baby martensite while without triggering a martensitic transformation, so as to circumvent the difficulty with the standard theory of martensite. These models take into account the role of inhomogeneities (or point defects) in stabilizing the baby martensite but without prematurely triggering a formal martensitic transformation. As a result, the models predict that there is a stable premartensitic tweed temperature region prior to the formal martensitic transformation and the tweed should eventually transform into martensite at low temperature. However, such models are challenged by experimental findings that similar "premartensitic tweed" occurs in non-transforming compositions [3] where the premartensitic microstructure does not end up into martensite.

In 2005 a critical experiment was performed for a non-transforming Ti-51.5Ni alloy with mottled premartensitic nanodomains [9]. It was a dynamical mechanical analysis (DMA) measurement, which measures the mechanical response of the system to small AC stresses, in analogous to the magnetic susceptibility measurement for magnetic systems. The experiment revealed that the hitherto unexplained non-martensitic premartensitic alloy exhibits a frequency-dependent modulus dip (Fig. 2 of Ref. 9) with the dip temperature following Vogel-Fulcher relation. This behavior very much resembles that found in spin glass and relaxor ferroelectrics (or electric dipole glass), and strongly suggests a glass transition of lattice strain. Thus it is named "strain glass." Mechanical ZFC/FC (zero-field-cool/field-cool) measurement [10] further shows characteristic branching of ZFC/FC curves, evidence of the non-ergodicity—another important signature of glass. Up to this point, the long-standing puzzle about the premartensitic tweed was solved, and it is a glass form of martensitic/ferroelastic systems, a mechanical analog of spin glass in magnetic systems and relaxor in ferroelectric systems [11, 12]. Glass features in spin glass and relaxor ferroelectrics can be referred to chapters by Sherrington and Kleemann in this volume [13, 14].

The experimental signatures of strain glass have been shown to be [11] (1) frequency-dependent modulus dip or loss peak following Vogel-Fulcher relation, (2) branching in mechanical ZFC/FC (zero-field-cool/field-cool) curves, (3) invariance of average structure during heating/cooling or absence of martensitic transformation, and (4) existence of nanosized strain domains or local strain ordering.

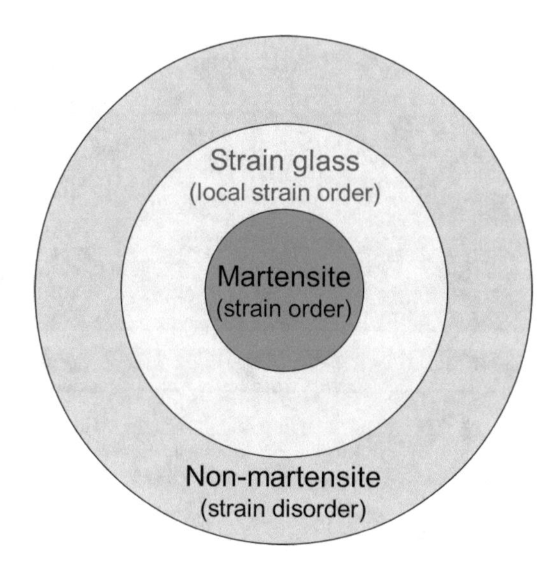

Fig. 7.1 Relationship among martensite, strain glass, and non-martensite revealed by the different strain ordering (i.e., long-range strain order, short-range strain order, and strain disorder, respectively) [11]

The relationship among martensite, strain glass, and non-martensite is demonstrated in Fig. 7.1. Martensite can be considered as "strain crystal"—a long-range ordered form of lattice strain. By contrast, strain glass is a "strain jelly"—a frozen local ordered form of lattice strain. Normal non-martensitic alloys can be considered as dynamically disordered strains without static local strain, in analogous to a normal paramagnetic material.

7.2 Origin of Strain Glass and a Generic Phase Diagram

Strain glass has been successfully modelled in phase-field models such as in Refs 15, 16; these models enable qualitative or quantitative explanation of the observed strain glass features. Here, we provide a tutorial "domino-and-stone model" (Fig. 7.2) that enables one to "feel" why and how strain glass is formed.

Figure 7.2a shows that martensitic transformation can be viewed as the long-range toppling of a domino chain when there are no "stones/defects" to disturb the process. Figure 7.2b shows that with doping "stones/defects" into the domino chain in the parent state, the long-range ordering is prohibited and instead a short-range toppling will appear, forming a locally ordered "strain glass (STG)" state. Such a local ordered domino phenomenon originates from different preference of each stone when the neighboring tile is toppled. This is a pedagogical picture of strain glass formation.

Wang et al. recently have provided a rigorous modelling and phase-field simulations on strain glass [16, 17] based on the above picture, and their results reproduce all known features of strain glass, including phase diagram. A detailed discussion about modelling can refer to another chapter of this book [17].

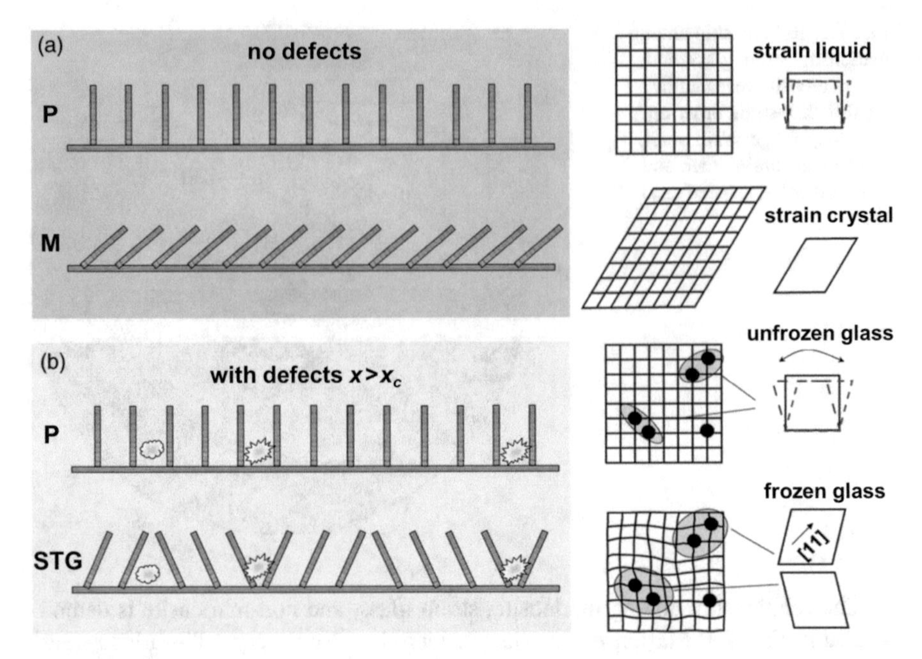

Fig. 7.2 A "domino-and-stone model" showing the microscopic formation mechanism of martensite and strain glass [11]. (**a**) Long-range toppling (strain ordering) without stones (defects). (**b**) Short-range toppling (strain ordering) with many stones (defects). P, M, and STG represent the parent phase (strain liquid), martensite (strain crystal), and strain glass, respectively

The above simple "domino-and-stone model" also shows how to produce strain glass, i.e., doping "stones" into the "domino chain." Up to now, experimental results indeed confirm that the strain glass can be generated through doping sufficient defects into a martensitic system [9, 18–22].

Figures 7.3 and 7.4 show a generic phase diagram of a strain glass system, which has been experimentally verified in all strain glass systems. The phase diagram contains four states of a ferroelastic system: normal parent phase (strain liquid), unfrozen strain glass (tweed), martensite (long-range strain order or strain crystal), and a strain glass (frozen local strain order or strain jelly) can be seen and the relationship among them are shown as follows. At high temperature, the system is in a normal parent phase, which can be viewed as an ideal strain liquid state with dynamically disordered lattice strain (i.e., lattice vibration). With temperature decreasing to T_{nd} (nanodomain formation temperature), the strain liquid becomes sticky, i.e., some quasi-static clusters of lattice strain appears. This state was previously named "precursor" or "premartensitic state." Upon further cooling, the strain liquid becomes stickier because of the formation of more quasi-static clusters. Finally, it either transforms into martensite or freezes into a strain glass depending on defect concentration. At low defect concentration ($x < x_c$), the system transforms into a martensite (strain crystal) due to the large thermodynamic driving force and small local energy barrier. However, when the defect concentration is beyond a

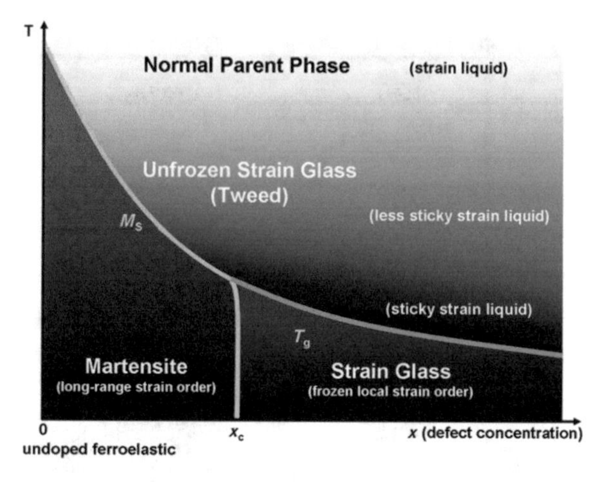

Fig. 7.3 A generic phase diagram of a strain glass system, where four different strain states (normal parent phase, martensite, tweed, and strain glass) exist [11]. A strain glass transition occurs when the defect concentration x exceeds a critical value x_c. M_s and T_g represent the martensitic transformation temperature and strain glass transition temperature, respectively

critical value ($x > x_c$), the local barrier becomes large enough to prevent a long-range strain ordering or martensite formation, and the system freezes into a strain glass (frozen strain liquid).

This generic strain glass phase diagram (Fig. 7.3) can be used as a guide to designing strain glasses. To discover or design a strain glass, one should first select a system exhibiting normal martensitic transformation and then dope defects to suppress the transition. When defect-doping level is above a critical concentration x_c, the martensitic transformation disappears and the system crossovers into strain glass.

So far, defects that make a strain glass have been found to be one of the following types according to their dimensions: point defects (0D), dislocations (1D), and nanosized precipitates (3D). Probably 2D defects (like boundaries) can also make a strain glass but this awaits future investigation. In the following sections we shall provide examples showing strain glass induced by point defects (0D), dislocations (1D), and nanosized precipitates (3D).

7.3 Strain Glass Induced by Point Defects

The first strain glass system reported is the famous Ti-Ni system in Ni-rich ($Ti_{50-x}Ni_{50+x}$) compositions, where excess Ni atoms are the point defects [9]. The transition behaviors as a function of defect concentration are shown in Fig. 7.5, which appears to be a common feature of all strain glass systems [19]. At the low defect concentration, a martensitic transformation occurs, characterized by a DSC peak/dip, the presence of hysteresis in the resistivity curve, and the frequency independence of elastic modulus dip temperature and of mechanical loss peak temperature. Above a critical concentration ($x > 1.3$), strain glass appears, which is characterized by the disappearance of DSC peak/dip, the absence of hysteresis

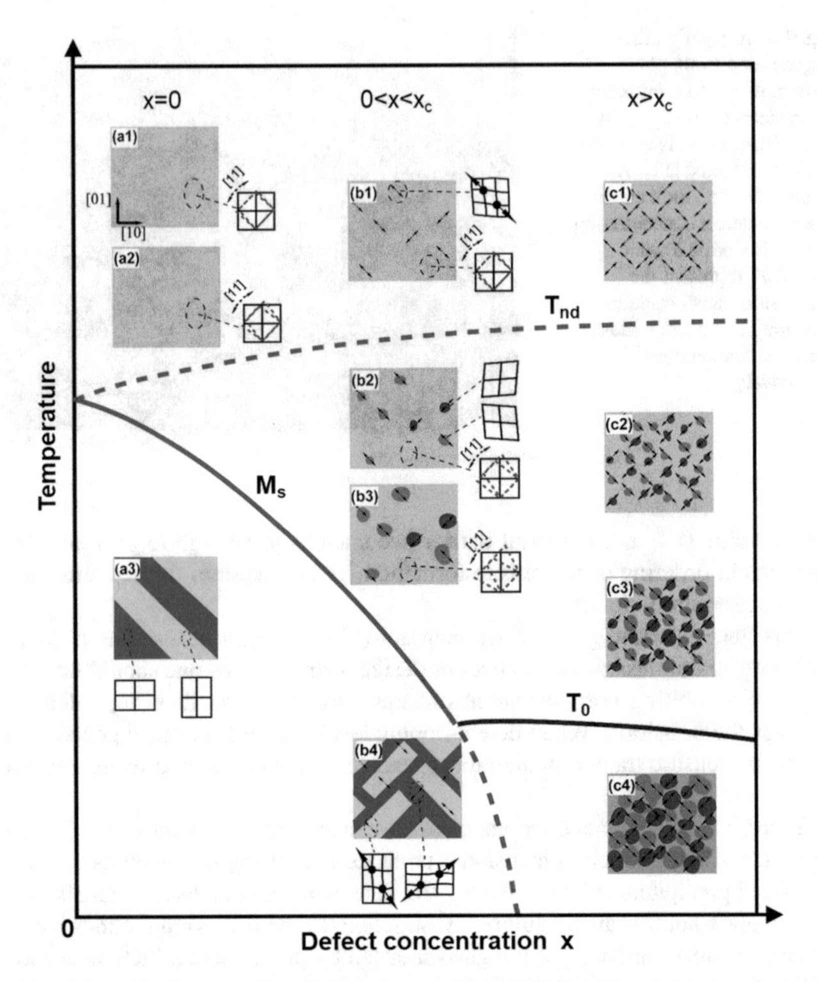

Fig. 7.4 Schematic demonstration of the microstructure evolution of a strain glass system as a function of temperature and defect concentration [19]. Arrows indicate the point-defects-induced random stress. T_0 and T_{nd} represent the ideal strain glass transition temperature and the nanodomain formation temperature, respectively

in the resistivity curve, and the frequency dependence of elastic modulus dip temperature and of mechanical loss peak temperature. The frequency dependence follows the Volgel-Fulcher relation ($\omega = \omega_0 \exp(-E_a/k_B(T_g - T_0))$, a key signature of strain glass transition. Other glass signatures such as branching of ZFC/FC curves and invariance of average structure, local strain ordering can be referred to Refs. [10–12].

Figure 7.6 shows the strain glass phase diagram of binary $Ti_{50-x}Ni_{50+x}$ ($0 < x < 3$) alloys [19]. Besides the well-known parent phase (B2) and martensite phase (B19′) in the previously reported phase diagram, two new states of precursor phase and strain glass are also shown. Although both states are characterized by a

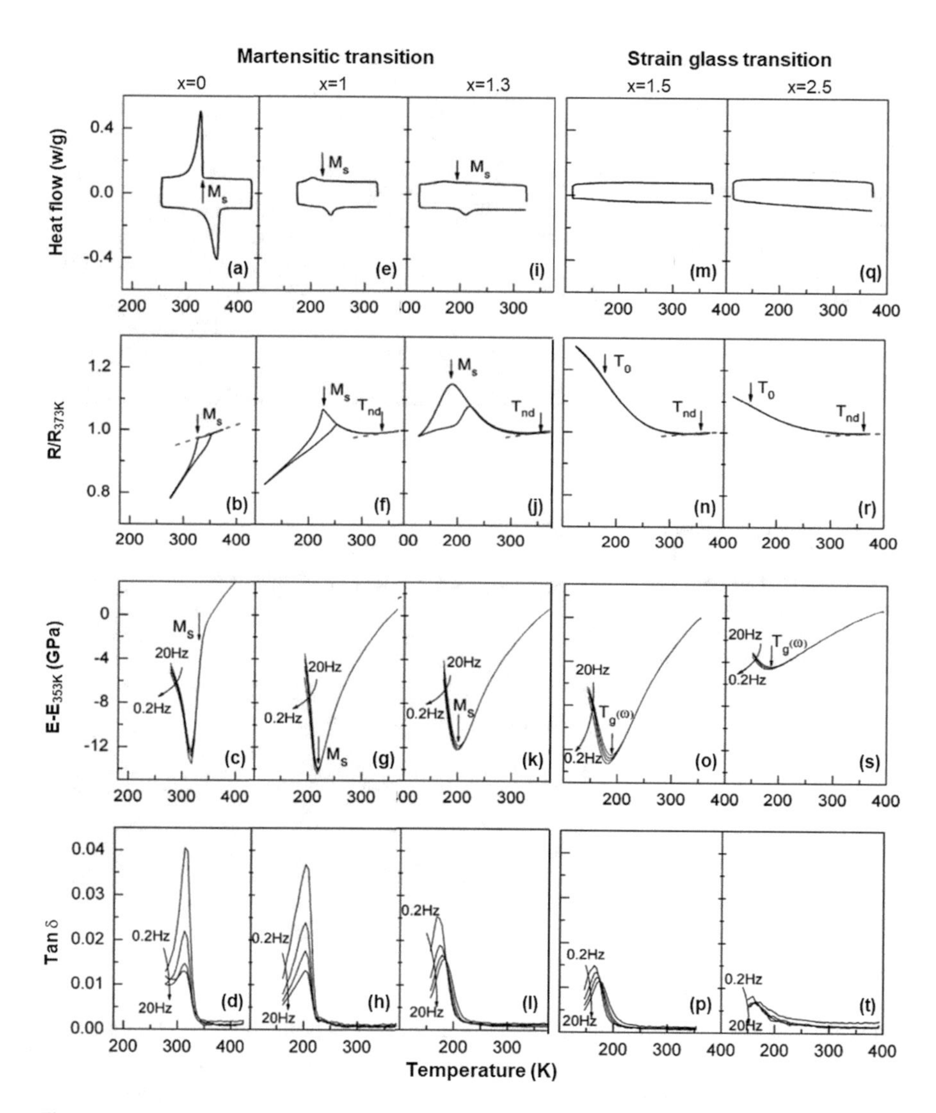

Fig. 7.5 Transition behavior of $Ti_{50-x}Ni_{50+x}$ as a function of defect concentration, which shows a crossover from martensitic transformation into strain glass transition [19]. (**a**), (**e**), (**i**), (**m**), and (**q**) show heat flow curves; (**b**), (**f**), (**j**), (**n**), and (**r**) show the normalized electrical resistivity; (**c**), (**g**), (**k**), (**o**), and (**s**) show the normalized storage modulus; (**d**), (**h**), (**l**), (**p**), and (**t**) show the internal friction. $T_g(\omega)$ represents the freezing temperature at different frequency

short-range strain ordering, the former is primarily a dynamically disordered strain state with some quasi-static strain nanodomains, the latter is a disordered strain state with frozen nano-sized strain domains.

After the discovery of strain glass in Ti-Ni system, more strain glasses were found in ternary Ti-Ni-X (X = Fe, Co, Cr, Mn) systems [20, 21] as shown in

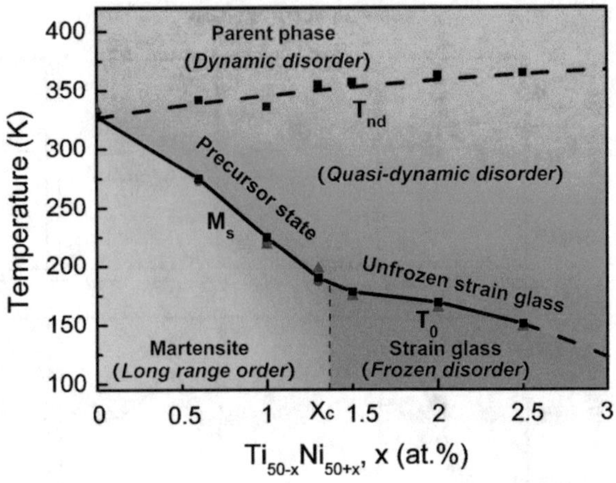

Fig. 7.6 Strain glass phase diagram of binary $Ti_{50-x}Ni_{50+x}$ alloys, where the excess Ni atoms act as point defects [19]. With doping more Ni atoms, the martensitic transformation gradually suppresses and instead the strain glass transition appears

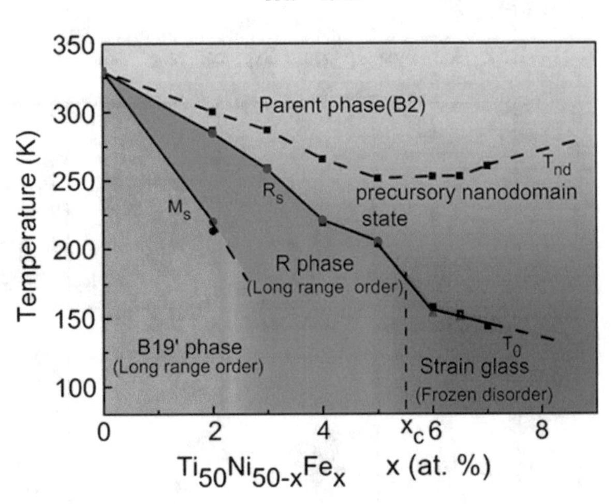

Fig. 7.7 Strain glass phase diagram of ternary Ti-Ni-Fe ($Ti_{50}Ni_{50-x}Fe_x$) alloys, where the Fe atoms act as point defects [20]. R_s represents the R-martensitic transformation temperature

Figs. 7.7 and 7.8. Here, the substitutional X atoms act as point defects. The transition behavior as a function of defect concentration is similar with that of Ti-Ni alloys as shown in Fig. 7.5. In these ternary Ti-Ni-X systems an R martensite is found to exist and locates between B19′ martensite and strain glass. The existence of R martensite in Ti-Ni-X ternary systems explains a puzzle why in Ti-Ni binary system the strain glass does not have a B19′ local symmetry but a R-like local symmetry. This is because R martensite is disfavored relative to B19′ for long-range strain ordering but favored for local strain ordering due to its small strain.

Strain glass in Ti-Ni-based systems appears well below room temperature, as shown above. Figure 7.9 shows an example that a room temperature strain glass can be achieved by doping point defects into high temperature martensitic systems Ti-Pd-X (X = Fe, Mn, Cr) [23–25]. This may enable potential application of strain glass at room temperature or even higher temperatures.

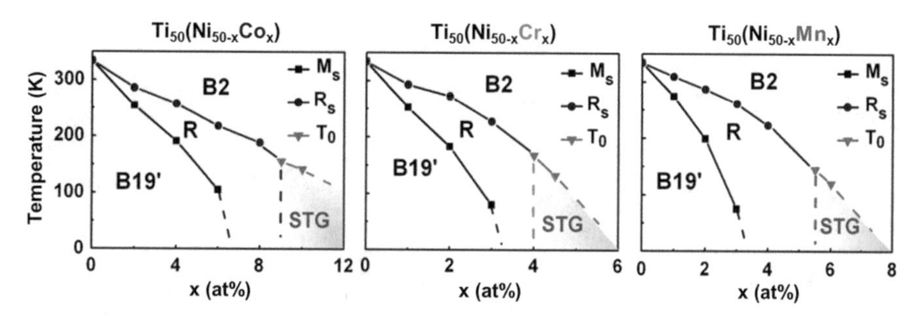

Fig. 7.8 Strain glass phase diagram of ternary Ti-Ni-X (X = Co, Cr, Mn) alloys, where the X atoms act as point defects [21]

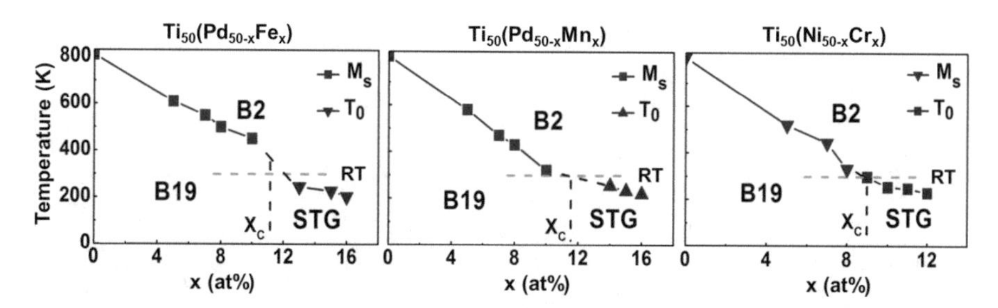

Fig. 7.9 Strain glass phase diagram of high temperature shape memory alloys Ti-Pd-X (X = Fe, Mn, Cr), where the X atoms act as point defects [24]

Figure 7.10 shows the strain glass phase diagram of Ti-xNb shape memory alloys [26]. With increasing Nb concentration, the martensitic transformation from the β-phase (BCC) to the α′′-phase (orthorhombic) is gradually suppressed and it finally disappears above a critical value (x_c~25). When $x > 25$, the system undergoes a strain glass transition. The GUM metal (Ti-23Nb-0.7Ta-2Zr-1.2O), which is known to exhibit interesting properties such as low modulus, superelasticity, and Invar and Elinvar effects after heavy plastic deformation, turned out to be a strain glass alloy [26]. Recently, more strain glass systems have been reported in β-Ti alloys such as Ti-26Nb-xO, Ti–xNb–2Zr–0.7Ta–1.2O, Ti–23Nb–2Zr–0.7Ta–xO, Ti–24Nb–4Zr–8Sn (Ti2448), and Ti-30Nb-1Mo-4Sn [27–29].

Figure 7.11 shows the strain glass phase diagram of $Ni_{55-x}Co_xMn_{20}Ga_{25}$ [30] and $Ni_{55-x}Co_xFe_{20}Ga_{25}$ [31] Heusler alloys, which are known as ferromagnetic shape memory alloys (FSMAs). Similar with the nonmagnetic strain glass systems, the martensitic transformation crossovers into strain glass transition by doping Co. In this magnetic system it is noted that the ferromagnetic transition deviates from the martensitic/strain glass transition with increasing the Co concentration, although two transitions coincide at the lower Co compositions.

Very recently, strain glass has been reported in another ferromagnetic shape memory alloy system Fe-Pd [32]. The strain glass phase diagram of $Fe_{100-x}Pd_x$

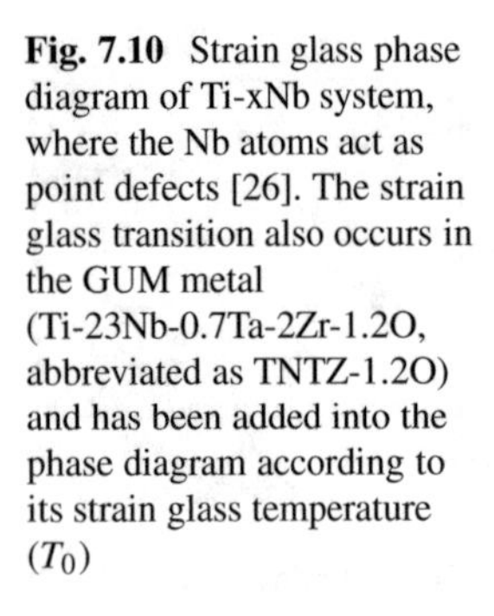

Fig. 7.10 Strain glass phase diagram of Ti-xNb system, where the Nb atoms act as point defects [26]. The strain glass transition also occurs in the GUM metal (Ti-23Nb-0.7Ta-2Zr-1.2O, abbreviated as TNTZ-1.2O) and has been added into the phase diagram according to its strain glass temperature (T_0)

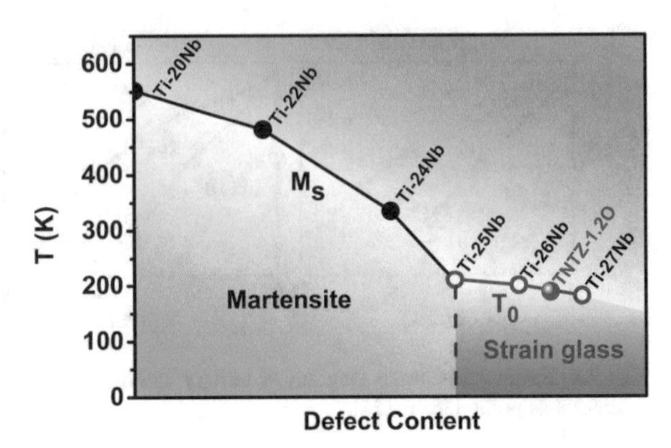

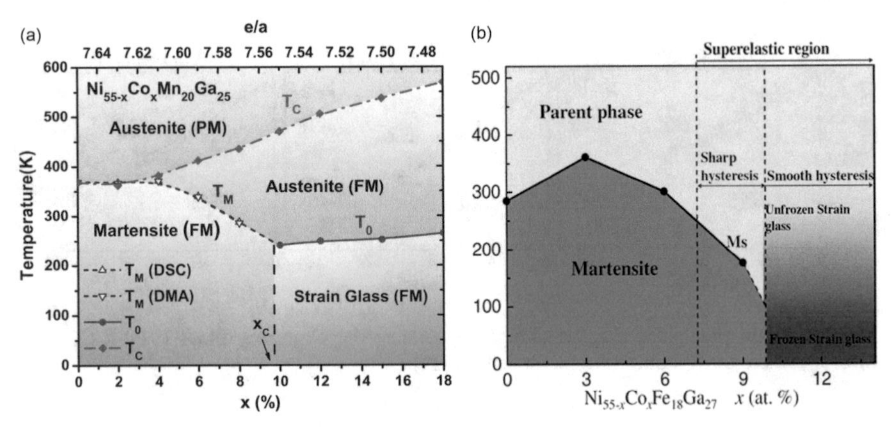

Fig. 7.11 Strain glass phase diagram of ferromagnetic shape memory alloys (**a**) $Ni_{55-x}Co_xMn_{20}Ga_{25}$ [30] and (**b**) $Ni_{55-x}Co_xFe_{18}Ga_{27}$ [31] Heusler alloys, where the Co atoms act as point defects

(Fig. 7.12a) resembles those of Heusler alloys as shown in Fig. 7.11. In the strain glass state of $Fe_{67.7}Pd_{32.3}$ a remarkable property has been found, i.e., a low-field-triggered large magnetostriction (Fig. 7.12b), which is advantageous to existing magnetostrictive materials like Terfenol-D and Ni-Mn-Ga. This finding indicates that ferromagnetic strain glass may be an effective approach to design high-performance magnetostrictive materials.

Strain glass is also reported in nonmetallic martensitic/ferroelastic ceramics. Figure 7.13 shows the strain glass phase diagram of $Bi_{1/2}Na_{1/2}TiO_3$-$xBaTiO_3$ (BNT-xBT) perovskite ceramics [33]. With doping BT into BNT, at low concentration (e.g., $x = 2$) two ferroelastic/ferroelectric transitions are observed, manifested by two dips in the modulus curves and two peaks in the mechanical loss curves (Fig. 7.14a). By contrast, at high concentration (e.g., $x = 6$) only strain glass transition occurs (Fig. 7.14c). Besides BNT-BT system, strain glass has also been reported in another ferroelastic ceramic system, La-doped $CaTiO_3$ [34]. The mechanical and functional properties of strain glass ceramics await future investigation.

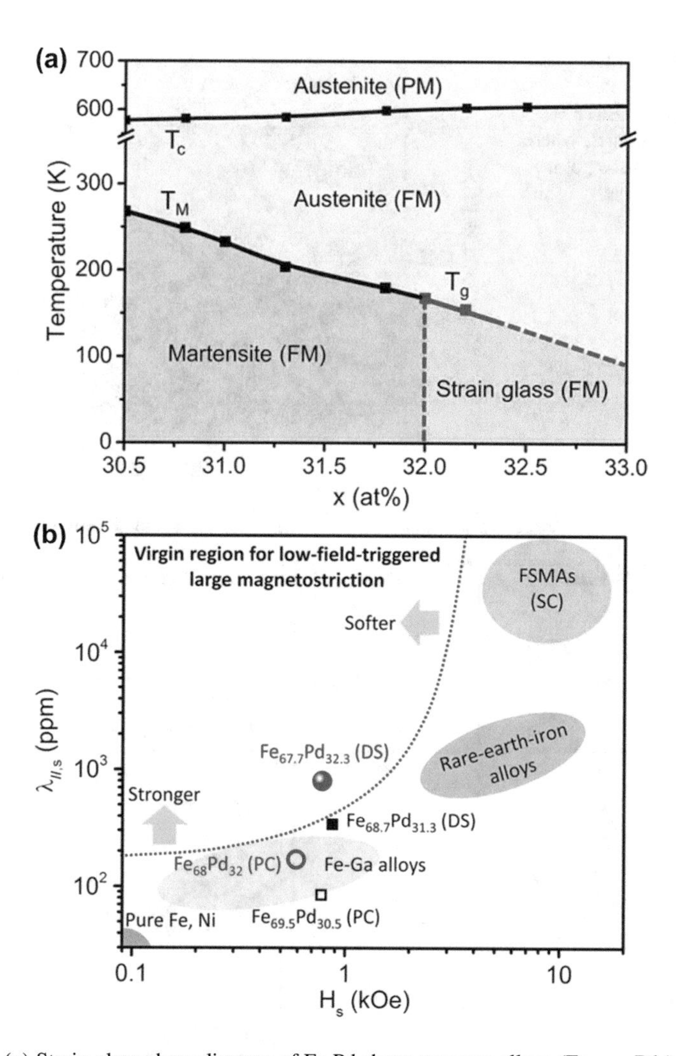

Fig. 7.12 (**a**) Strain glass phase diagram of Fe-Pd shape memory alloys ($Fe_{100-x}Pd_x$), where the Pd atoms act as point defects [32]. (**b**) Comparison of the magnetostrictive benchmark ($\lambda_{//,s}$ vs H_s) between Fe-Pd alloys and main magnetostrictive families, including single crystal (SC) of FSMAs, rare-earth-iron Laves phase alloys, and Fe-Ga alloys [32]. DS and PC represent directionally solidfied alloys and polycrystalline alloys, respectively

$Au_7Cu_5Al_4$ alloy has been reported to exhibit features of the strain glass transition, but upon further cooling a martensitic transformation occurs [35, 36]. A $Ni_{45}Co_5Mn_{36.6}In_{13.4}$ alloy has been reported to show the absence of a thermally induced martensitic transformation but the existence of a stress-induced martensitic transformation (i.e., superelasticity) [37]. It is likely that this alloy is also a strain glass and further investigation is necessary.

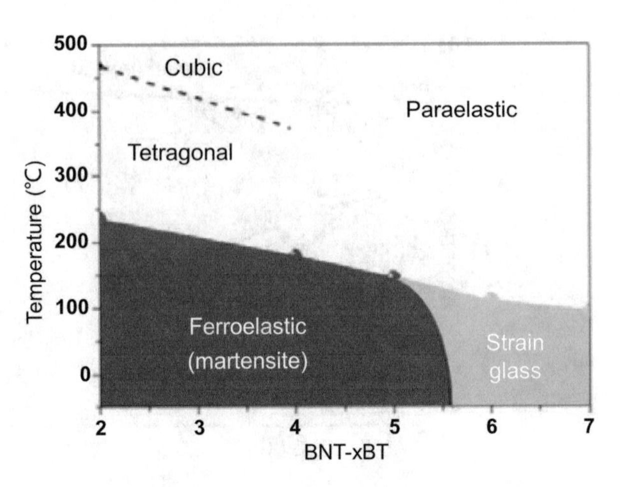

Fig. 7.13 Strain glass phase diagram of $Bi_{1/2}Na_{1/2}TiO_3$-$xBaTiO_3$ (BNT-xBT) ceramics, where the Ba^{2+} ions act as point defects [33]

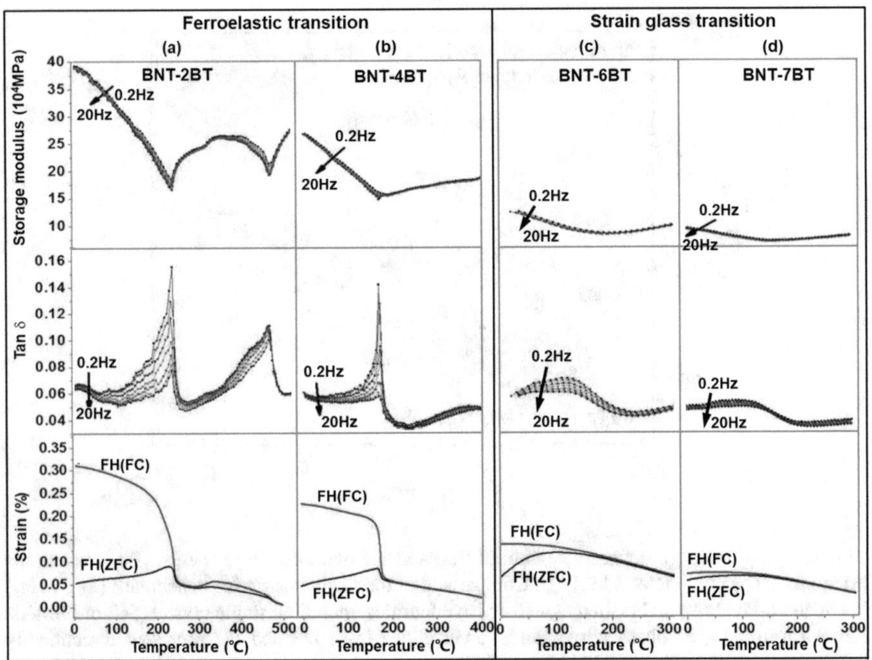

Fig. 7.14 (**a–d**) Transition evolution of BNT-xBT from the ferroelastic/martensitic transition ($x = 2, 4$) to the strain glass transition ($x = 6, 7$) with increasing the Ba^{2+} concentration [33]

7.4 Strain Glass Induced by Dislocations and Nano-Precipitates

Strain glasses introduced so far are induced by point defects that can be considered as 0D defects. In the following we shall show that strain glass can also be induced by introducing dislocations (1D defects) or nanoparticles (3D defects) [38–40].

Figure 7.15 shows a temperature vs. dislocation-density phase diagram of $Ti_{50}Ni_{45}Fe_5$ alloys, where dislocations are produced through cold-rolling [38]. Being similar with point-defect-induced strain glass phase diagram, strain glass appears when dislocation density is above a critical value ($\rho > \rho_c$). The nanodomains of strain glass exhibit an R-like local structure, which is a reminiscence of the B2-R martensitic transition.

Figure 7.16 shows another example of dislocation-induced strain glass in a $Ti_{49.2}Ni_{50.8}$ alloy [39]. With increasing plastic deformation (thickness reduction, ε_p), the B2-B19′ martensitic transformation gradually weakens and eventually vanishes above a critical value ($\varepsilon_p >$ 25%). Figures 7.16b, c show the existence of strain glass transition for the $\varepsilon_p = 27\%$ sample. Interestingly, this strain glass exhibits a quasi-linear slim-hysteretic superelasticity with a large recoverable strain of ~4% over a temperature range as wide as 200K (from 323 to 123K) (Fig. 7.16d).

Figure 7.17 shows the microstructure of $\varepsilon_p = 27\%$ strain glass sample at different temperatures [39]. From the diffraction pattern, two sets of satellite spots can be observed and are identified to be R-like spots (locating around $1/3(011)_{B2}$) and B19′-like spots (locating around $1/2(011)_{B2}$). In situ dark-field imaging shows that the density of R-like nanodomains remains unchanged while the density of B19′-like

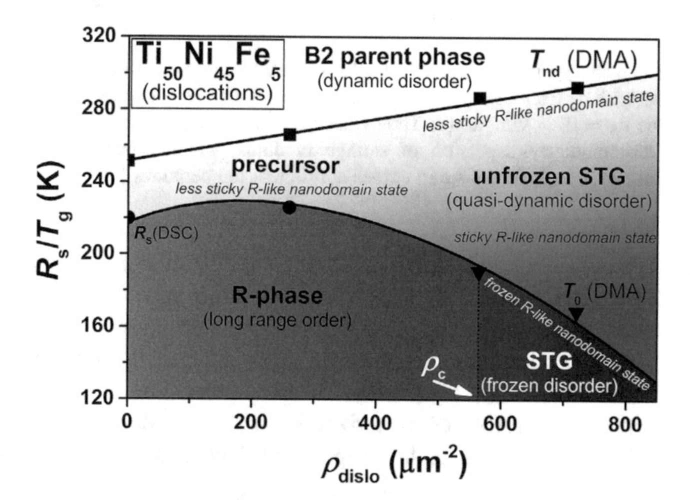

Fig. 7.15 Temperature vs. dislocation-density strain glass phase diagram of $Ti_{50}Ni_{45}Fe_5$ alloys, where dislocations act as the defects [38]. With increasing dislocation density, the B2-R martensite transition crossovers into a strain glass transition

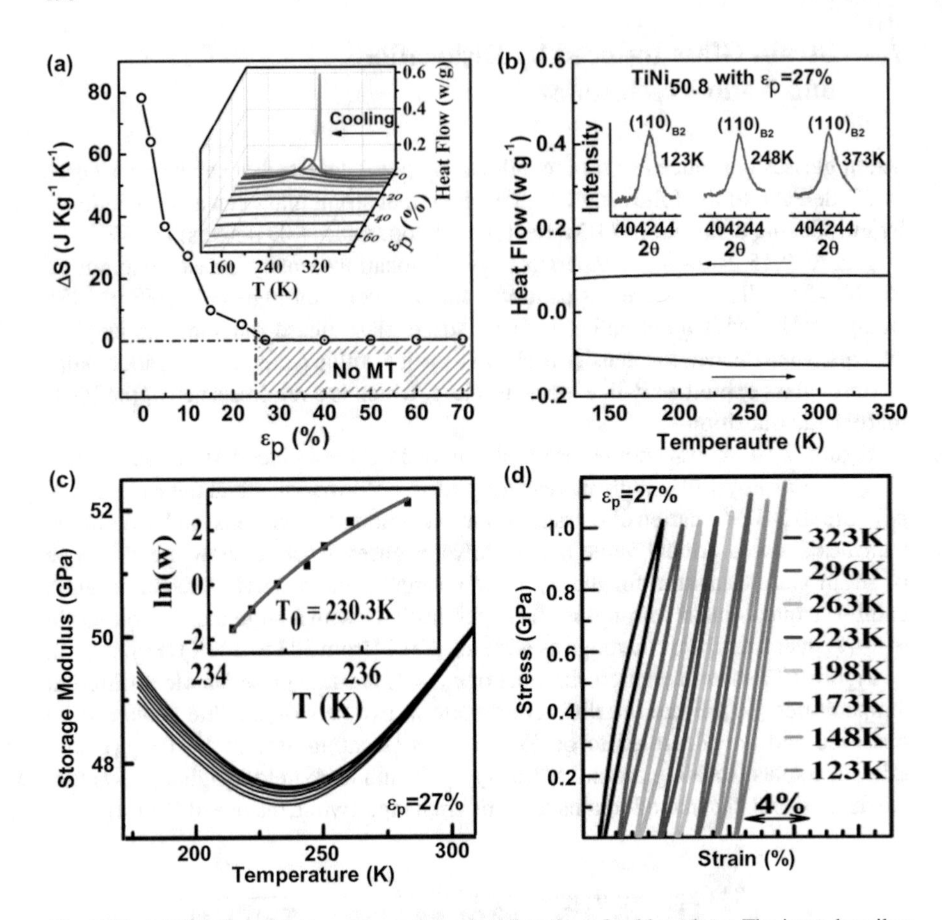

Fig. 7.16 (**a**) The change in transition entropy as a function of cold work ε_p. The inset describes the DSC results with ε_p from 0% to 70% upon cooling [39]. (**b**) The disappearance of martensitic transition in the $\varepsilon_p = 27\%$ sample. (**c**) The evidence for strain glass transition in the $\varepsilon_p = 27\%$ sample, i.e., the frequency-dispersion of storage modulus, which follows the Vogel-Fulcher relationship (Inset) [39]. (**d**) Stress-strain curves at different temperatures [39]

nanodomains increases upon cooling, indicating the B19′-like nanodomains playing a major role in the properties. Therefore, the large superelasticity with slim hysteresis can be explained by the growth of B19′-like nanodomains.

Figure 7.18 shows an example of strain glass induced by 3D defects (nano-precipitates) in a $Ti_{48.7}Ni_{51.3}$ alloy [40]. The precipitate-free sample undergoes a normal martensitic transition, characterized by a DSC peak/dip, the presence of hysteresis in the resistivity curve, the temperature-dependent structural change, and the frequency independence of elastic modulus dip temperature (Fig. 7.18a). By contrast, in the nano-precipitate-bearing sample, a strain glass transition occurs, being characterized by the absence of DSC peak/dip, the absence of hysteresis in the resistivity curve, no temperature-dependent structural change, and the frequency

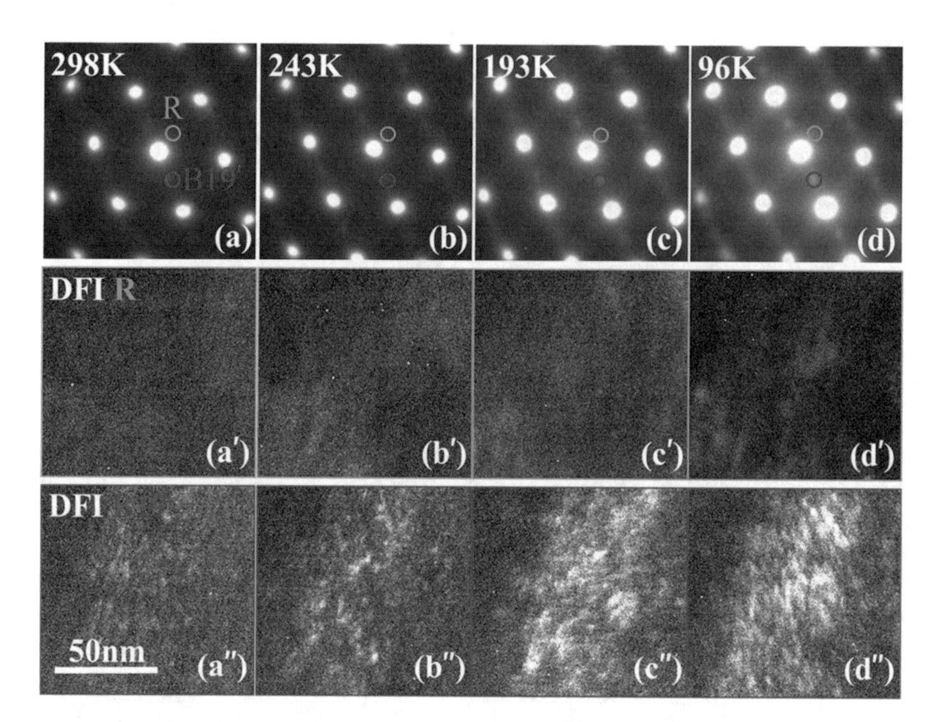

Fig. 7.17 (**a–d**) TEM dark-field images and corresponding diffraction patterns of $Ti_{49.2}Ni_{50.8}$ strain glass alloy cold rolled down to a thickness reduction of 27% at 298 K (>T_g), 243 K (~T_g), 193 K (<T_g), and 96 K, respectively [39]. The zone axis is $[111]_{B2}$

dependence of modulus dip temperature (Fig. 7.18b). This nano-precipitate-induced strain glass has R-like nanodomains, being the same as that of the point-defect-induced strain glass in the $Ti_{48}Ni_{52}$ alloy (Fig. 7.19). Another example of nano-precipitate-induced strain glass can be found in a Fe-based shape memory alloy system [41].

7.5 Competing Consequences of Defect-Doping in Ferroelastic/Martensite Systems: New Martensite vs. Strain Glass

It should be noted that doping defects into a ferroelastic/martensitic system does not always lead to strain glass; in many cases it leads to a new martensite. Examples include doping Pd or Cu into Ti-Ni alloys eventually lead to a new martensite B19 in Ti-Ni-Pd/Cu alloys, in contrast with the B19′ martensite in the undoped alloy. Therefore, there exists a competition between a new martensite and strain glass in some defect-doped systems.

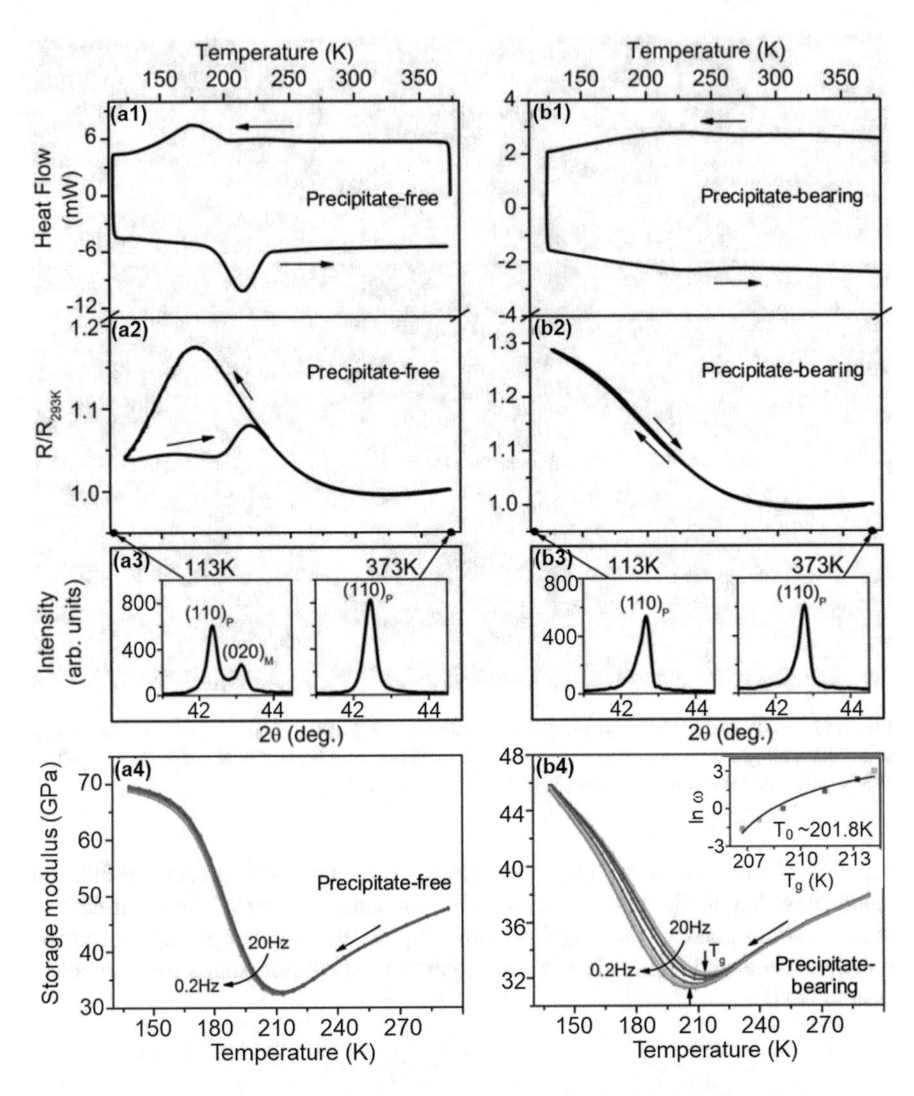

Fig. 7.18 (**a**, **b**) Evidence for martensitic transformation and strain glass transition in precipitate-free and precipitate-bearing samples, respectively [49]

Figure 7.20 shows a 3D phase diagram of $Ti_{50-y}Ni_{50+y-x}Pd_x$ alloys. At y = 0 (i.e., no anti-site defects in Ti-site), the $Ti_{50}Ni_{50-x}Pd_x$ phase diagram (Fig. 7.20a) shows a crossover from B19′ martensite into a new martensite B19 by doping Pd [42]. On the other hand, at $y = 1$ the $Ti_{49}Ni_{51-x}Pd_x$ phase diagram in Fig. 7.20b shows an interesting "sandwich-like" shape, characterized by a crossover from B19′ martensite into strain glass first and then into B19 martensite. The DMA data corresponding to this unique phase diagram are given in Fig. 7.21.

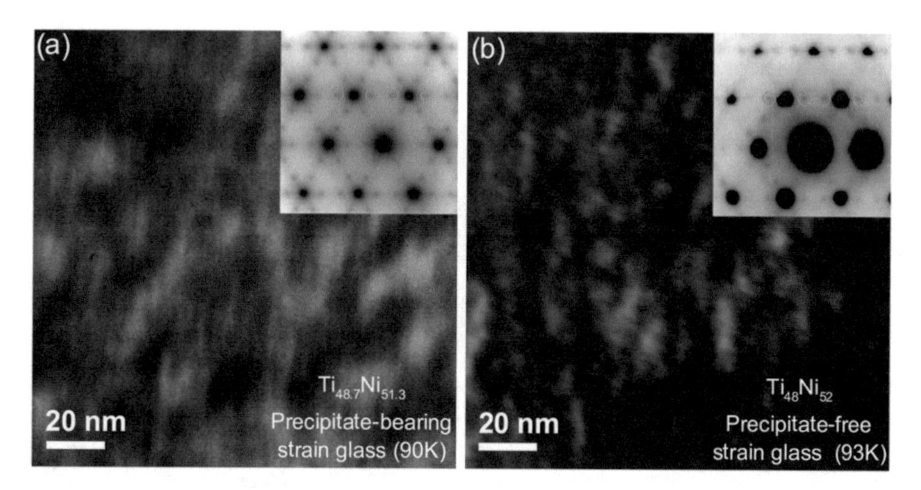

Fig. 7.19 TEM images and corresponding diffraction patterns of (**a**) nano-precipitates induced strain glass and (**b**) point-defects induced strain glass. Both have the same R-like nano-domains (indicated by the incommensurate 1/3 diffuse spots in the inset) [49]

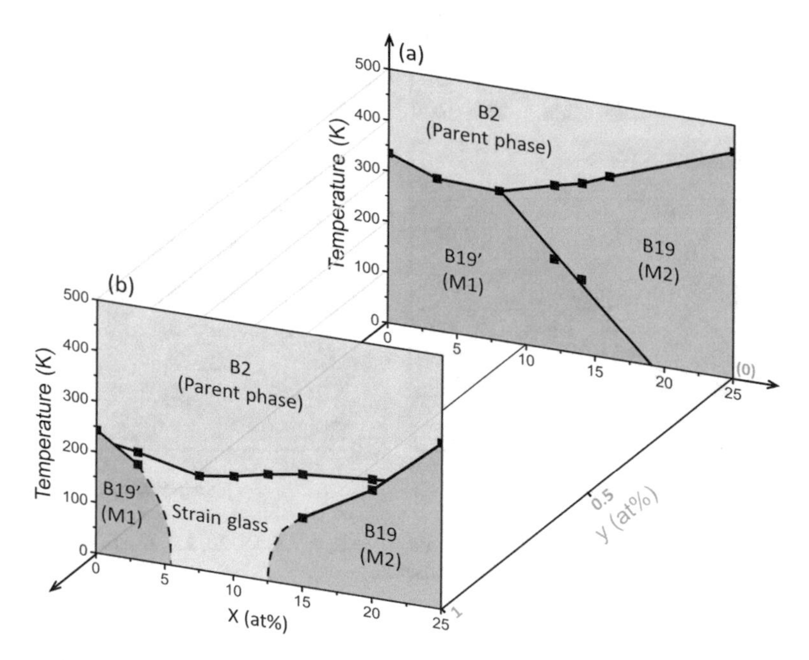

Fig. 7.20 3D phase diagram of $Ti_{50-y}Ni_{50+y-x}Pd_x$ ($x = 0$–25, $y = 0$ and 1) [42]. (**a**) Phase diagram of $Ti_{50}Ni_{50-x}Pd_x$ ($x = 0$–25). (**b**) Sandwich-like strain glass phase diagram of $Ti_{49}Ni_{51-x}Pd_x$ ($x = 0$–25)

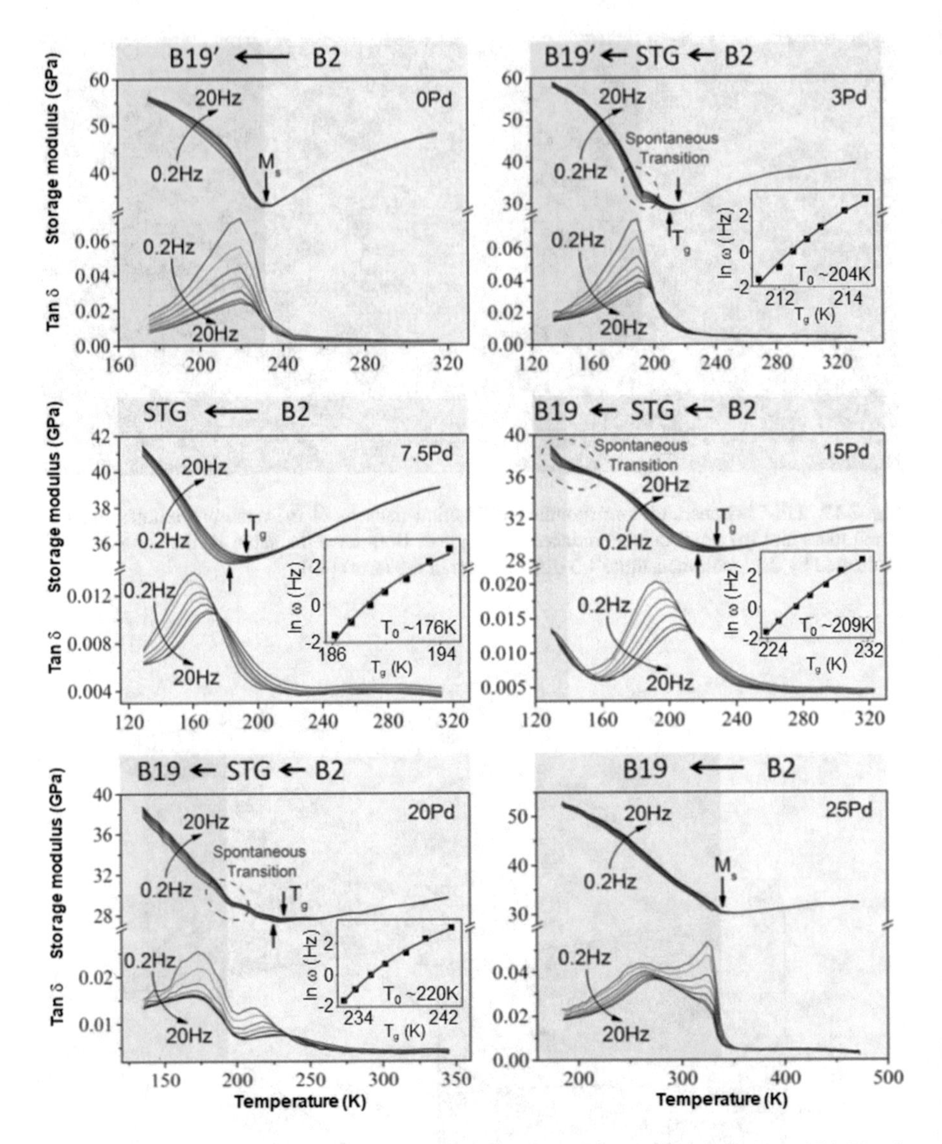

Fig. 7.21 Transition behaviors of $Ti_{50}Ni_{50-x}Pd_x$ ($x = 0, 3, 7.5, 15, 20, 25$). Insets exhibit Vogel-Fulcher relation fitting of each strain glass transition

This sandwich-like strain glass phase diagram can be understood from the dual role of Pd. On one hand, it destabilizes B19′ martensite and stabilizes B19 martensite, thus resulting in a crossover from B19′ martensite into B19 martensite in $Ti_{50}Ni_{50-x}Pd_x$ (Fig. 7.20a). On the other hand, in $Ti_{49}Ni_{51-x}Pd_x$ the system is close to strain glass due to excess Ni doping. In such a situation the randomness caused by Pd doping changes the B19′ martensite into a strain glass first and then into a new martensite B19 due to the stabilization effect of Pd on B19 martensite.

7.6 Summary and Outlook

In this chapter we have reviewed the state of the art of strain glass research since 2005. It is shown that strain glass can be viewed as a "strain jelly" with frozen local strains, and can be generally produced by doping into a martensitic/ferroelastic system with sufficient amount of defects, such as dopant atoms (0D), dislocations (1D), and nano-precipitates (3D). It is likely that planar defects (2D) may also produce strain glass, and this speculation awaits future experiment for verification.

Strain glass exhibits unusual properties unexpected for a martensitic alloy, including superelasticity with slim hysteresis over a wide temperature range, Invar, Elinvar effect, low modulus, and high damping. Ferromagnetic strain glass alloy Fe-Pd is shown to exhibit a remarkable low-field-triggered large magnetostriction. These unique properties suggest strain glass may become a new class of functional/structural materials.

Strain glass exhibits all glass features, being physically parallel with cluster spin glass and relaxor ferroelectrics. These three types of glasses can be generalized by a new name "ferroic glass" and they share very similar glass features.

Besides the typical ferroic glasses (i.e., without long-range ferroic order, Fig. 7.22 left), it is in principle possible to have another two classes of ferroic-glass-derived states/materials. One is glass-ferroic composite [43], a mixture of nanosized ferroic domains with large ferroic domains (Fig. 7.22 center). This state occurs at the crossover composition between ferroic phase and ferroic glass. Another is LRO-matrix ferroic glass, nanosized ferroic glass domains embedded

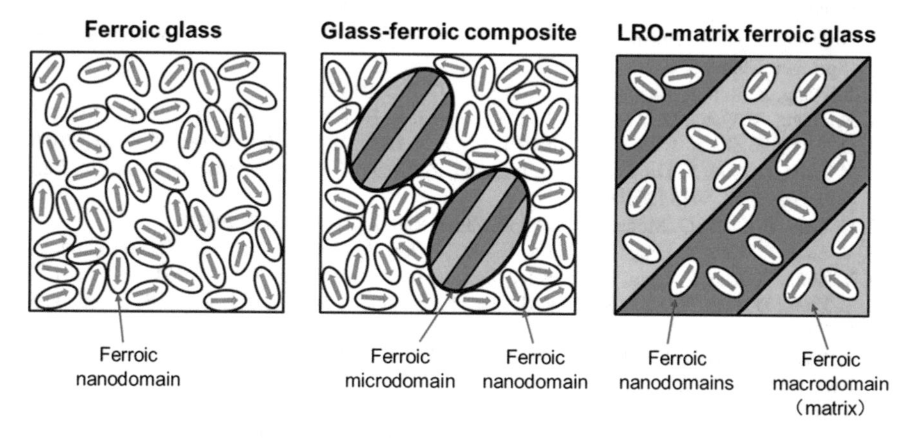

Fig. 7.22 Schematic microstructure of ferroic glass, glass-ferroic composite, and LRO-matrix ferroic glass

into large ferroic domain (Fig. 7.22 right). These new states are expected to show unusual properties absent in both ferroic glass and ferroic phases. Investigations along these directions are highly desired, and they may provide new opportunities for ferroic materials.

Acknowledgments The work is supported by the National Natural Science Foundation of China (51831006, 51701150, 51621063, 51431007), Program for Changjiang Scholars, and Innovative Research Team in University (IRT_17R85), China Postdoctoral Science Foundation (2017M610637), and the Fundamental Research Funds for the Central Universities.

References

1. D. Schryvers, L. Tanner, Ultramicroscopy **32**, 241 (1990)
2. B.C. Muddle, editor. Martensitic transformations (Proc. ICOMAT-89), Mater. Sci. Forum. **56–58** (1990)
3. K. Otsuka, X. Ren, Prog. Mater. Sci. **50**, 511–678 (2005)
4. A. Planes, L. Manosa, Solid State Phys. **55**, 159 (2001)
5. G.B. Olson, M. Cohen, Metall. Trans. A. **6**, 791 (1975)
6. S. Kartha, T. Castan, J.A. Krumhansl, J.P. Sethna, Phys. Rev. Lett. **67**, 3630 (1991)
7. S. Kartha, J.A. Krumhansl, J.P. Sethna, L.K. Wickham, Phys. Rev. B **52**, 803 (1995)
8. S. Semenovskaya, A.G. Khachaturyan, Acta Mater. **45**, 4367 (1997)
9. S. Sarkar, X. Ren, K. Otsuka, Phys. Rev. Lett. **95**, 205702 (2005)
10. Y. Wang, X. Ren, K. Otsuka, A. Saxena, Phys. Rev. B **76**, 132201 (2007)
11. X. Ren, Phys. Status Solidi B **251**, 1982 (2014)
12. Y. Ji, D. Wang, Y. Wang, Y. Zhou, D. Xue, K. Otsuka, Y. Wang, X. Ren, npj Comput. Mater. **3**, 43 (2017)
13. D. Sherrington, in *Frustrated Materials and Ferroic Glasses*, ed. by T. Lookman, X. Ren. (Springer, Cham, 2018)
14. W. Kleemann, J. Dec, in *Frustrated Materials and Ferroic Glasses*, ed. by T. Lookman, X. Ren. (Springer, Cham, 2018)
15. Y. Wang, X. Ren, K. Otsuka, A. Saxena, Acta Mater. **56**, 2885 (2008)
16. D. Wang, Y. Wang, Z. Zhang, X. Ren, Phys. Rev. Lett. **105**, 205702 (2010)
17. D. Wang et al., in *Frustrated Materials and Ferroic Glasses*, ed. by T. Lookman, X. Ren. (Springer, Cham, 2018)
18. Y. Wang, X. Ren, K. Otsuka, Phys. Rev. Lett. **97**, 225703 (2006)
19. Z. Zhang, Y. Wang, D. Wang, Y. Zhou, K. Otsuka, X. Ren, Phys. Rev. B **81**, 224102 (2010)
20. D. Wang, Z. Zhang, J. Zhang, Y. Zhou, Y. Wang, X. Ding, X. Ren, Acta Mater. **58**, 6206 (2010)
21. Y. Zhou, D. Xue, X. Ding, Y. Wang, J. Zhang, Z. Zhang, D. Wang, K. Otsuka, J. Sun, X. Ren, Acta Mater. **58**, 5433 (2010)
22. G. Liao, B. Chen, Q. Meng, M. Wang, X. Zhao, Rare Metals **34**, 829 (2015)
23. Y. Zhou, D. Xue, X. Ding, K. Otsuka, J. Sun, X. Ren, Appl. Phys. Lett. **95**, 151906 (2009)
24. Y. Zhou, D. Xue, X. Ding, K. Otsuka, J. Sun, X. Ren, Phys. Status Solidi B **251**, 2027 (2014)
25. Y. Zhou, D. Xue, Y. Tian, X. Ding, S. Guo, K. Otsuka, J. Sun, X. Ren, Phys. Rev. Lett. **112**, 025701 (2014)
26. Y. Wang, J. Gao, H. Wu, S. Yang, X. Ding, D. Wang, X. Ren, Y. Wang, X. Song, J. Gao, Sci. Rep. **4**, 3995 (2014)
27. Y. Nii, T. Arima, H.Y. Kim, S. Miyazaki, Phys. Rev. B **82**, 214104 (2010)
28. H.Y. Kim, L. Wei, S. Kobayashi, M. Tahara, S. Miyazaki, Acta Mater. **61**, 4874 (2013)
29. L.S. Wei, H.Y. Kim, S. Miyazaki, Acta Mater. **100**, 313 (2015)

30. Y. Wang, C. Huang, J. Gao, S. Yang, X. Ding, X. Song, X. Ren, Appl. Phys. Lett. **101**, 101913 (2012)
31. D.P. Wang, X. Chen, Z.H. Nie, N. Li, Z.L. Wang, Y. Ren, Y.D. Wang, EPL **98**, 46004 (2012)
32. S. Ren, D. Xue, Y. Ji, X. Liu, S. Yang, X. Ren, Phys. Rev. Lett. **119**, 125701 (2017)
33. Y. Yao, Y. Yang, S. Ren, C. Zhou, L. Li, X. Ren, EPL **100**, 17004 (2012)
34. Y. Ni, Z. Zhang, D. Wang, Y. Wang, X. Ren, J. Alloys Compd. **577S**, S468 (2013)
35. J. Liu, M. Jin, C. Ni, Y. Shen, G. Fan, Z. Wang, Y. Zhang, C. Li, Z. Liu, X. Jin, Phys. Rev. B **84**, 104102 (2011)
36. X. Jin, J. Liu, M. Jin, J. Alloys Compd. **577S**, S155 (2013)
37. J.A. Monroe, J.E. Raymond, X. Xu, M. Nagasako, R. Kainuma, Y.I. Chumlyakov, R. Arroyave, I. Karaman, Acta Mater. **101**, 107 (2015)
38. J. Zhang, D. Xue, X. Cai, X. Ding, X. Ren, J. Sun, Acta Mater. **120**, 130 (2016)
39. Q. Liang, D. Wang, J. Zhang, Y. Ji, X. Ding, Y. Wang, X. Ren, Y. Wang, Phys. Rev. Mater. **1**, 033608 (2017)
40. Y. Ji, X. Ding, T. Lookman, K. Otsuka, X. Ren, Phys. Rev. B **87**, 104110 (2013)
41. Z. Zhou, J. Cui, X. Ren, AIP Adv. **7**, 045019 (2017)
42. S. Ren, C. Zhou, D. Xue, D. Wang, J. Zhang, X. Ding, K. Otsuka, X. Ren, Phys. Rev. B **94**, 214112 (2016)
43. Y. Ji, X. Ding, D. Wang, K. Otsuka, X. Ren, Phys. Rev. B **92**, 241114 (2015)

Chapter 8
Discrete Pseudo Spin and Continuum Models for Strain Glass

Dezhen Xue and Turab Lookman

Abstract Strain glass refers to a frozen disordered state of lattice strain, conjugate to the long-range ordering in ferroelastics. A number of descriptions have been invoked over the last few years to model strain glasses. These include the continuum Landau free energy approach with added disorder, and the discrete pseudo spin model. We review these and focus on the discrete pseudo spin model, which is derived from the continuum Landau model in the sharp interface limit. We also show how the pseudo spin model leads to predictions that can be confirmed experimentally. We conclude by briefly discussing how the continuum model can be coupled with machine learning to provide a basis for rapidly finding material attributes that optimize alloy response.

8.1 Introduction

Ferroic materials form an essential subgroup of functional materials whose physical properties are sensitive to the changes in external conditions such as temperature, pressure, electric, and magnetic fields. A phase transition with symmetry-breaking usually occurs in ferroic materials, leading to two or more orientation states of a physical property (or an order parameter) corresponding to the same energy. Moreover, the states can be switched by the application of an external field (electric, magnetic or stress depending on the ferroic property). The most important three classes of primary ferroic materials are ferroelectrics with long-range ordering of electric dipoles, ferromagnets with long-range ordering of magnetic moments, and ferroelastics with long-range ordering of lattice strains below their critical

D. Xue (✉)
State Key Laboratory for Mechanical Behavior of Materials, Xi'an Jiaotong University, Xi'an, China
e-mail: xuedezhen@xjtu.edu.cn; xuedezhen@mail.xjtu.edu.cn

T. Lookman
Theoretical Division, Los Alamos National Laboratory, Los Alamos, NM, USA
e-mail: txl@lanl.gov

T. Lookman, X. Ren (eds.), *Frustrated Materials and Ferroic Glasses*, Springer Series in Materials Science 275, https://doi.org/10.1007/978-3-319-96914-5_8

temperatures. These ferroic materials have an anisotropic long-range interaction encoding the various symmetries at play due to magnetic dipole interaction, electric dipole interaction and strain compatibility, respectively, and consequently possess a multi-orientation or domain configuration [1]. Improper ferroelectrics can also show multi-orientation states due to the breaking of symmetry due to short wavelength modes. For example, $YMnO_3$, a magnetoelectric, shows the characteristic six orientation states associated with the rotation of octahedra in the unit cell [2]. The hysteretic response of ferroics to external stimuli, and cross-coupling between different responses, enables them to find a variety of applications as memory devices, actuators, sensors, and transducers.

However, in the presence of quenched-in disorder, the long-range ordering associated with the order parameters can be perturbed resulting in a frustrated state. Examples include certain relaxor ferroelectrics and the usual spin glasses or cluster-spin glass phases discussed in the previous chapters in this book. The state usually manifests in its non-ergodic response, slow dynamics and nanoscale heterogeneities, and certain types of experiments are performed to characterize these aspects. The state exhibiting such behavior is described as *ferroic glass*. This suggests the need to exercise care in merely relying on these characteristics to identify a glassy state as there are alloys that can display such signatures but are poly-twinned rather than glassy. Recent experiments on shape memory alloys, as discussed by Ji et al. in this book, have shown the existence of a strain glass phase in which localized random configurations of lattice distortions are kinetically frozen below a glass transition temperature, T_g. This glass phase was initially observed in non-martensitic Ni-rich $Ti_{50-x}Ni_{50+x}$ alloys with B2 structure above a compositional threshold $x \sim 1.3$, below which the B2 to B19' martensitic transformation takes place [3]. It can be produced by introducing either point defects, compositional variations, precipitates or dislocations into the host martensitic alloys, and has been identified in numerous alloys including Ti-Ni-X (X = Fe, Co, Cr, Mn), Ti-Pd-X (X = Fe, Co, Cr, Mn) systems [4–6], as well as ferromagnetic systems [7]. The strain glass shows typical features of "dynamic or kinetic freezing" measured by dynamic mechanical analysis (DMA) experiments, and the breaking down of ergodicity measured by zero-field-cooling and field-cooling (ZFC/FC) experiments [8, 9]. The phenomena observed in strain glass are analogous to those observed in other ferroic glasses, including spin glass and relaxor ferroelectrics. The concept of strain glass is not only of interest in terms of distill physics purely based on strain, but also provides its unique aspects relevant to applications such as the shape-memory effect and pseudoelasticity, which are typical of alloys undergoing martensitic transformations [10]. Moreover, the pretransitional "tweed" phase, which can exist over a wide temperature range 100 K above the martensitic transformation, shows no glass-like responses within the current experimental measurement capabilities. However, the tweed phase shows short-range strain order and has previously been suggested and modeled as a glassy state [11–13].

Motivated by the experiments referred to above, there have been a number of studies recently focused on describing the glass phase. These have attempted to elucidate the nature of strain glass in analogy with spin glass and relaxor ferroelectrics [14–17]. The focus has been to recognize the importance of the elastic long-range interaction and disorder via chemical inhomogeneity or compositional fluctuations in the context of martensites, where the interplay of short range and long range order gives rise to an ordered phase with a characteristic twin width (see Chaps. 7, 9, 10). These studies have largely been of two kinds, namely (a) those that utilize a continuum Landau free energy approach with elastic energy and disorder and solve a relaxation dynamics equation [18–20], and (b) those that derive a discrete model from the continuum Landau description and utilize the tools of statistical mechanics to obtain predictions of the glass behavior [14, 15, 21–23]. Our purpose here is to review these mesoscopic models for strain glass. Thus, the work we describe makes connections with Chap. 1 (Sherrington) as well as Chaps. 7 (Ji et al.), 9 (Lloveras et al.) and 10 (Wang et al.). We will also show that the discrete formulation or pseudo spin model leads to predictions that can be confirmed experimentally. We conclude this chapter with some suggestions of how in the context of Landau continuum models, informatics tools can aid to accelerate the discovery of alloys with given parameters encoding defect concentration and potency.

8.2 A Continuum Landau Model with Elastic Interactions and Defects

A continuum Landau model and its variations have been used to simulate the glassy features, the microstructure evolution and especially how intrinsic inhomogeneities arise, including when coupled to magnetization and charge [24]. Here we will focus on the origin of the elastic interactions in a purely strain based picture as it is the competition of the long-range and short range interactions within the Landau description that creates heterogeneities, such as the ordered twins in martensites. The disorder is an additional factor that essentially perturbs the ordered state and forms a glass. We will consider a two-dimensional (2D) square to rectangle transformation driven by the deviatoric shear, which serves as an order parameter for the shear driven martensitic transformation in an alloy such as FePd. This transition is a 2D analog of a cubic to tetragonal or tetragonal to orthorhombic transformation and is one of the simplest that illustrates the salient physics. To describe elastic effects, the linearized strain tensor in a global reference frame is defined as $\varepsilon_{ij} = (\partial u_i/\partial r_j + \partial u_j/\partial r_i)/2$ ($i = 1, 2$; $j = 1, 2$), where u_i is the lattice displacement. The strain tensor components, ε_{xx} is the longitudinal strain, ε_{yy} is the transverse strain and ε_{xy} is the simple shear strain. The symmetry adapted strains [25] e_1, e_2, and e_3 representing the dilatational, deviatoric, and shear modes, are defined by $e_1 = \frac{1}{\sqrt{2}}(\varepsilon_{xx} + \varepsilon_{yy})$, $e_2 = \frac{1}{\sqrt{2}}(\varepsilon_{xx} - \varepsilon_{yy})$ and $e_3 = \varepsilon_{xy}$. And the three adapted strains are not independent but related through the compatibility relation,

$$\nabla^2 e_1 - \left(\frac{\partial^2}{\partial x^2} - \frac{\partial^2}{\partial y^2}\right) e_2 - \sqrt{8}\frac{\partial^2}{\partial x \partial y} e_3 = 0 \tag{8.1}$$

as they are derivatives of the same underlying displacement field. From a computational point, a key difference between the approach using the compatibility equation and that typically used in the phase field approach is that we eliminate the displacements using this relationship and only solve for the strains. The usual phase field approach eliminates strains and solves for the displacements. The free energy density is written as the sum of three contributions, namely

$$f(e_1, e_2, e_3) = f_h(e_2) + f_{\text{grad}}(\nabla e_2) + f_{\text{non-OP}}(e_1, e_3), \tag{8.2}$$

where $f_h(e_2)$ is the homogeneous Landau part accounting for the required non-linearities in the order parameters, $f_{\text{grad}}(\nabla e_2)$ is the gradient (Ginzburg) term responsible for the interface energy in the order parameters, and $f_{\text{non-OP}}(e_1, e_3)$ is the contribution from the non-order parameter components of the strain which is assumed to be harmonic and gives the long-range elastic interaction. The forms of those contributions are chosen as

$$f_h(e_2) = \frac{1}{2}A_2[T]{e_2}^2 + \frac{1}{4}\beta {e_2}^4 + \frac{1}{6}\gamma {e_2}^6 \tag{8.3}$$

$$f_{\text{grad}}(\nabla e_2) = \frac{1}{2}g|\nabla e_2|^2 \tag{8.4}$$

$$f_{\text{non-OP}}(e_1, e_3) = \frac{1}{2}A_1{e_1}^2 + \frac{1}{2}A_3{e_3}^2, \tag{8.5}$$

where $A_1 = C_{11} + C_{12}$ is the bulk modulus and $A_2 = C_{11} - C_{12} = 2C'$ and $A_3 = 4C_{44}$ are elastic modulus associated with deviatoric and shear modes, respectively [26]. The C_{11}, C_{12}, and C_{44} are the elastic constant tensor components for a crystal with square symmetry. The time-dependent Ginzburg-Landau equation is used for the time-evolution of the order parameter e_2,

$$\frac{\partial e_2(\mathbf{r}, t)}{\partial t} = \Gamma \frac{\delta F}{\delta e_2(\mathbf{r}, t)}, \tag{8.6}$$

where Γ is a kinetic coefficient which controls the rate of free energy evolution. The essential Landau contribution to the free energy for a first-order transition leads to a single energy well corresponding to the parent phase at high temperature, and a double energy well corresponding to the two martensite variants below the martensitic transformation temperature.

Disorder may be added to drive the system to a strain glass or a premartensite tweed [18]. One way to consider the effects of disorder is via a spatial fluctuation in the martensitic transformation temperature, through $A_2[T, \eta(r)] = \alpha_T(T - T_c) + \alpha_\eta \eta(r)$, where T_c is the lower stability limit of the high temperature parent phase in the clean limit and $\eta(r)$ is a random variable that can gaussian distributed around zero and with spatial correlations that can delta correlated or described by an exponential pair correlation function that Lloveras et al. (Chap. 9) have used.

Disorder may also be added by introducing a global variation of the martensite stability together with a local breaking of the symmetry of the Landau potential.

The former changes the stability of martensite through $A_2[T, c] = \alpha[T - T_c - \alpha_c c]$, where again T_c is the lower stability limit of the high temperature parent phase, c is the defect concentration, and α_c is the strength of defect necessary to change the transformation temperature [20]. The latter incorporates local random deviatoric stress field that interacts with the strain order parameters directly and breaks the symmetry of the Landau potential through an extra contribution to the free energy density of the form $-e_2\zeta(r)$, where $\zeta(r)$ is a spatially distributed random field coupled directly to the order parameter e_2 [20].

A third approach, which we will consider in the next section, is to only introduce the random dilatational stress field that couples to the non-order parameter strain e_1. The time-dependent simulations in all cases reproduce microstructure similar to the premartensite tweed, strain glass state, and experimentally observed phase diagram of a ferroelastic system in the presence of defects. The glass behavior is monitored largely by the deviation in ZFC/FC curves (as in experiment) and/or presence of glass like morphology. The difficulty with this is that it is not so straightforward to identify a glass in the sense of a spin glass as an appropriate order parameter is not defined. Thus, such models tend not to be particularly predictive. Hence, we will consider in Sect. 8.4 a discrete version of the continuum strain model that allows us to evaluate a spin glass like OP for strain glass.

8.3 A Strain Glass with Randomly Distributed Dopants

As mentioned, a random-field model for strain glass can be formulated with a deviatoric strain/stress field directly coupled to the order parameter e_2. Intuitively, the dopants with different atomic sizes effectively replacing the host atom would cause a local volume change of the lattice. In general, the dopants have different atomic sizes compared to the host alloy and the size mismatch gives rise to a local dilatational strain or stress. Thus, the effect of dopants can be modeled as a randomly distributed dilatational stress in the system and the concentrations can be varied by changing the number of dopants. A dilatational internal stresses coupled to a volumetric strain ($e_1 = \frac{1}{\sqrt{2}}(\varepsilon_{xx} + \varepsilon_{yy})$) can be introduced.

We thus consider a 2D Ginzburg-Landau free energy that includes point defect doping for a square to rectangle martensitic transition. The free energy density is written as the summation of four contributions:

$$f(e_1, e_2, e_3, \rho, \sigma_{11}) = f_h(e_2) + f_{\text{grad}}(\nabla e_2) + f_{\text{non-OP}}(e_1, e_3) + f_{\text{defect}}(e_1, \rho), \quad (8.7)$$

where $f_h(e_2)$, $f_{\text{grad}}(\nabla e_2)$, and $f_{\text{non-OP}}(e_1, e_3)$ are defined as before and are the homogeneous Landau contribution, the gradient (Ginzburg) term and the contribution due to the non-order parameter strain components, respectively. We couple randomly distributed dilatational stresses, ρ, to the volume strain e_1, that is,

$$f_{\text{defect}}(e_1, \rho) = -e_1\rho \quad (8.8)$$

The local dilatation stress will not only affect the dopant crystallographic site but also its nearest neighbors and beyond because of long-range elastic forces. Therefore, the dilatation stress will acquire a certain distribution rather than a Delta function. We assume a Gaussian distribution of the form

$$\rho(r) = h\frac{1}{\sigma\sqrt{2\pi}}e^{\frac{r^2}{2\sigma^2}}, \tag{8.9}$$

such that different chemical dopants will have varied strengths in the host lattice affected surrounding regions heterogeneously. Therefore, the combination of parameters h (strength) and σ (range of stress disturbance) in Eq. (8.9) characterize the dopants. By relaxing the free energy $F(e_2, e_1, e_3, \rho(r))$ through solving the time-dependent Ginzburg-Landau evolution equation (8.6), we can reproduce the strain glass phase diagram. Figure 8.1 shows our preliminary simulation results of the microstructure change with number of dopants or defect concentration. It has been known for decades that the martensitic transformation temperature change depends differently on the concentration for different types of point defects. For Cr dopant, one percent in concentration can change the transformation temperature by more than 150 K, where as for Co one needs more than 7% to decrease the martensitic transformation temperature by 150 K. Certain dopants such as Fe, Cr, V, Mn, Co can result in a strain glass state in TiNi, but dopants of Zr, Hf, Cu do not give give to a strain glass. By varying the range (σ) and the strength or potency (h) within this model, the behavior of different dopants can be reproduced.

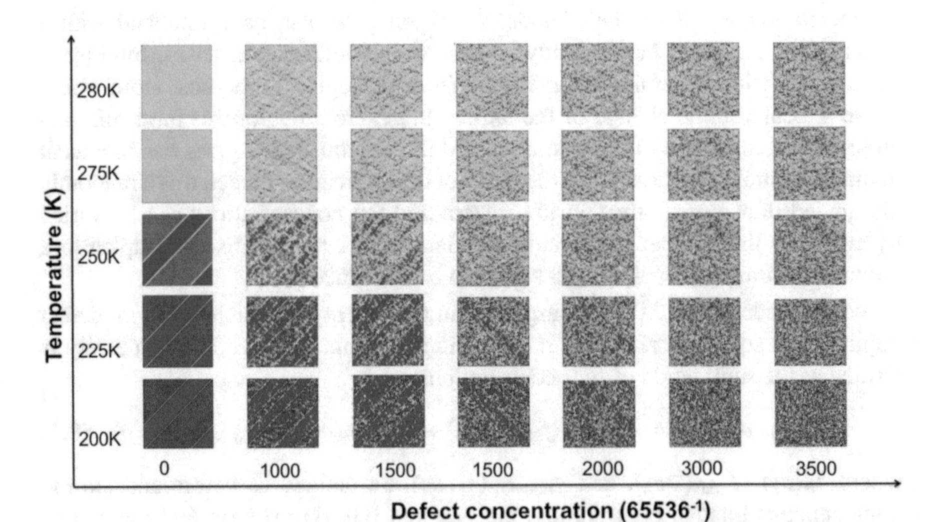

Fig. 8.1 Strain states with different defect concentrations at different temperatures. Green describes the parent phase; Blue and red colors describe the two martensitic variants

8.4 A Discrete Pseudo Spin Model for Strain Glass

An alternative, yet complementary approach to strain glass we discuss here is a discrete pseudo spin model for martensites, which we obtain from the previous continuum model, to which we add quenched disorder of varying strengths. The methods of statistical mechanics then allow us to identify and distinguish the different phases using the analog of a spin glass order parameter. The Renormalization Group (RG) approach, as well as mean field and Monte Carlo approaches, can then be used to study the pseudo spin model. RG allows us to integrate out microscopic degrees of freedom so that the attractive basins characterize the physics at large scales. The values of the interaction strength and strength of disorder uniquely characterize the different phases, including the glassy state, thereby allowing the phase diagram in terms of temperature and disorder to be predicted. Thus, our approach makes contact with the discussion in Chap. 1 (Sherrington). The continuum to discrete limit is achieved by replacing the OP strain, e_2, for the square to rectangle transformation by the discrete variable (pseudo spin), S, using $e_2 \to |e_2 S|$; where $S = 0, +1, -1$ are the minima of the free energy representing austenite and the two martensite variants. If we perform this mapping on the nonlinear Landau free energy, it collapses to the linear crystal-field form $w(T)S^2$ as $S^2 = S^4 = S^6$, where the coefficient, $w(T)$, is some function of temperature. Similarly, the gradient or interaction energy term $\frac{g}{2}(\nabla e_2)^2$ may be written in terms of the product $S_i S_j$ involving discrete S values on neighboring sites $< i, j >$ by using the finite difference definition of gradient. Thus, the energy of the original square to rectangle problem now transforms to

$$H = -\sum_{<i,j>} J_{ij}(T) S_i S_j + w(T) \sum_i S_i^2 + A_1 \sum_{i,j} S_i U_{ij} S_j, \tag{8.10}$$

with $S_i = 0, +1, -1$, and where U is the long-range anisotropic elastic interaction with strength A_1. This is the well-known spin-1 or Blume-Capel model with long-range interactions [21]. The interaction J is related to the interface energy, g, and the minimum energy or ground state of H gives the well-known twin microstructure in both 2D and 3D for a cubic to tetragonal transformation. Models such as these, even though they are quite simplified, encode the main physics of the continuum description, which is phenomenological anyway, and have the advantage of being studied by well-known methods from spin glass theory. It is important to recognize that such models, including the Landau phenomenological description, are effective or mean field which can predict universal features, such as the phase diagram rather than quantitative comparisons to experiments.

In analogy with usual spin glasses, we can add disorder to mimic the effects of changing composition or defects (e.g., point defects, dislocations, precipitates) in the strain alloys. The disorder-free hamiltonian of Eq. (8.10) does not consider the long-range interaction. Based on Eq. (8.10), we take the nearest neighbor couplings to be quenched independent random variables J_{ij}. It is drawn from the distribution

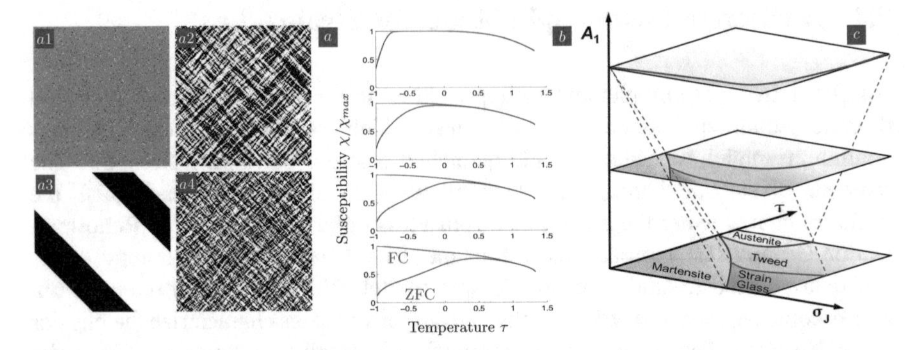

Fig. 8.2 Monte Carlo simulation results. (**a**) Typical microstructures on a 256 × 256 lattice in the different phases of the phase diagram. (**a1**) austenite, (**a2**) tweed, (**a3**) martensite, and (**a4**) strain glass. (**b**) Example of FC and ZFC curves for different levels of disorder. (**c**) Qualitative phase diagram showing the influence of the long-range interaction and disorder on the various phase transitions. Four different phases are shown: austenite, martensite, tweed, and strain glass. Reprinted figure with permission from [23] Copyright (2018) by the American Physical Society

$P(J_{ij})$ with mean given by $J(T)$ and variance s_J, which is a measure of the quenched disorder in the system. The form of the distribution is irrelevant to the geometry of the phase diagram, indicating that this approach tries to capture salient, universal features. A real-space RG approach can be applied to obtain the experimentally observed phase diagram. In addition to the glass phase, this approach predicts a tweed precursor phase, consistent with Monte Carlo simulations shown in Fig. 8.2. Therefore, such an approach appears simpler and more reliable than the replica/mean-field approach for this model.

The martensite (ferroelastic) phase is the analog of the ordered ferromagnetic phase, and in terms of order parameters (OPs) used in mean field and replica theory, this phase is identified by a non-zero magnetization $m = \overline{< S_i >} \neq 0$, where the averaging over the disorder is represented by the bar and the angle brackets correspond to an average with respect to Boltzmann weights. There are two high symmetry or paraelastic phases which are disordered. One favors the state $S = 0$, and the other is identified as tweed with OP given by the martensite volume fraction $p = \overline{S_i^2}$ that separates the two phases. The tweed precursor is found to be ergodic and non-glassy, in agreement with recent experiments.

At large scales, the effective hamiltonian favors variants $S = \pm 1$, and this phase is also identified by the Edward-Anderson order parameter $q = \overline{S_i^2}$, which corresponds to the overlap between two replicas $q =< S_i^1 S_i^2 >$ of the system in the replica formulation. We find that a first order phase transition occurs between the austenite and martensite phase with $\tau \approx 0$ without disorder ($\sigma_J = 0$), as expected. As the disorder increases, an intermediate tweed phase exists before it transforms into a low temperature phase (either martensite or glass). In the limit of large disorder and low temperatures, there is a spin glass phase that we interpret as strain glass. If the disorder is intermediate ($1.3 < \sigma_J < 2.3$ in our model), we predict a spontaneous phase transition from glass to martensite for a given concentration.

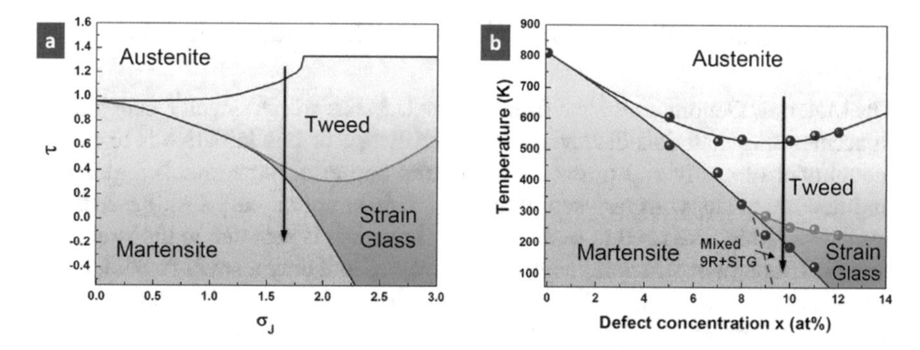

Fig. 8.3 Comparison between theoretical and experimental phase diagrams. (**a**) Phase diagram in the temperature-disorder (τ, σ_J) plane for our spin model, obtained within the RG projection approximation. τ is the normalized temperature and σ_J characterizes the amount of quenched disorder in the system. (**b**) Experimental phase diagram of the ternary ferroelastic $Ti_{50}(Pd_{50-x}Cr_x)$). STG refers to strain glass. 9R is a new martensite phase formed, other than B19, if TiPd-based alloys are doped with certain defects. It is a martensite with long periodic stacking structure. Reprinted figure with permission from [23] Copyright (2018) by the American Physical Society

To check the predictions of the RG calculations, the temperature-composition phase diagram of a modeled strain glass system given by $Ti_{50}(Pd_{50-x}Cr_x)$ was experimentally established, shown in Fig. 8.3b. The transformation behavior was systematically investigated as a function of defect concentration x. In the low doping concentration regime ($x < 8$), the system undergoes a normal B2 $\rightarrow$ B19 martensitic transformation with a sharp increase in electrical resistivity, a frequency-independent peak in internal friction and a frequency-independent dip in storage modulus. In the high doping concentration regime ($x > 12$), the alloy transforms upon cooling from austenite through tweed to strain glass. A frequency-dependent storage modulus dip and an internal friction peak can be observed for these strain glass alloys, demonstrating a dynamic freezing strain-glass transition in these alloys. Such behavior is in contrast with the frequency-independent response of the low doping concentration alloys in the course of the martensitic transformation. The most interesting phenomenon occurs within the crossover regime ($9 < x < 12$) between martensite and strain glass, where the alloys experience all four strain states of parent phase, tweed, strain glass, and martensite phase upon cooling. In particular, a spontaneous transformation from strain glass to the martensite phase (9R) takes place at the crossover regime. The $Ti_{50}(Pd_{40}Cr_{10})$ alloy possesses a frequency-dispersive internal friction peak and a storage modulus dip, followed by a frequency-independent internal friction peak, which shows a similar feature to the martensitic transformation. This indicates a spontaneous phase transformation from strain glass to martensite phase occurs. According to the experimental results, a phase diagram for $Ti_{50}(Pd_{50-x}Cr_x)$ alloys, where a crossover composition regime is included, is shown in Fig. 8.3b. This phase diagram is in good agreement with the phase diagram from the RG approach as shown in Fig. 8.3a.

8.5 Coupling Information Sciences with Landau Models

The Materials Genome Initiative (MGI) in the U.S. has created much recent interest in accelerating materials discovery. A key challenge of this initiative is to reduce the number of costly and time-consuming trial and error experiments required to find new materials with targeted properties. This is not an easy task because the space over which we need to search for new materials is vast due to the structural, chemical, and microstructural complexity involved and only a small fraction of the space has been experimentally investigated. Data-driven machine learning tools have created much interest as they are very efficient in optimally guiding new experiments or calculations to find materials with desired properties. However, applying some of these tools that rely merely on data can be a problem because they can yield suboptimal results, as the available training data are often limited compared to the number of features (or material descriptors) and size of the space over which one is searching for new compounds [27]. An advantage of materials science is that knowledge in the form of scaling relations or constitutive laws and functional relationships are often available from theory or known. Such prior knowledge can be used with data to accelerate the discovery of new materials with targeted properties [28]. The Landau model we have discussed applied to shape memory alloys (SMAs) has been shown to capture reasonably well the underlying physics of the shape memory effect (SME) and superelasticity (SE). Such a model with dopants, which we have discussed in the context of strain glass, provides a prototype example of how we can couple the results from simulations of the model with a data-driven optimization method to find desired attributes or descriptors encoding specific alloys that give rise to small energy dissipation [29].

The SE effect arises as a result of a stress induced martensitic transformation and it appears in the parent phase above the transformation temperature. When the high symmetry parent phase is stress loaded beyond a critical value, it transforms to the low symmetry martensite phase; upon unloading, the martensite reverts back to the parent phase which is the stable phase in energy. The martensitic transformation gives rise to the large, non-linear, but recoverable strain, leading to a variety of applications for SE. However, the martensitic transformation is typically a first-order phase transition accompanied by large hysteresis in the stress-strain curve. The enclosed area between loading and unloading curves is the amount of energy dissipated during the stress-strain cycle and a measure of hysteresis. For practical applications, a large energy dissipation or hysteresis is undesirable because it results in serious fatigue problems of SMAs in devices (such as cardiovascular stents) that require high sensitivity, precision, and durability. Finding new SMAs with low energy dissipation accompanying SE is critical for realizing SMAs in practical applications. Therefore, it serves as one of the design targets in SMAs. One can modify the chemistry of SMAs experimentally, for example, by doping alloying elements in the host alloy. The chemical modification would vary the stress-strain

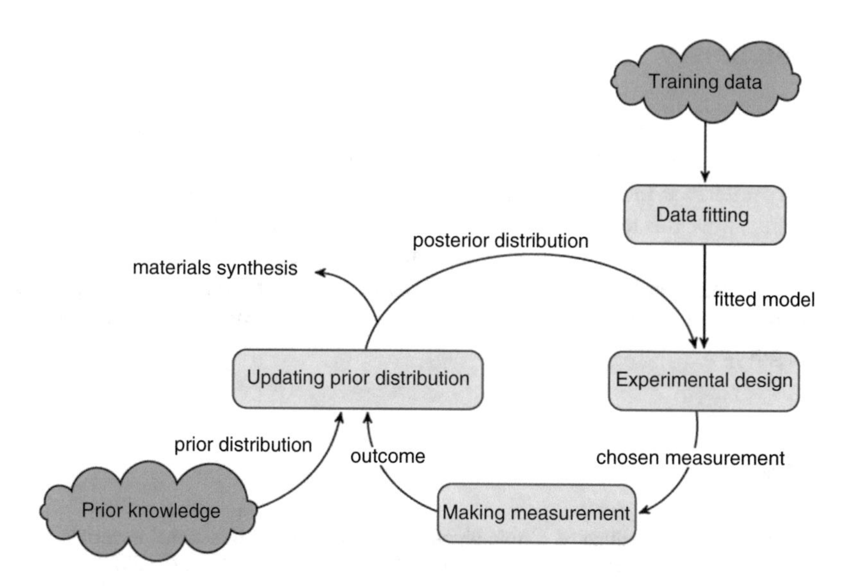

Fig. 8.4 An illustrative view of the experimental design process when used recursively. Reprinted figure from [29], Copyright (2018), with permission from Elsevier

response of SE and consequently, the energy dissipation. Our dilatational random-field model for strain glass described in Sect. 8.3 can "mimic" the doping effects by varying the model parameters. We thus integrated the Landau model (which computes the stress-strain curve for prototypical SMAs at different temperatures and doping concentrations) with the experimental design algorithms to rapidly optimize the material specific model parameters (c, h, and σ) that minimize the energy dissipation associated with SE.

We assume a virtual set of different dopants that can be used. The range (σ) and strength or potency (h) of the dopants, and the associated SE properties are unknown. The design strategy is shown in Fig. 8.4. We build an inference model for the energy dissipation as a function of dopant concentration, dopant potency, and dopant range based on a small training dataset which was established beforehand. The trained model is used together with the optimization algorithms in the experimental design step to find the best dopant and concentration for the next measurement. After the outcome of the chosen measurement is obtained by using the Landau model, the prior distribution that encodes the prior knowledge of unknown dopant parameters is updated to the posterior distribution. One can make further measurements so that the posterior distribution serves as the new prior distribution for the next design loop. In case no further measurements are needed, the posterior distribution can be used to design the low energy dissipation material. The design loop provides a potential strategy for designing materials with targeted properties given that some parameters or features in the problem are unknown.

8.6 Summary

We review the continuum Landau free energy approach with added disorder and show how the discrete pseudo spin model is derived from the continuum model in the sharp interface limit. We examine the predictions of this model, in particular, we discuss how the temperature-composition phase diagram of a modeled strain glass system given by $Ti_{50}(Pd_{50-x}Cr_x)$ can be established (Fig. 8.3b). The use of machine learning and a phenomenological model, such as Landau theory, provides a new paradigm for accelerating how we can find new materials, including new ferroic glasses.

References

1. A. Saxena, Broken symmetry, ferroic phase transitions and multifunctional materials. Integr. Ferroelectr. **131**(1), 3–24 (2011). https://doi.org/10.1080/10584587.2011.616380
2. B.B. Van Aken, T.T.M. Palstra, A. Filippetti, N.A. Spaldin, The origin of ferroelectricity in magnetoelectric $YMnO_3$. Nat. Mater. **3**, 164–170 (2004). https://doi.org/10.1038/nmat1080
3. S. Sarkar, X. Ren, K. Otsuka, Evidence for strain glass in the ferroelastic-martensitic system $Ti_{50-x}Ni_{50+x}$. Phys. Rev. Lett. **95**, 205702 (2005). https://link.aps.org/doi/10.1103/PhysRevLett.95.205702
4. Y. Zhou, D. Xue, X. Ding, K. Otsuka, J. Sun, X. Ren, High temperature strain glass transition in defect doped Ti–Pd martensitic alloys. Phys. Status Solidi B **251**(10), 2027–2033 (2014). https://onlinelibrary.wiley.com/doi/abs/10.1002/pssb.201350360
5. Y. Zhou, D. Xue, X. Ding, K. Otsuka, J. Sun, X. Ren, High temperature strain glass in $Ti_{50}(Pd_{50-x}Cr_x)$ alloy and the associated shape memory effect and superelasticity. Appl. Phys. Lett. **95**(15), 151906 (2009). https://doi.org/10.1063/1.3249580
6. Y. Zhou, D. Xue, X. Ding, Y. Wang, J. Zhang, Z. Zhang, D. Wang, K. Otsuka, J. Sun, X. Ren, Strain glass in doped $Ti_{50}(Ni_{50-x}D_x)$ (D=Co, Cr, Mn) alloys: implication for the generality of strain glass in defect-containing ferroelastic systems. Acta Mater. **58**(16), 5433 – 5442 (2010). http://www.sciencedirect.com/science/article/pii/S1359645410003733
7. J. Monroe, J. Raymond, X. Xu, M. Nagasako, R. Kainuma, Y. Chumlyakov, R. Arroyave, I. Karaman, Multiple ferroic glasses via ordering. Acta Mater. **101**, 107–115 (2015). https://doi.org/10.1016/j.actamat.2015.08.049. http://www.sciencedirect.com/science/article/pii/S1359645415006266
8. Y. Wang, X. Ren, K. Otsuka, A. Saxena, Evidence for broken ergodicity in strain glass. Phys. Rev. B **76**, 132201 (2007). https://link.aps.org/doi/10.1103/PhysRevB.76.132201
9. Y. Zhou, D. Xue, Y. Tian, X. Ding, S. Guo, K. Otsuka, J. Sun, X. Ren, Direct evidence for local symmetry breaking during a strain glass transition. Phys. Rev. Lett. **112**, 025701 (2014). https://link.aps.org/doi/10.1103/PhysRevLett.112.025701
10. Y. Wang, X. Ren, K. Otsuka, Shape memory effect and superelasticity in a strain glass alloy. Phys. Rev. Lett. **97**, 225703 (2006). https://link.aps.org/doi/10.1103/PhysRevLett.97.225703
11. S. Kartha, T. Castán, J.A. Krumhansl, J.P. Sethna, Spin-glass nature of tweed precursors in martensitic transformations. Phys. Rev. Lett. **67**, 3630–3633 (1991). https://link.aps.org/doi/10.1103/PhysRevLett.67.3630
12. S. Kartha, J.A. Krumhansl, J.P. Sethna, L.K. Wickham, Disorder-driven pretransitional tweed pattern in martensitic transformations. Phys. Rev. B **52**, 803–822 (1995). https://link.aps.org/doi/10.1103/PhysRevB.52.803

13. J.P. Sethna, S. Kartha, T. Castán, J.A. Krumhansl, Tweed in martensites: a potential new spin glass. Phys. Scr. **1992**(T42), 214 (1992). http://stacks.iop.org/1402-4896/1992/i=T42/a=034
14. D. Sherrington, A simple spin glass perspective on martensitic shape-memory alloys. J. Phys. Condens. Matter **20**(30), 304213 (2008). http://stacks.iop.org/0953-8984/20/i=30/a=304213
15. D. Sherrington, A spin glass perspective on ferroic glasses. Phys. Status Solidi B **251**(10), 1967–1981 (2014). https://doi.org/10.1002/pssb.201350391
16. D. Sherrington, S. Kirkpatrick, Solvable model of a spin-glass. Phys. Rev. Lett. **35**, 1792–1796 (1975). https://link.aps.org/doi/10.1103/PhysRevLett.35.1792
17. K. Wolfgang, Relaxor ferroelectrics: cluster glass ground state via random fields and random bonds. Phys. Status Solidi B **251**(10), 1993–2002 (2014). https://onlinelibrary.wiley.com/doi/abs/10.1002/pssb.201350310
18. P. Lloveras, T. Castán, M. Porta, A. Planes, A. Saxena, Influence of elastic anisotropy on structural nanoscale textures. Phys. Rev. Lett. **100**, 165707 (2008). https://link.aps.org/doi/10.1103/PhysRevLett.100.165707
19. P. Lloveras, T. Castán, M. Porta, A. Planes, A. Saxena, Glassy behavior in martensites: interplay between elastic anisotropy and disorder in zero-field-cooling/field-cooling simulation experiments. Phys. Rev. B **80**, 054107 (2009). https://link.aps.org/doi/10.1103/PhysRevB.80.054107
20. D. Wang, Y. Wang, Z. Zhang, X. Ren, Modeling abnormal strain states in ferroelastic systems: the role of point defects. Phys. Rev. Lett. **105**, 205702 (2010). https://link.aps.org/doi/10.1103/PhysRevLett.105.205702
21. S.R. Shenoy, T. Lookman, Strain pseudospins with power-law interactions: glassy textures of a cooled coupled-map lattice. Phys. Rev. B **78**, 144103 (2008). https://link.aps.org/doi/10.1103/PhysRevB.78.144103
22. R. Vasseur, T. Lookman, Effects of disorder in ferroelastics: a spin model for strain glass. Phys. Rev. B **81**, 094107 (2010). https://link.aps.org/doi/10.1103/PhysRevB.81.094107
23. R. Vasseur, D. Xue, Y. Zhou, W. Ettoumi, X. Ding, X. Ren, T. Lookman, Phase diagram of ferroelastic systems in the presence of disorder: analytical model and experimental verification. Phys. Rev. B **86**, 184103 (2012). https://link.aps.org/doi/10.1103/PhysRevB.86.184103
24. A.R. Bishop, T. Lookman, A. Saxena, S.R. Shenoy, Elasticity-driven nanoscale texturing in complex electronic materials. Europhys. Lett. **63**(2), 289 (2003). http://stacks.iop.org/0295-5075/63/i=2/a=289
25. T. Lookman, S.R. Shenoy, K.O. Rasmussen, A. Saxena, A.R. Bishop, Ferroelastic dynamics and strain compatibility. Phys. Rev. B **67**, 024114 (2003). https://link.aps.org/doi/10.1103/PhysRevB.67.024114
26. D. Xue, Y. Zhou, X. Ding, T. Lookman, J. Sun, X. Ren, Aging and deaging effects in shape memory alloys. Phys. Rev. B **86**, 184109 (2012). https://link.aps.org/doi/10.1103/PhysRevB.86.184109
27. D. Xue, P. V. Balachandran, J. Hogden, J. Theiler, D. Xue, T. Lookman, Accelerated search for materials with targeted properties by adaptive design. Nat. Commun. **7**, 11241 (2016). https://doi.org/10.1038/ncomms11241
28. D. Xue, P.V. Balachandran, R. Yuan, T. Hu, X. Qian, E.R. Dougherty, T. Lookman, Accelerated search for $BaTiO_3$-based piezoelectrics with vertical morphotropic phase boundary using Bayesian learning. Proc. Natl. Acad. Sci. **113**(47), 13301–313306 (2016). http://www.pnas.org/content/113/47/13301
29. R. Dehghannasiri, D. Xue, P.V. Balachandran, M.R. Yousefi, L.A. Dalton, T. Lookman, E.R. Dougherty, Optimal experimental design for materials discovery. Comput. Mater. Sci. **129**, 311–322 (2017). https://doi.org/10.1016/j.commatsci.2016.11.041. http://www.sciencedirect.com/science/article/pii/S0927025616306024

Chapter 9
Mesoscopic Modelling of Strain Glass

P. Lloveras, T. Castán, M. Porta, A. Saxena, and A. Planes

Abstract Glassiness is ubiquitous in nature but it still keeps many fascinating phenomena hidden. The discovery about a decade ago of glassy behavior in strain nanoclusters (the strain glass) has extended ferroic glasses to include the ferroelastic property. Here, by means of numerical modelling and comparison with experimental data in the literature, we identify disorder and anisotropy as key parameters whose interplay determines the ferroelastic behavior in alloys: While anisotropy-driven systems exhibit a normal ferroelastic transition, disorder-driven systems may result in the strain glass state. Interestingly, strain glass preserves functional properties such as the shape memory effect (SME) and superelasticity. Moreover, it exhibits hysteresis reduction and widening of operational temperature-stress range, which enhances its technological appeal. Precisely based on the occurrence of the SME, the relevance of geometrical frustration in strain glass is called into question as it might play a minor role in the freezing process. In magnetostructural systems, the multiferroic coupling could yield strain-mediated magnetic glass.

P. Lloveras (✉)
Grup de Caracterització de Materials, Departament de Física, EEBE, Universitat Politècnica de Catalunya, Barcelona, Catalonia, Spain

Barcelona Research Center in Multiscale Science and Engineering, Barcelona, Catalonia, Spain
e-mail: pol.lloveras@upc.edu

T. Castán · A. Planes
Departament de Física de la Matèria Condensada, Facultat de Física, Universitat de Barcelona, Barcelona, Catalonia, Spain
e-mail: teresa.castan@fmc.ub.edu; antoni.planes@fmc.ub.edu

M. Porta
Departament de Física Quàntica i Astrofísica, Facultat de Física, Universitat de Barcelona, Barcelona, Catalonia, Spain
e-mail: marcel@fqa.ub.edu

A. Saxena
Theoretical Division, Los Alamos National Laboratory, Los Alamos, NM, USA
e-mail: avadh@lanl.gov

T. Lookman, X. Ren (eds.), *Frustrated Materials and Ferroic Glasses*, Springer Series in Materials Science 275, https://doi.org/10.1007/978-3-319-96914-5_9

9.1 Introduction

Glassiness is one of the most intriguing phenomenon in condensed matter physics [1]. It basically refers to long-lived out-of-equilibrium states characterized by the presence of quasistatic disorder, that replace the long-range anisotropic order arising in thermodynamic solid phases. The freezing of these states results from a slowing-down and subsequent arrest of the relaxation dynamics that is initiated by the drop of thermal fluctuations across a phase instability within a (quasi-)degenerate, multiwell free-energy landscape. This prevents the system from spontaneously finding the path towards the thermodynamic equilibrium and thus blocks the occurrence of a thermodynamic phase transition. The temperature below which the thermally activated dynamically-disordered phases become effectively frozen (i.e., when typical relaxation times reach the arbitrary threshold of $\sim$ 100 s) is denoted the *glass transition* temperature.

The (extremely) slow yet continuous nonequilibrium relaxation process towards the equilibrium is called aging and renders the glass properties to depend on time and on thermal and external field history. While the time dependence entails frequency dependent peaks in the susceptibility obtained by broadband spectroscopy, the history dependence leads to a loss of ergodicity (in contrast to equilibrium states which are ergodic [2]), which is revealed by the splitting of the curves in Zero-Field-Cooling/Field-Cooling (ZFC/FC) protocols. Other traces characteristic of the glass transition are the absence of signatures associated with the suppressed thermodynamic phase transition, namely the lack of a calorimetric peak and anomalies in the susceptibility.

Glassy characteristics are met in a wide variety of systems, involving one or more physical quantities (such as translational, orientational, magnetic [3], vortex [4], polar [5], orbital degrees of freedom [6], etc.) and exhibiting different correlation lengths (canonical, clusterized). Moreover, depending on the particulars of the system, different additional requirements for the vitrification must accompany the thermal deactivation, such as fast cooling, geometrical frustration, and/or quenched-in disorder. Therefore, glassiness does not refer to a unified framework accounting for the underlying physics, but it rather corresponds to the aforementioned description of the phenomenology associated with a certain relaxation dynamics.

Beyond the fundamental interest in condensed matter physics, glasses exhibit unique features related to structure, magnetism, and electricity that are absent in equilibrium phases, and that makes them useful in a wide range of applications [7]: windows, optical components, containers, construction materials, medicine (bioactive implants), electronic components, recording heads, transformer cores, diffraction gratings, planar channel waveguides, optical fiber amplifiers, lasers and optical switches, and also in the development of new approaches to studying problems in computer science, neural networks, biology and economics, and a variety of other topics [8]. Hence, gaining insight into the underlying physics is also important in order to control the emergence and stability of glassiness which is crucial before considering any application.

This chapter is devoted to modelling cluster glasses involving strain as the primary frozen disordered physical quantity. This will include *strain glass* and *strain-mediated magnetic glass*. Prior to going into the details of modelling a brief overview on glasses will be presented, sketching the concepts of frustration, non-ergodicity, and cluster glasses with respect to aspects that are relevant for strain glasses. More importantly, we will particularly focus on the key role of anisotropy and intrinsic disorder, which will be supported by our simulations. This will provide insight into the common trends and requirements for the emergence of glassiness in ferroelastics and ferroic systems in general. Multiferroic couplings and their relationship to strain glass will be discussed as well.

9.1.1 Canonical Structural Glasses

Prototypical structural glasses refer to certain amorphous solids, and are obtained by *fast* cooling from a supercooled liquid. The consequent sudden drop of the thermal fluctuations traps the system in local minima between free-energy barriers such that the relaxation towards the equilibrium crystalline phase is prevented. Instead, atoms (or molecules) arrange in a disordered quasistatic network, with random positions and orientations with no specific symmetry. These systems lack translational and orientational invariance so that a Bravais lattice cannot be defined. Long range order is absent, resulting in diffraction patterns with only one or very few diffuse peaks, resembling that of liquids. However, they behave like solids, with diverging viscosity at finite temperature (which is proportional to the relaxation time), and often display useful technological properties.

Some liquids undergo thermodynamic phase transitions towards intermediate phases where some type of dynamical disorder is still present, before ordering completely across a lower-temperature transition towards the crystalline phase. This is the case, for instance, of liquid crystals (nematic, smectic, discotic, cholesteric, etc.), which are also known as anisotropic liquids due to the anisometric shape of their molecules. This confers on them orientational order but (partial) translational disorder. In contrast, plastic crystals are formed by globular molecules, yielding orientational disorder but translational order. Like in supercooled liquids, upon fast cooling these mesophases may undergo a glass transition involving the freezing of the disordered state, giving rise to nematic, smectic [9, 10], and orientational glasses [11].

It is worth emphasizing that the *unique* condition for vitrification of all these systems is fast cooling, in contrast to other glass-forming systems that originate from frustration and/or intrinsic disorder, which will be reviewed later. The minimum cooling rate needed to obtain glass depends on the system [12]; in fact, all ranges have been reported: while all systems are suitable for undergoing the glass transition provided that the cooling rate is high enough, some systems can easily form glassy compounds, so that glassiness can hardly be avoided upon cooling. Signatures of the glass transition may resemble second-order phase transitions (such as a continuous change of volume versus temperature with slope change and a step in calorimetric signal) but it cannot be designated as a true thermodynamic transition since the glass transition temperature depends on the cooling rate [12]. Slowing-down dynamics may be very intricate and include multiple relaxations (primary, secondary, etc.) associated with different degrees of freedom (intermolecular, intramolecular) in such a way that susceptibilities may exhibit more than one peak associated with glassiness.

9.1.2 Geometrical Frustration

Paramagnetic phases consist of spins arranged on a lattice with dynamically disordered orientations, i.e. dominated by thermal fluctuations. This gives rise to magnetically isotropic systems. Upon cooling below the Curie temperature, typically the spins spontaneously evolve to an ordered configuration with parallel orientations, establishing an anisotropic magnetic phase called ferromagnetic. Nevertheless, under certain circumstances the ferromagnetic transition may be suppressed, with the spins frozen in a static orientationally disordered configuration which is called spin glass. Glassiness in magnetic systems may result from fast cooling as in the case of structural glasses, but also from an incompatibility between the magnetic interactions and the underlying topology of the lattice. The latter situation is referred to as geometrical frustration [13], and may take place when antiferromagnetic exchange interactions arise. The simplest, prototypical example is an antiferromagnetic order on a triangular lattice. There, it becomes clear that energy minimization of all bonds cannot be satisfied at the same time. This leads to multiplicity of ground states with residual entropy at $T = 0$. Notice that this is essentially different from the case of the structural glasses discussed in the preceding section, where the global thermodynamic minimum does exist, despite being not reached due to kinetic reasons. Although in both classes of glasses the existence of multiple quasi-degenerate minima causes the trapping of the system in a limited area of the phase space, leading to non-ergodicity and a lack of long-range order, in purely frustrated systems such as the antiferromagnetic triangular lattice, glassiness is an intrinsic feature of the system, as it appears inevitably regardless of the cooling rate.

Other examples of geometrically frustrated magnets are found in the Kagome lattice [14, 15], pyrochlore oxides [16], and artificial spin ice [17], where the coexistence of ferromagnetic and antiferromagnetic interactions resembles the configurational disorder associated with the doubly degenerate position of hydrogen in water ice. Notice that water ice is a frustrated system, with a residual entropy at $T = 0$ but, interestingly, it must be considered a thermodynamic phase, as it is reached on cooling across a first-order phase transition with a finite volume change and associated latent heat. Therefore, one can deduce that frustration is not exclusive in glassy systems. Some magnetic systems with competing ferromagnetic and antiferromagnetic interactions may exhibit exotic frustrated magnetic phases, although they cannot be designated as glasses [18]. Even liquids have been proposed as frustrated systems as well [19, 20], although this is still under discussion. From this it can be inferred that the set of experimental evidence shared by all glasses, regardless of the diverse underlying physical causes, must be met to establish rigorously the existence of vitrification. That is to say, systems exhibiting only a few of them cannot be claimed as glasses.

9.2 Anisotropy and Intrinsic Disorder

In addition to fast cooling and frustration, a third independent cause for glassiness is the presence of quenched-in disorder, which is particularly relevant in strain glasses. We remark here that these three triggering factors are independent, although the two latter cases, frustration and quenched-in disorder, often appear together. A deeper discussion concerning this issue will be presented in Sect. 9.3.2. In the present section, we will focus on the role of intrinsic disorder and anisotropy. They will be proposed as quite general parameters that can be used for characterization of thermodynamic and glassy phases, in particular those concerning ferroic systems and their corresponding cluster glasses.

9.2.1 Anisotropy

We have seen that the thermodynamic phase transitions that are susceptible to be replaced by vitrification are accompanied by a local symmetry breaking with a consequent increase in anisotropy in the low-temperature phase, and the establishment or increase of long-range order. The latter allows to define the anisotropy as an intensive thermodynamic variable, setting up a link between long-range order and the anisotropy of the thermodynamic phases [21–23]. A simple example is the lattice periodicity and related space group resulting from crystallization. Instead, the crossover to the glassy state occurs only if the symmetry loss is inhibited, and the long-range order cannot be achieved or increased. In this case, the anisotropy is kept rather constant throughout the freezing process.

In addition, in many systems such as ferroics, the local symmetry breaking is degenerate, and the existing symmetry-related domains propagate long distances in an orderly manner mediated by dipolar-like long-range interactions and subjected to specific boundary conditions, such as phase boundaries or free surfaces. This *self-accommodation* process yields the emergence of long-range, low-symmetry domains with specific morphologies that can also be considered as an additional thermodynamic signature of the anisotropy of the system. Let us recall here that this feature is common to all ferroics, including ferromagnets, ferroelectrics, and ferroelastics. It is however apparent that the underlying physics is different: Whereas the electric and magnetic degrees of freedom are vectors, elastic strain is a rank-2 tensor. Moreover, while electric and magnetic fields are long ranged in nature, elastic interactions are propagated by a "knock-on effect." In fact, they emerge from compatibility constraints in the strain field arising from the underlying displacement field.

In the case of ferromagnets, for instance, there is a shape anisotropy arising from dipolar interactions (i.e. the demagnetizing factor) and a magnetocrystalline anisotropy coming from coupling of the magnetic degrees of freedom to the underlying lattice. Precisely, within ferroic variables, the strain is of particular interest because very often it plays an important role in multiferroic couplings, as it defines the lattice where magnetic and electric dipoles reside. With respect to the former, lattice (or sublattice) spacing determines the sign of the exchange parameter, thus setting the ferromagnetic or antiferromagnetic character of magnetic interactions. In turn, electric dipoles emerge from off-center ions as a consequence of a lattice distortion and magnetoelectric coupling is often mediated by strain [24]. Therefore, to gain insight into ferroics it is useful to start with the analysis of pure ferroelastic systems, where strain is the unique ferroic property that fully characterizes the system in terms of the free energy.

Ferroelastic transitions are usually triggered by the softening of certain phonon modes that stabilize the low-symmetry phase. Such soft directions, and the subsequent low elastic constants provide the lattice with easy channels that rule the long-range elastic interactions. This results in highly anisotropic patterns that usually organize in the form of twin related domains (or variants). In cubic systems, for instance, the elastic anisotropy factor is defined as $\mathscr{A} = C_{44}/C'$ with C_{44} and C' being elastic constants associated with shear and deviatoric modes, respectively. In ferroelastic transitions, $\mathscr{A}$ gains relevance as a result of the softening of C' while other elastic constants maintain their magnitude to a good approximation. Indeed, it can be inferred that, in general, the anisotropy is directly involved in the strength of the forces giving rise to long-range order. In the next section, mathematical explanation supporting this argument will be given for a square-to-rectangle transition. The role of long-range anisotropic elastic interactions is indeed prominent. In most of the systems, they suppress the otherwise decisive role of the critical fluctuations, thus rendering the transition athermal. In this case, thermal fluctuations are not the triggering factor leading up to the transition but temperature acts as a scalar control parameter like an external applied field. Hence, it is reasonable that some ferroelastic (and other ferroic) phase transitions are referred to as anisotropy-driven transformations.

Now it is worth considering two aspects related to twinning: First, twinning is not an *inherent fact* of the phase transition but the anisotropic long-range response of the ferroelastic phase to the coexistence with the high-symmetry phase taking advantage of the degenerate multi-well structure of the free energy. Second, the specific transformation path taken by the self-accommodation process, and consequently the microstructure of a given ferroelastic material depend on a number of additional factors: the initial nucleation conditions, the specimen size, grain size, history, external conditions, coupling with other entities like magnetic fields, the presence of impurities and defects, the specific composition, etc., and there may be a high degree of complexity. In fact, defects may also act as pinning centers for twin boundary motion, eventually leading to unique paths for forward and backward transformations, which results in small hysteresis in temperature [25]. The length scale of twins may range from few nanometers to tenths of millimeters [26], with twin boundary mobilities that may be very high or essentially zero [27]. Self-similar patterns—twins within twins and hierarchical patterns—have also been observed. Polycrystals also show a coexistence of variants with multiple length scales, etc. Hence, it is apparent that the specific stabilized configuration of the twin interfaces is different from case to case. Within this framework, it has been suggested that, while anisotropy underpins the long-range interactions, when combined with real heterogeneous nucleation, it may also cause that some ferroics are unable to go over the phase space but they get trapped into a certain region once a configuration is stabilized, entailing a loss of ergodicity [28–30]. Also, the role of energy and entropy barriers, and the possibility of glass-like behavior in ferroelastic models without quenched disorder have been discussed elsewhere [31]. Hence, some systems exhibiting multidomain patterns resulting from quasi-degenerate multiwell energy profile may also be described as inhomogeneous, non-ergodic and/or frustrated anisotropy-driven systems. However, they lack disorder and display long-range order resulting from a first-order phase transition, thus excluding the glassy character.

9.2.2 *Intrinsic Disorder*

The presence of intrinsic inhomogeneities has been already mentioned because it leaves significant marks on the specific stabilized structures, but actually their origin is diverse. Examples are point defects from intrinsic compositional fluctuations, vacancies, interstitial and substitutional atoms from doping or self-doping, line defects like dislocations, among others. Unavoidable or intentional disorder can in general be characterized as random and local. While the local strain field arising from impurities tries to propagate long distances through the knock-on effect, its statistically random character may prevent any self-organized global process, and the strain field finally decays at a short range.

Disorder has been observed to cause a number of effects on materials [32]. First, it may cause rounding of phase transitions [33, 34], moving away from the sharp case in the ideal clean limit and giving rise to multiphase coexistence well above and below the transition point. Second, and intimately related to the first, disorder may induce local free-energy barriers in such a way that the total free energy of the system can adopt a bumpy profile with many degenerate and nearly degenerate low-energy states that, in general, do not correspond to the global minimum of energy, if it still exists [32].

In general, it is observed that the rounding of the transition results in anomalies in the specific heat such as softening and shift of the peak towards lower temperatures, changes in the baseline, power-law singularities, etc. Indeed, a more accurate sample treatment may result in a decrease and even removing of the anomalies. With respect to this, anomalies in C_p have been proposed to be an indicator of the level of dopant and lattice imperfections in a material [35]. Also, impurities have been observed to modify the elastic constants of the material and, subsequently, its elastic anisotropy. Consequently, the phase diagrams strongly depend on the level of doping. Slight changes in the alloy composition can result in a large shift of the transition temperature or even inhibition of the transformation.

It is significant, and may also be paradoxical, that ferroic systems in general and ferroelastics in particular may exhibit frustration and non-ergodicity in two extreme and opposite cases—from anisotropy-driven to disorder-driven—where physics is dominated by either long-range or local interactions, respectively. Here it is worth mentioning that frustration in ferroelastics may give rise to precursor nanoscale textures, whose origin lies in the interplay between anisotropy and disorder. The magnitude of the elastic anisotropy factor has been suggested to determine the morphology of these patterns [36]. Pretransitional tweed consisting of cross-hatched modulations of small strain of the low-temperature phase at temperatures above the ferroelastic transition arise as the natural response of long-range interactions to local coupling to disorder. Kartha et al. [37] showed that in the limit of infinite anisotropy, tweed can be considered as a frustrated system and formally identified as a spin glass. In finite-anisotropy real materials, however, tweed seems to lack glassy signatures that do appear in strain glass [38].

9.2.3 Cluster Glasses

It is hence clear that both anisotropy and quenched-in disorder are intrinsic to ferroic materials and play key roles in determining many of their properties. Both high-anisotropy and high-disorder limits may lead to anisotropy-driven and disorder-driven non-ergodicity and frustration [39], yet undergoing normal ferroic transitions. However, above a certain threshold of disorder the growth of the domains is interrupted at the nanoscale and long-range order cannot be achieved. Hence, the thermodynamic transition throughout the system is suppressed and both the symmetry breaking and consequent gain of anisotropy occur within these locally transformed nanoregions only, losing their thermodynamic character. Frustration may emerge as the long-range interactions cannot be satisfied because of the random fields created by impurities [40, 41].

The slowing down and final arrest of the domain growth results in a vitrified nanostructured state that is denoted *cluster glass*, as it is evidenced by the usual frequency dispersion peak and non-ergodic ZFC/FC measurements. Cluster glasses corresponding to all types of ferroic systems have been reported: cluster-spin glasses [42–45], ferroelectric relaxors [5, 46], and strain glasses. Cluster glasses were first interpreted as superparamagnetic and superferroelectric states but experimental signatures pointing to glass behavior discarded such hypothesis.

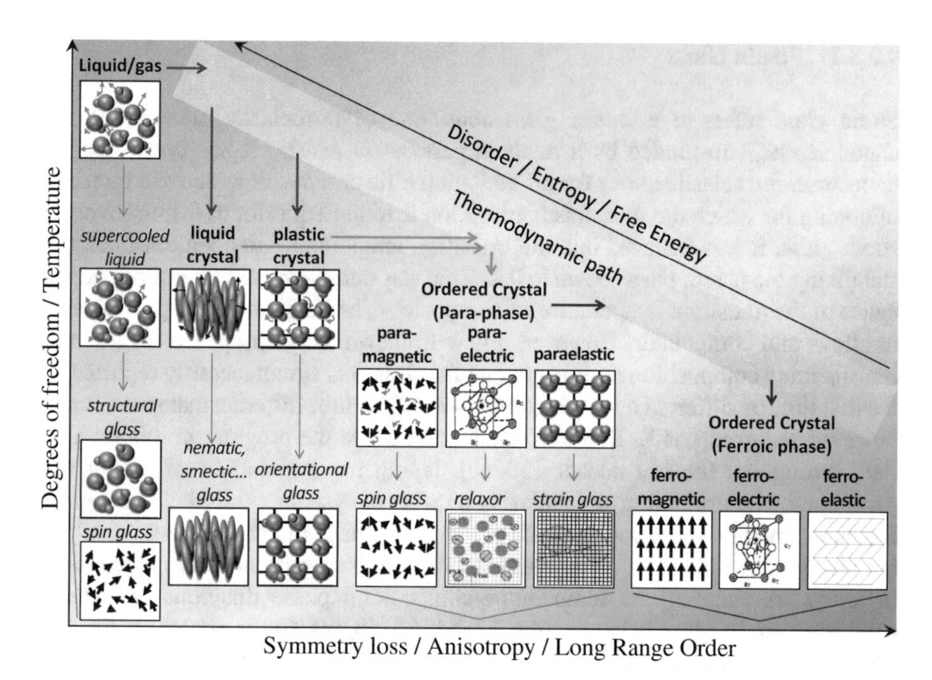

Fig. 9.1 Routes to glassiness: qualitative temperature-anisotropy phase diagram. Thermodynamic (glass) phases are indicated by bold (italic) letters and the corresponding thermodynamic (glass) transitions by thick (thin) arrows. Each phase is accompanied by a schematic representative configuration

As a summary, Fig. 9.1 displays some typical thermodynamic and glass phases linked by their corresponding transitions in a temperature-anisotropy diagram. They are represented by bold and italic letters and thick and thin lines respectively, and are accompanied by qualitative representative snapshots of configurations. The direction of entropy and free energy increase is also shown.

Interestingly, these nanostructured glasses also display functional properties, which in the past were thought to be associated with the occurrence of ferroic transitions. This widens the operational range of alloy compositions and temperatures that can be used for technological applications. In addition, even some properties may be enhanced with respect to normal ferroics and other new ones may emerge. For instance, the rounding of the susceptibility around the glass crossover spreads the temperature interval where high response to an applied field is obtained. Hereafter we will focus on ferroelastic systems and strain glasses, as they are the very last ferroic cluster glasses discovered so far, although of crucial importance for both an understanding of fundamental physics and technological applications.

9.2.3.1 Strain Glass

Strain glass refers to a cluster glass consisting of ferroelastic (non-zero strain) nanodomains, surrounded by a retained paraelastic matrix. It has been observed to occur in ferroelastic alloys for off-stoichiometric compositions above a threshold of doping for which the ferroelastic transition is inhibited. Prior to the discovery of strain glass, it was believed that the transition temperature dropped drastically by stabilizing the parent phase down to 0 K. This was consistent with the strong dependence of the transition temperature on composition, a widely observed phenomenon in alloys and compounds. However, a new framework was proposed for the non-transforming composition regime after strain glass was simultaneously reported for the first time by different research groups [47–49] in three different materials. It was, however, the group of X. Ren in Tsukuba that drove the progress in this research field throughout the last decade [50–60], laying the foundations of a firm new paradigm by incorporating an increasing number of systems exhibiting strain glass, including mainly shape memory alloys and ferromagnetic shape memory alloys, but also gum metals and others, thus establishing this phenomenon as a generality in alloys. A summary of temperature-composition phase diagrams of systems exhibiting strain glass behavior [49, 52–54, 56–58, 60, 61] is shown in Fig. 9.2. The critical composition x_c above which the ferroelastic transition is inhibited may in general depend on multiple features of the particular system. However, there is experimental and theoretical evidence indicating the relevance of the elastic anisotropy $\mathscr{A}$ as a key factor influencing x_c. Although there are no systematic experimental studies on the influence of $\mathscr{A}$ on the existence of the strain glass phase, available data [52, 53, 62–66] is compiled in Fig. 9.3, showing indeed a correlation between $\mathscr{A}$ and x_c. More specifically, the higher the anisotropy, the higher the critical concentration. While in Ti-Ni- and Ti-Pd-based alloys, strain glass behavior has been observed above x_c, in Fe-Pd, Cu-Al-Mn, and Ni-Al alloys the existence of strain glass has not been studied yet. Close to x_c, mixed phases containing both normal ferroelastic and strain glass regions, and spontaneous strain glass-to-ferroelastic transition have also been reported. This is analogous to observations in ferromagnetic and ferroelectric systems, and extends the common framework shared by ferroic systems.

As the transition temperature strongly depends on the composition, it is natural to expect that other characteristics such as the elastic anisotropy may also be modified by doping. For instance, ab initio density functional theory (DFT) calculations predict a strong dependence of $\mathscr{A}$ on composition in Ti-Nb alloys: $\mathscr{A} = 3.2$ for Ti-18.75 at.%Nb, $\mathscr{A} = 2.4$ for Ti-25 at.%Nb and $\mathscr{A} = 1.1$ for Ti-31.25 at.%Nb [67].

Strain glasses have been reported to meet all the required signatures to claim the existence of glassy behavior: absence of a peak in calorimetry measurements, frequency dependent peak in the real part (ac storage modulus) and the imaginary part (loss tangent), non-ergodicity in Zero-Stress-Cooling/Stress-Cooling (ZSC/SC) protocols, and evidence in X-Ray diffraction and high-resolution transmission electron microscope (HRTEM) imaging [59].

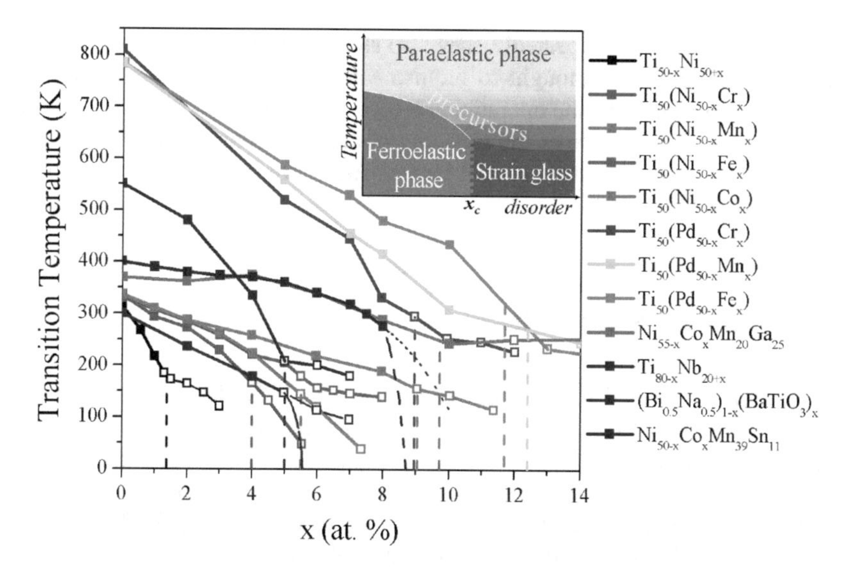

Fig. 9.2 Experimental temperature-composition phase diagrams of shape-memory alloys that are known to display strain glass behavior in the non-transforming regime. References for the materials are the following: $Ti_{50-x}Ni_{50+x}$ [49], $Ti_{50}Ni_{50-x}Cr_x$ [53], $Ti_{50}Ni_{50-x}Mn_x$ [53], $Ti_{50}Ni_{50-x}Fe_x$ [54], $Ti_{50}Ni_{50-x}Co_x$ [53], $Ti_{50}(Pd_{50-x}Cr_x)$ [52], $Ti_{50}(Pd_{50-x}Mn_x)$ [60], $Ti_{50}(Pd_{50-x}Fe_x)$ [60], $Ni_{55-x}Co_xMn_{20}Ga_{25}$ [57], $Ti_{80-x}Nb_{20+x}$ [58], $(1-x)(Bi_{0.5}Na_{0.5})TiO_3 - xBaTiO_3$ [56], $Ni_{50-x}Co_xMn_{39}Sn_{11}$ [61] Solid symbols stand for the normal ferroelastic transition, empty symbols refer to the crossover towards the strain glass and dashed lines indicate the critical composition x_c separating both regimes. Dotted line in $Ni_{50-x}Co_xMn_{39}Sn_{11}$ stands for a proposed strain glass crossover. The inset is a schematic qualitative picture of the phase diagram, applicable to all cases in general. Precursors refer to either pretransitional textures anticipating the ferroelastic transition or unfrozen nanodomains above the strain glass

Among ferroelastics exhibiting a strain glass composition regime, it is specially worth focusing on shape memory alloys because they display two outstanding features related to the ferroelastic transition: The shape memory effect (SME) and superelasticity [68], which consist of the shape recovery after severe stress-induced deformations. This behavior confers upon them functional properties that are intensively exploited in a broad variety of technological applications, ranging from medical devices to sensors and actuators, muscular wires in robotics, mechanical aeronautic and underwater couplings, and others.

It is well known that the operational characteristics of both superelasticity and shape memory effect (i.e. onset temperature, hysteresis, etc.) are crucially affected by the specific alloy composition [69–74]. To date, stoichiometric Ti-Ni alloy is the most used shape-memory alloy due to its lasting durability (wear and corrosion resistance), repeatability, and biocompatibility. However, Ti-Ni exhibits large hysteresis in stress and temperature, and a narrow operating temperature interval, which challenges a further technological development. The discovery of strain glass, first observed precisely in off-stoichiometric Ti-Ni alloy, may overcome these obstacles since, unexpectedly, they also show both functional properties [50], which in principle were thought to require a normal ferroelastic behavior. This results in the widening of the operational temperature window and may reduce the hysteresis as a consequence of the rounding of the stress-induced transition. Other appealing properties absent in normal ferroelastics have been observed to arise in the strain glass [75].

9.3 Modelling Strain Glass

This section is devoted to the analysis of numerical simulations obtained from a mesoscopic model for ferroelastics. Modelling ferroelastic systems has been extensively addressed by many different groups. Mesoscopic approaches can be basically summarized in the sharp-interface minimizers introduced by Ball and James [26, 76], phase fields based on Kachaturyan's elasticity [77–79] or similar theories [80], and models derived from Ginzburg-Landau (GL) theory [25, 81–83]. In turn, the latter, either in the usual continuum version [51, 83–85] or combined with a discrete pseudo-spin mapping approach [31], has been also employed to investigate non-ergodicity, frustration and glassy behavior in ferroelastic systems.

9.3.1 The Model

Here we present a GL free-energy functional to perform simulations of a square-to-rectangle structural transition. This can be considered as the 2-dimensional analogue of a 3-d cubic-to-tetragonal transition, commonly occurring among ferroelastics. This becomes more meaningful since the former can be conceived as the cross-section of the latter, where some strain textures of interest take place in certain 2-d planes.

In a square lattice, the corresponding symmetry adapted strains e_1, e_2, and e_3 are hydrostatic, deviatoric, and shear modes, with associated harmonic elastic constants $A_1 = (C_{11} + C_{12})$, $A_2 = (C_{11} - C_{12}) = 2C'$ and $A_3 = 4C_{44}$, respectively. Since the tetragonal distortion e_2 is the order-parameter (OP) strain of the transition, the structural free-energy density f_s contains a GL sixth-order polynomial expansion f_{GL} in terms of e_2 as allowed by symmetry:

$$f_{\mathrm{GL}}(e_2) = \frac{A_2(T)}{2} e_2^2(\mathbf{r}) + \frac{\beta}{4} e_2^4(\mathbf{r}) + \frac{\gamma}{6} e_2^6(\mathbf{r}) + \frac{\kappa}{2} |\nabla e_2(\mathbf{r})|^2 \tag{9.1}$$

where the nonlinear elastic parameters β and γ together with $A_2(T) = \alpha_T(T - T_c)$ account for the structural transition and the Ginzburg coefficient κ captures the interfacial energy cost. Here, T_c is the lower stability limit of the high-temperature phase and defines the equilibrium transition temperature T_t and the higher stability limit of the low-temperature phase T_i. Indeed, above T_t, Eq. (9.1) renders $e_2 = 0$, which corresponds to the undistorted paraelastic (or square) phase. Below T_t, $e_2 = \pm e$, that corresponds to the two twin rectangular orientation variants. However, to properly describe heterogeneous structures in both ferroelastic systems and strain glasses, anisotropy and disorder must also be included. In the following we will see how both parameters can naturally be incorporated in the model.

9.3.1.1 Anisotropy

In addition to the OP expansion, the secondary strains e_1 and e_3 are taken into account up to the harmonic contribution only, since they are expected to be small. Hence, the resulting non-OP free-energy density $f_{\text{non-OP}}$ is written as:

$$f_{\text{non-OP}}(e_1, e_3) = \frac{A_1}{2}e_1^2 + \frac{A_3}{2}e_3^2, \tag{9.2}$$

so that the total free-energy density is expressed as $f_s(e_1, e_2, e_3) = f_{\text{GL}}(e_2) + f_{\text{non-OP}}(e_1, e_3)$. To ensure lattice integrity, the St. Vénant compatibility relation that links the three strains derived from the 2-d underlying displacement field must be taken into account [86]. It reduces f_s to only two independent strains, e_2 and e_3. Here, linear elasticity is assumed. After further minimization of the total free energy with respect to e_3, it can be expressed in terms of only e_2. The resulting expression for the non-OP free energy can be expressed as

$$F_{\text{non-OP}} = \int f_{\text{non-OP}} d\mathbf{r} = \int \frac{A_3}{2} \frac{(k_x^2 - k_y^2)^2}{(A_3/A_1)k^4 + 8(k_x k_y)^2} |\tilde{e}_2(\mathbf{k})|^2 d\mathbf{k}, \tag{9.3}$$

where $\mathbf{k}$ is the wavevector of the reciprocal space and $\tilde{e}_2(\mathbf{k})$ stands for the Fourier transform of $e_2(\mathbf{r})$. This term is crucial to understand two characteristic features of ferroelastic systems. On the one hand, it has a dipolar-like expression (decaying as $1/r^2$ in real space) that accounts for long-range elastic interactions. On the other hand, it reveals that it is minimized for $k_x = \pm k_y$, thus providing an explanation for the directionality of the [11] and [1$\bar{1}$] twin interfaces. Notice that, since the elastic anisotropy factor can be expressed as $\mathscr{A} = A_3/2A_2$, then at constant temperature $\mathscr{A} \sim A_3$ from which it follows that $\mathscr{A}$ is directly related to the strength of the dipolar interaction in Eq. (9.3). This is consistent with the highly anisotropic character of twinning and explains why large values of anisotropy favor long-range patterns. Here, variations in $\mathscr{A}$ will be carried out by modifying the value of A_3, while the ratio A_3/A_1 (in the denominator of the Fourier-space kernel) is kept constant, making clear that we change the weight of the long-range interactions. Actually, we have checked that variations of this ratio do not lead to qualitatively new physics.

9.3.1.2 Disorder

As here we mainly focus on shape memory alloys, it is appropriate to take into account that they contain intrinsic disorder associated with inherent compositional fluctuations and/or intentional doping that, in turn, is known to highly influence their transition temperature. Hence, to naturally account for disorder we introduce a random static field coupled to the harmonic term by replacing T_c by a distribution of local stability limits (that in turn entail a distribution of local transition temperatures) $\tilde{T}_c(\mathbf{r}) = T_c + \eta(\mathbf{r})$, where $\eta(\mathbf{r})$ is a random variable exponentially (i.e., short-range) correlated in space and gaussian distributed with zero mean and variance ζ^2. Variations of disorder will be taken into account by changing the value of standard deviation ζ.

It is worth noting here that results are qualitatively independent of this specific form of disorder. In fact, other forms of disorder have been considered in similar models [31], but they do not lead to new physics as the key feature lies in the local character of disorder competing against long-range anisotropic order. In particular, our disorder is fully characterized by the correlation length and the standard deviation, that determine the effective density and intensity of local phase instabilities, respectively. Precisely, the comparison with experiments will indicate that these characteristics have a physically relevant correspondence to experimental disorder features at the mesoscale, which, in principle, could be rigorously approached from a combination of experiments (specific doping, level of off-stoichiometry) and ab initio calculations.

9.3.1.3 Numerical Simulations

Model parameters we use correspond to $Fe_{70}Pd_{30}$ and are given in Ref. [81]. The free energy F_s is discretized by means of the finite differences scheme onto a square mesh, typically of 512×512 mesoscopic unit cells. Boundary and initial conditions will be imposed according to the needs of the particular simulation experiments. The system will evolve following a purely relaxational dynamics until reaching a stable configuration. Exhaustive numerical details of the simulations can be found elsewhere [83–85]. Having said that, however, it is not our aim here to focus on a quantitative description of a particular system but instead to provide a general theoretical framework for strain glass from which qualitative behavior can be inferred. Following this guideline, simulation results will be presented with neither numerical scales nor specific anisotropy and disorder values, but only general trends will be indicated. This will make it easier for a qualitative comparison with a number of experimental observations in different systems.

First, a preliminary analysis of the model behavior will be carried out to discuss and establish the physical grounds ruling ferroelastics and strain glasses. Next sections will be devoted to the simulations of diverse aspects that are relevant for their characterization, namely the structural morphology, thermodynamics, and thermomechanics.

9.3.2 Preliminary Analysis: Origin of Glassy Behavior

First, it is important to notice that the local transition temperatures do not correspond to any thermodynamic transition temperature T_t as the latter must be unique across a thermodynamic transition. It will only occur at T_t if long-range anisotropic interactions are strong enough to correlate distant sites with different local phase stability, thus unifying the transformation temperature. Instead, if long-range interactions are weak with respect to disorder, the thermodynamic behavior may be broken, with the subsequent suppression of the global transition.

From this point of view we can anticipate the prime achievement of this model, which is twofold: (1) the ferroelastic transition is suppressed above a critical value of disorder, and (2) the disorder threshold depends on the elastic anisotropy $\mathscr{A}$: The higher the anisotropy, the higher the disorder threshold, and it is found that the latter goes approximately as $\sim \sqrt{\mathscr{A}}$. This is shown in Fig. 9.3, in agreement with experimental data from shape-memory alloys [87], and confirms the picture set out in the introduction that points to the anisotropy and the disorder as the two key parameters whose balance determines either normal ferroelastic or glassy behavior.

Also, it is interesting to explore the limit of zero anisotropy, that entails the removal of the long-range interactions. In this case the model still displays glassy behavior, and from the resultant local character of the free energy, the hypothesis of geometrical frustration can be excluded. This is consistent with the idea that the

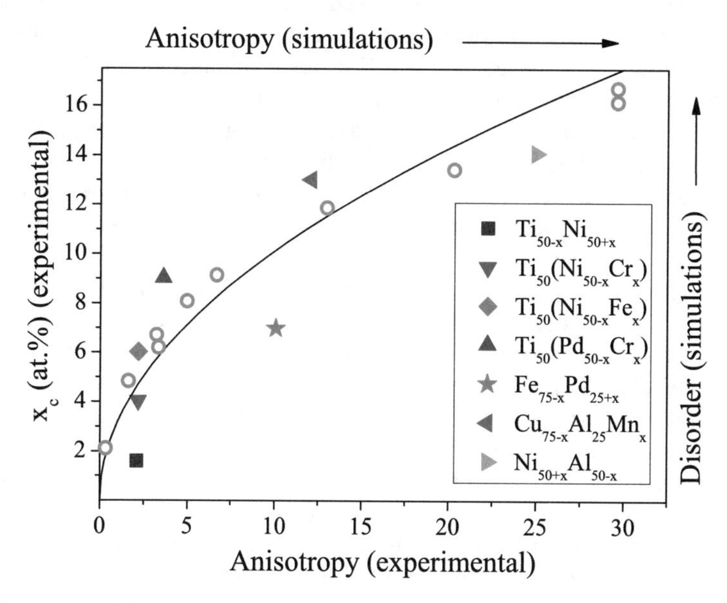

Fig. 9.3 Crossover between twinned and non-transforming systems. Solid symbols have been obtained from experiments on different shape-memory alloys: $Ti_{50-x}Ni_{50+x}$ [62], $Ti_{50}Ni_{50-x}Cr_x$ [53], $Ti_{50}Ni_{50-x}Fe_x$ [62, 63], $Ti_{50}Pd_{50-x}Cr_x$ [52], $Fe_{75-x}Pd_{25+x}$ [64], $Cu_{50-x}Al_{25}Mn_{25+x}$ [65] and $Ni_{50+x}Al_{50-x}$ [66]. Empty circles are simulation results. Figure adopted from Ref. [87]

origin of the vitrification may lie in the kinetic arrest of the domain growth due to the encounter between domains that belong to different variants. The process can be summarized as follows: In the early stages of nucleation and growth of ferroelastic nanodomains, the strain fields can be absorbed by the surrounding paraelastic matrix. However, when the interfaces of different growing domains approach each other, the domains either will percolate if they are of the same variant or they will stop growing if they correspond to different variants. It is yet another example showing that quenched disorder can exist without frustration (one of the best known examples being the Mattis model [88]) and frustration without disorder (for example, the antiferromagnetic triangle; see also Ref. [31]): These are not inseparable concepts and can exist independently [8].

The absence or minor role of frustration in strain glass is consistent with the occurrence of the SME in the strain glass phase of $Ti_{48.5}Ni_{51.5}$ [50] and $Ti_{50}(Pd_{50-x}Cr_x)$ [52], that reveals that the stress-induced ferroelastic phase (i.e., the macroscopic strain) is mostly preserved after the unloading process. This would indicate that: either (1) geometrical frustration does not occur in strain glass or (2) the stress field is able to rearrange or overcome lattice imperfections. In the latter case, frustration would either be removed or not be strong enough to bring the system back to the strain glass phase, respectively. Notice that this is in contrast to the case of significant geometrically frustrated systems, where the removal of the external field entails the reestablishment of a frustrated configuration. A similar situation takes place in complex spin glasses whose behavior is known to lie in the combination of both frustration and quenched disorder: There, while the application of an external magnetic field may induce the ferromagnetic phase as in the strain-glass mechanical analogue, the removal of magnetic field results in that case in the partial or total demagnetization of the system [89–91]. This is probably due to the fact that the role of frustration is in general more important in spin glasses than in strain glasses: While in the former case the antiferromagnetic interactions are essential for the existence of magnetic frustration, in the latter case, antiferroelastic interactions, if existent, would play a much less relevant role in the dynamics of the systems analyzed so far.

In real ferroelastic materials, which have finite anisotropy, the disorder consists of topological defects of the lattice that create a strain field that may have long-range character. Thus, in addition to the local phase instabilities considered by the model disorder, real disorder may render geometric incompatibilities that effectively give rise to geometric frustration preventing the development of long-range order. Hence, the presence of some degree of frustration cannot be discarded. In fact, experiments in O-doped Ti-Nb strain glass [34] do show strain below the glass crossover upon unloading that have some analogy to the situation in spin glasses described above, where frustration plays a more relevant role. In O-doped Ti-Nb alloy, SME is not obtained from the strain glass phase but superelasticity is observed instead. This could be a consequence of the fact that temperature is still too high for SME to occur or the fact that frustration cannot be avoided at all. Maybe the slightly larger anisotropy for this system compared to Ti-based alloys might increase the degree of frustration, although the difference in their nominal values is really small and there is no available exhaustive data concerning the anisotropy and range of SME in these off-stoichiometric and/or doped systems to carry out a deeper analysis.

In summary, neither experiments nor simulations are conclusive so far, so that the relevance of frustration is an open question deserving further investigation and might depend on the specific system and doping. Broadband spectroscopy measurements in loaded-unloaded configurations, and cooling the system from the high-T phase after the shape memory protocol could probably give more insight into the existence of frustration in these systems. Strain glasses are likely complex systems dominated, as in spin glasses, by the combination of frustration and quenched disorder, although with different relative weights for each quantity with respect to their magnetic counterpart.

9.3.3 Structural Morphology

In this section we focus on the morphology of the structures to identify characteristic features that distinguish strain glasses from normal ferroelastics. As the embryos of ferroelastic distortions (either twinning or nanodomains) are often incubated well above the transition in the form of precursor textures that anticipate the oncoming (thermodynamic or local) symmetry breaking, it is interesting to begin with the analysis of the strains arising in this regime. Moreover, since precursors are well accepted to emerge as a consequence of compositional disorder cooperating with anisotropy, they are particularly suited to be investigated by the present model.

Following the suggestion of Murakami et al. [36], who associated the shape of such pretransitional structures with the particular value of $\mathscr{A}$ in some shape-memory alloys, in Fig. 9.4 we show the dependence of high-temperature configurations on the anisotropy [panels (a, c–e)], accompanied by the structure factor and experimental snapshots [panels (b, f–i)] for comparison. The agreement is excellent and confirms that, indeed, large values of $\mathscr{A}$ modulate the strain according to well-defined directions (cross-hatched textures), whereas low $\mathscr{A}$ results in uniform strain droplets of almost spherical shape. Also, in this last case, both experimental and simulation observations lead to an increasing number of such droplets when the transition is approached on cooling.

In Fig. 9.5 the evolution of the structures on cooling yields evidence of the differences between normal ferroelastic and strain glass behavior. In normal ferroelastics, precursors transform to twinned textures on cooling across the ferroelastic transition. In contrast, the inhibition of the ferroelastic transition in strain glass is accompanied by the preservation of the precursor nanostructures down to low temperatures. During this process, domain growth and coarsening slow down basically occur because the retained paraelastic matrix is progressively distorted and hence incorporated into strain glassy nanodomains. Secondly, percolative processes may take place, but domain boundary mobility is very low at these low temperatures. This mechanism makes the glassy nanostructures to be irregular in shape, from almost spherical to ramified droplets, with no preferential directions. Simulations are depicted along with experimental configurations for comparison.

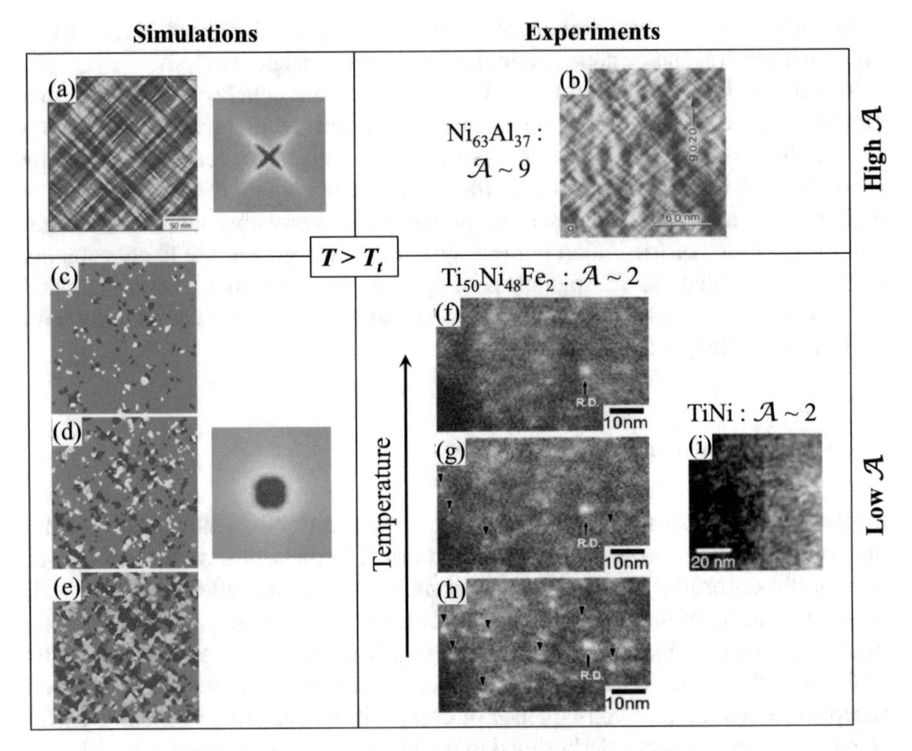

Fig. 9.4 Influence of anisotropy on the morphology of precursor strain textures: high anisotropy favors cross-hatched striations named tweed (**a**), in agreement with experiments (**b**), taken from Ref. [92]. Instead, low anisotropy configurations (**c**)–(**e**) do not exhibit preferential directions but consist of rather isotropic short-range domains. The number of domains increases when the transition is approached, in agreement with experiments (**f**)–(**i**), taken from Refs. [36, 49]. Both high- and low-anisotropy configurations are accompanied by the corresponding structure factor, which are X-shaped and almost spherical, respectively. Experimental values of $\mathcal{A}$ are indicated in each case

It is worth noting that retained paraelastic phase surrounds the glassy strain nanodomains and, due to averaging, this results in X-ray patterns that resemble those at the high-T phase (with broader peaks) [49], which is yet another evidence of the suppression of the ferroelastic transition.

The interfacial dynamics can be inferred from the evolution on cooling of the strain profiles along [11] direction and quantified by computing the corresponding domain size distributions. In Fig. 9.6a, it is clearly seen that in ferroelastic systems, the high-temperature pretransitional strain configuration is uncorrelated with the low-temperature twinning interfaces due to the ferroelastic transition. Moreover, they exhibit different typical widths as indicated by the distinct peak positions (dashed lines) of the domain size distributions, in agreement with experimental observations [25, 93–95]. Instead, in strain glass (Fig. 9.6b), the local strains arising at high temperatures survive on cooling. This is further supported by the invariance

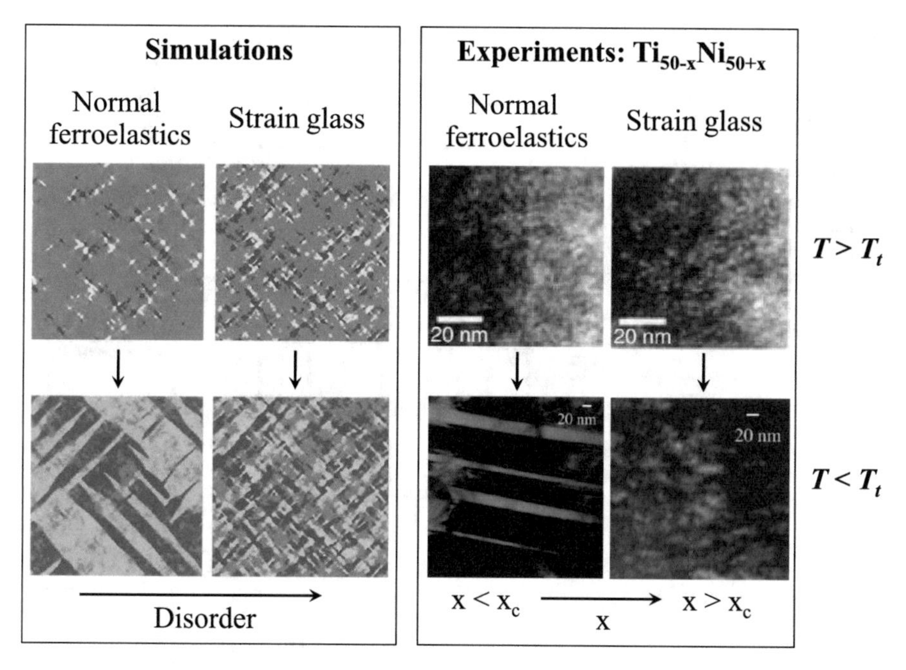

Fig. 9.5 Effect of disorder on ferroelastic systems: on cooling, low-disorder systems undergo ferroelastic transitions with consequent development of twinning, whereas above a certain threshold of disorder, the transition is suppressed, preventing the formation of long-range domains, with strain glass characteristics. The agreement with experiments [51] indicates that disorder can be associated with point defects arising from off-stoichiometric composition

of the peak position of the distributions, confirming the stabilization of the high-temperature nanostructural domains on cooling. The increasing tails indicate a slight domain coarsening. These strong correlations are a signature of history dependence, which limits the region of the phase space accessible to the system, therefore leading to non-ergodic behavior.

9.3.4 *Thermodynamics*

The effective kinetic arrest in glasses leaves well-established marks on the thermodynamic quantities that differentiate the former from the thermodynamic phase. On the one hand, the peak in calorimetry associated with the first-order transition progressively softens and shifts to lower temperatures as the transition weakens, and disappears when the transition is suppressed (see Fig. 9.7a–c).

On the other hand, as the anisotropy is introduced in the model as a local parameter, in general it may not correspond to the thermodynamic anisotropy accessible through elastic constants measurements. Actually, $\mathscr{A}$ may couple to

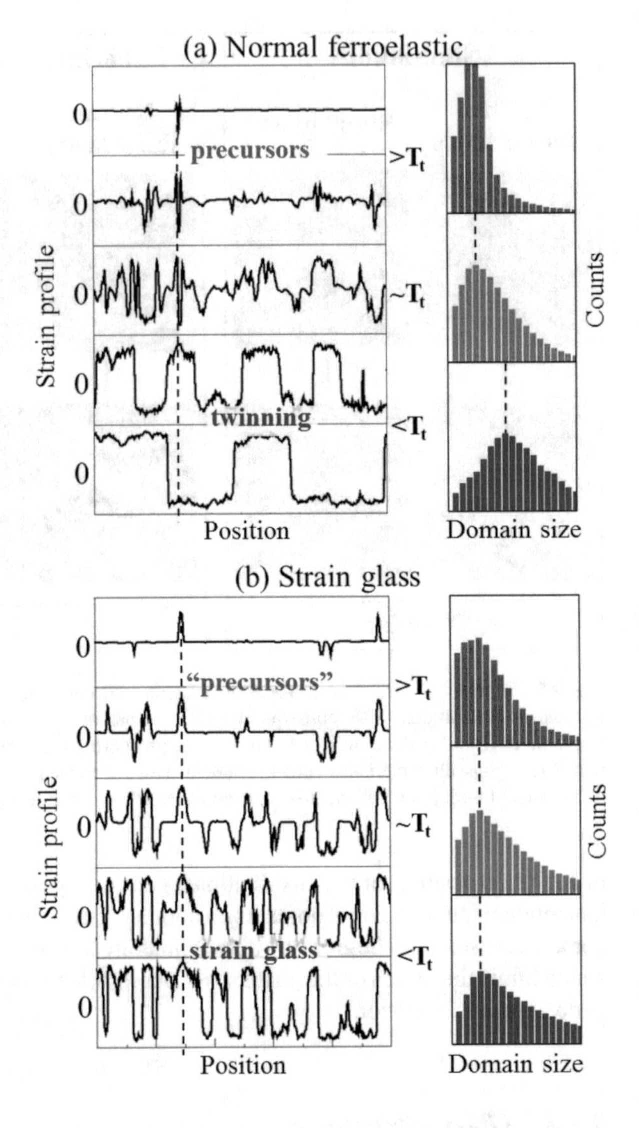

Fig. 9.6 Temperature evolution of representative strain profiles and typical domain size distributions on cooling towards (**a**) the ferroelastic phase and (**b**) the strain glass state. Vertical dashed lines are depicted to highlight differences or correlations between high-temperature and low-temperature structures

disorder as the latter influences the overall response of the system to external stimuli, such as the softening of the elastic constant underlying the ferroelastic transition. Hence, it is important to carry out simulation experiments to obtain the thermodynamic elastic response when local anisotropy and disorder are modified. Results are shown in Fig. 9.7d–f, where it can be observed that the dip in the dc-stress field response flattens compared to normal ferroelastics.

As a nonequilibrium system, the thermodynamic instability entails that its properties continuously evolve in time. In turn, this time dependence of glass properties makes them depend (1) on the measurement time scale, giving rise to frequency dependent susceptibility, and (2) on thermal and external field history.

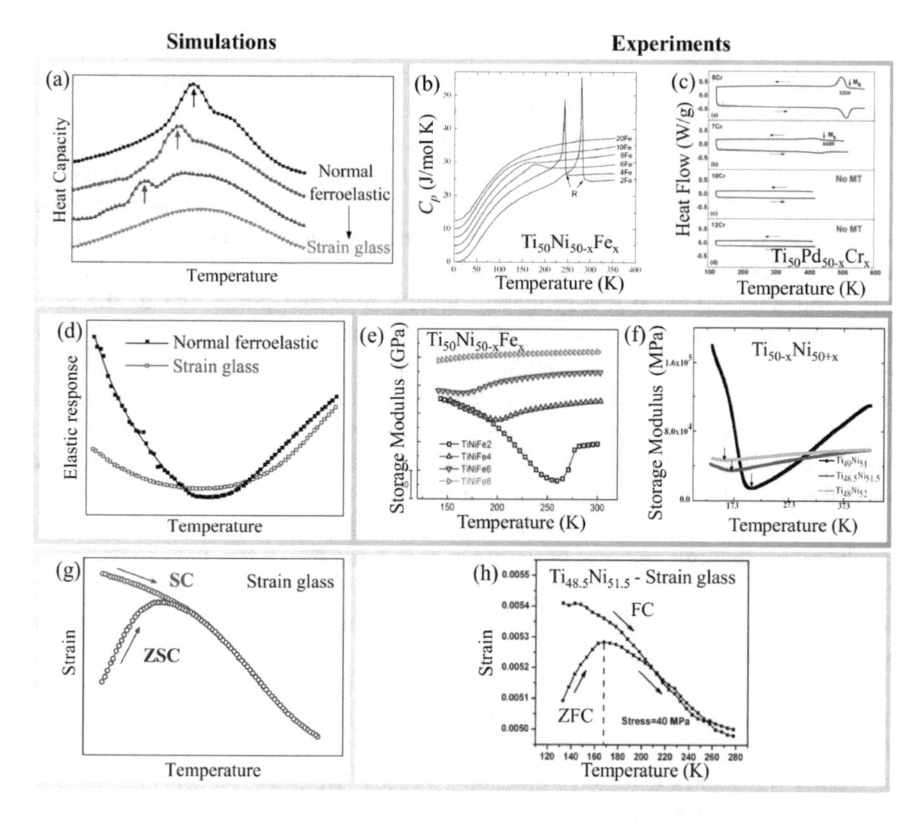

Fig. 9.7 Simulations and experiments of some thermodynamic signatures, from normal ferroelastics to strain glass: (**a**)–(**c**) Calorimetry, (**d**)–(**f**) elastic response and (**g**) and (**h**) zero-stress-cooling/stress-cooling results. Experimental measurements taken from Refs: (**b**) [96], (**c**) [52], (**e**) [97], (**f**) [49] and (**h**) [98]

These two signatures are inferred by means of low-ac-field broadband spectroscopy and ZSC/SC experiments, respectively. While the former cannot be simulated with the relaxation dynamics used here, ZSC/SC simulations are shown in Fig. 9.7g–h. The splitting between the ZSC and SC curves indicates the glassy behavior.

This set of simulations is in qualitative agreement with experimental measurements and reveals that the ferroelastic nanodomains replacing twinning in the disorder-anisotropy diagram (see Fig. 9.3) meet the requirements for glassiness.

9.3.5 *Thermomechanics*

We have seen that ferroic cluster glasses in general, and strain glasses in particular, exhibit all the thermodynamic evidence for vitrification. Therefore, it is seemingly surprising that they also exhibit a number of functional properties that in non-glass systems are intimately associated with a first-order transition. Remarkably, experiments have revealed that thermomechanical phenomena such as the SME and superelasticity [35] also take place in strain glasses. This is of primary technological interest because it broadens the compositional range (and consequently the temperature intervals) that can be exploitable for applications.

The SME refers to macroscopic shape recovery upon heating across the reverse ferroelastic phase transition after being deformed at low temperature by means of a stress field. It occurs because twinning preserves the macroscopic shape in the temperature-induced low-symmetry locally distorted phase due to the self-accommodation process. Then, application of stress causes twin reorientation resulting in macroscopic deformation, while heating to the parent phase removes any strain, thus restoring the initial undeformed shape. Superelasticity stands for a nonlinear reversible deformation beyond the elastic limit, originating through the stress-induced ferroelastic transition from the paraelastic phase. The application of stress leads to the nucleation and growth of a single variant, thus preventing twinning and giving rise to deformations as big as 10% in some cases. Depending on the phase stability, removal of the stress may bring the system back to the initial state (by undergoing the backward transition with hysteresis) or remain in the single variant state if the ferroelastic phase is still metastable. In this latter case, SME is obtained again on heating.

The effect of a stress field can be introduced in the model by performing a Legendre transform $G = F - \sigma e_2$ of the free-energy F, where σ is the stress field conjugated to the OP strain e_2.

Figure 9.8 shows a loading–unloading process from the paraelastic phase at different temperatures for three different sets of anisotropy and disorder: (a) the anisotropy-driven system exhibits normal ferroelastic behavior. By increasing enough the amount of disorder, it evolves towards (b) strain glass, corresponding to disorder-driven systems. For comparison with a reference framework, the homogeneous Landau paths for equilibrium and metastability limits are displayed with dashed and dotted lines, respectively. Nonlinear behavior is obtained in all cases, indicating the stress-induced transformation. Again, disorder causes the rounding of the transition. Interestingly, for technological implications, superelasticity is observed in non-transforming regimes, and it is accompanied by a hysteresis reduction, enhancing the range of temperatures with reversible effect. Residual strain remains, however, to higher temperatures compared to the other cases. The resemblance with experimental observations in O-doped Ti-Nb [34] shown in Fig. 9.8c is notable. Nevertheless, the phenomenology observed in experiments is diverse: For instance, the SME in Ti-Ni strain glass [50] does not exhibit such a pronounced rounding nor a hysteresis reduction, but preserves the essential characteristics of the SME in ferroelastic Ti-Ni (see Fig. 9.8d and Sect. 9.3.2 for discussion). In all cases, arrows originating at the end of the unloading process indicate the recovery of the initial shape after a heating process, entailing the SME.

In our simulations, the critical stress field above which nonlinear behavior occurs decreases when the glassy regime is approached (Fig. 9.8b), compared to normal ferroelastics (Fig. 9.8a). This disagreement with the experiments shown in

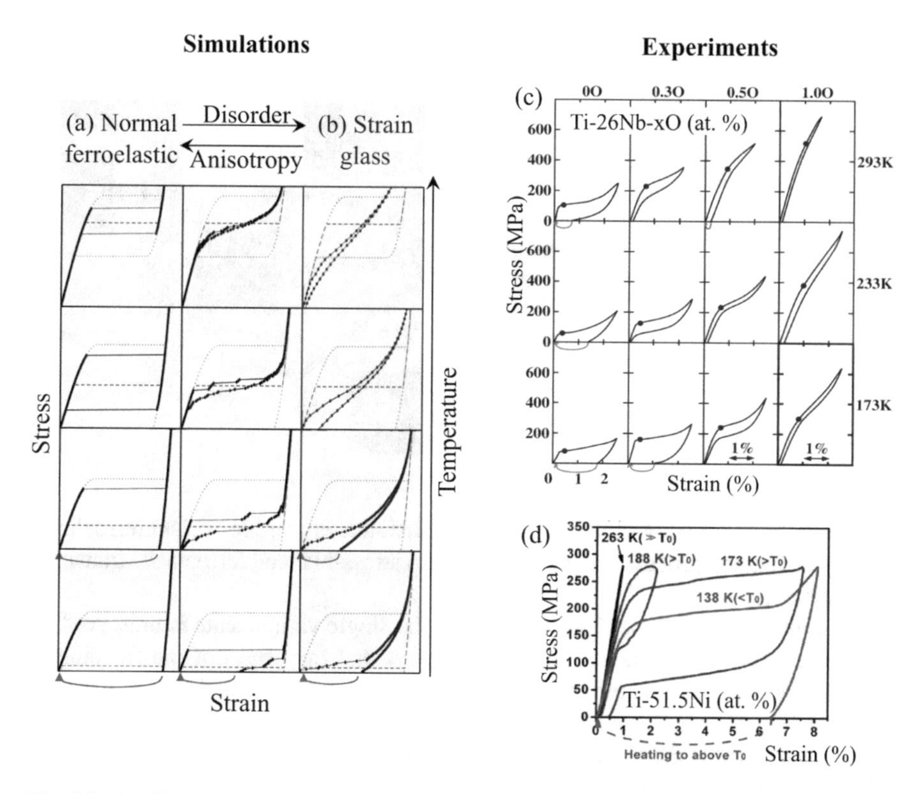

Fig. 9.8 (**a**) Simulations of stress-strain behavior as a function of temperature, from (**a**) anisotropy-driven (normal ferroelastics) to (**b**) disorder-driven (strain glass) systems. Dashed and dotted red lines correspond to the equilibrium and metastable limits of the Landau homogeneous free energy respectively, and are displayed to highlight the transition rounding caused by disorder. (**c**) Stress-strain experiments on O-doped Ti-Nb as a function of defect concentration, adopted from Ref. [34]. (**d**) Stress-strain behavior in Ti-Ni strain glass, adopted from Ref. [50]. In all cases, arrows after unloading indicate shape recovery on heating above the zero-stress transition temperature, leading to the shape-memory effect

Fig. 9.8c,d may come from the fact that impurities may act as pinning sites for strain propagation and the domain walls may have lower mobility than twin boundaries, thus requiring a larger stress to trigger the transformation.

Simulations of the shape-memory effect in the strain glass are displayed in Fig. 9.9. The $\sigma - e_2 - T$ curves illustrating the SME are averaged over 50 curves to approach the thermodynamic limit. Snapshots accompanying the curves show representative configurations at intermediate stages of the loading–unloading–heating process. When loading from the glass state (steps (i)→(ii)→(iii)), there is a combination of local variant switching and transformation from retained unstrained phase. These mechanisms cause a gradual growth of the domains of the variant favored by the stress field, evidenced by a rounded loading curve, in contrast to the sharp variant reorientation in the normal ferroelastic case occurring from a global

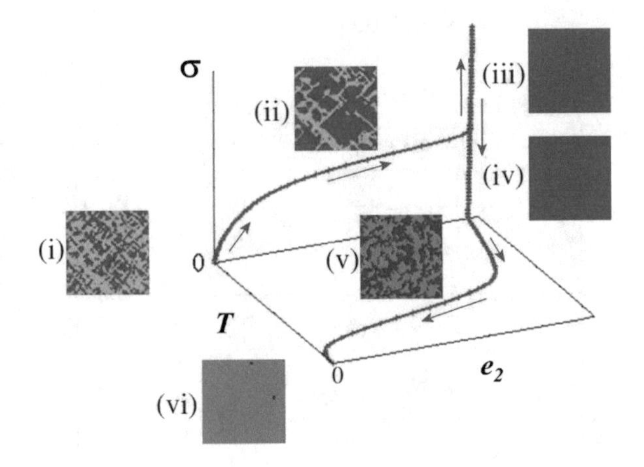

Fig. 9.9 Shape memory effect occurring in the strain glass. (i)–(iii): loading from the strain glass. (iii)–(iv) Unloading. (iv)–(vi) Heating above the ferroelastic transition

high twin-boundary mobility. As the final loaded state is a single-variant ferroelastic phase, the loading process can be described as a stress-induced ferroelastic transition from the glass phase.

On removing the stress field ((iii)→(iv)) the single variant state is preserved, in agreement with experiments. This may suggest that the origin of the freezing in strain glass is not dominated by topological frustration but a kinetic arrest imposed by a random distribution of energy barriers that impede the correct relaxation of the system towards the thermodynamic equilibrium; however, the latter does exist. This is similar to canonical structural glasses: once crystallization is achieved, it must be melted to freeze again, in contrast to geometrically frustrated systems, where the removal of the external field restores the system to a disordered state corresponding to one of the degenerate minima (e.g., the antiferromagnetic triangle). A deeper discussion regarding the presence of frustration in strain glass is contained in Sect. 9.3.2.

Finally, the process (iv→v→vi) corresponds to the heating process across the ferroelastic transition towards the paraelastic phase, recovering its initial undistorted shape. From this state, in our model the system will evolve towards the glass state if it is cooled down again.

It is worth remarking that the origin of these thermomechanical functional properties lies in the fact that in strain glass forming systems the application of stress can induce the formation of the ferroelastic phase (consisting of a single variant state), in contrast to the fact that the temperature-induced process leads to vitrification. This is yet another common feature of ferroic glasses where the transition towards a thermodynamic phase (ferromagnetic, ferroelectric) can be driven from the glass phase by the corresponding conjugated field. Moreover, the field-induced transition establishes a key difference with structural glasses where there is no external field that is able to induce crystallization from an amorphous state. Instead, pressure has been observed to favor amorphization in these systems [99, 100].

9.3.5.1 Elastocaloric Response

Caloric effects refer to isothermal entropy changes or adiabatic changes in temperature induced by application of an external field. Giant caloric effects near first-order phase transitions in the solid-state are of increasing interest in functional materials due to their application in novel cooling technologies [101]. Ferroic materials are being intensively studied as caloric materials, since they exhibit reversible transitions that can be driven by magnetic, electric, and/or mechanical fields (anisotropic stress or hydrostatic pressure), depending on the nature of the OP associated with the transition.

Interestingly, relaxor ferroelectrics have attracted particular attention in this field [102]. This is particularly intriguing as they are the cluster-glass counterpart of ferroelectric systems. Hence, relaxors do not exhibit the ferroelectric first-order phase transition, and thus they lack the corresponding latent heat from which giant caloric effects are expected. Indeed, experimental results report a slight softening of the maximum isothermal entropy change achievable, as the entropy change on cooling across the relaxor transition is not as high as in the corresponding ferroelectric transition. However, the former entails a temperature range and a hysteresis reduction with respect to the latter that results in significant improvement of electrocaloric performance.

On the other hand, we have seen that, as strain glass can be transformed to an ordered, normal ferroelastic phase by application of stress, it is able to display functional stress-induced properties such as shape-memory effect and superelasticity, and they are of the same order of magnitude as in its normal ferroelastic analogue. For these two reasons, it is natural to also expect interesting elastocaloric effect in strain glass derived from the stress-induced ferroelastic transition. Our model permits the evaluation of the elastocaloric effect from the stress-strain curve. Taking into account the Maxwell relation $(\partial S/\partial\sigma)_T = (\partial e_2/\partial T)_\sigma$, the elastocaloric effect can be obtained as

$$\Delta S(T; 0 \to \sigma) = \int_0^\sigma \left(\frac{\partial e_2}{\partial T}\right) d\sigma. \tag{9.4}$$

Indeed, simulations of the present model predicted this phenomenon to occur [85], which recently has been experimentally confirmed [103]. Both simulations and experiments of the elastocaloric effect in strain glass are shown in Fig. 9.10 for different stress values. Basic features are comparable to those observed in relaxors, with a tendency towards a rounding and slight flattening of the effect in the strain glass case which would entail higher temperature range and lower hysteresis compared to the elastocaloric effect in normal ferroelastics. Interestingly, for a given stress, the area below the curve remains essentially constant (not shown), which is associated with the refrigerant capacity [101]. We can then conclude that caloric effects are yet another example of functional properties that arise unexpectedly in non-transforming, cluster-glass ferroics, which enhances the spectrum of materials and temperature range that are appealing for novel refrigeration techniques.

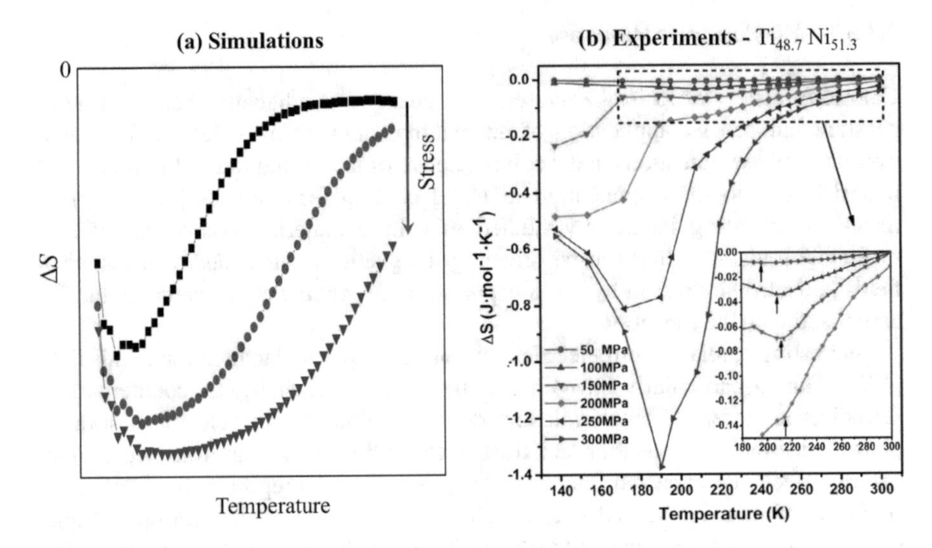

Fig. 9.10 Elastocaloric effect of strain glass: (**a**) Simulations and (**b**) experimental results on Ti-Ni, taken from Ref. [103]

9.4 Modelling Strain-Mediated Magnetic Glass

As the underlying physical forces ruling the different ferroic properties are of the same nature, namely electromagnetic, they normally couple each other. Interestingly, in some cases this coupling emerges at a significant level. This multiferroic interplay is at the origin of the magnetostructural, magnetoelectric, and piezoelectric cross-variable responses observed in many ferroics and yields mixed-variable patterns with correlated domains such as magnetic twins, magnetic stripes within twins [104–106], magnetoelastic tweed [107], polar tweed [108], and strong correlations between magnetic and electric dipoles [109].

Also, the possibility of controlling macroscopic physical properties by different external fields has led to fascinating functional properties: Magnetostriction refers to a volume change driven by a magnetic field [110], while the magnetic shape memory effect results from large strains arising due to magnetic field-induced twin reorientation [111]. Giant electrocaloric and magnetocaloric effects are enhanced when they are accompanied by a structural transition [101]. Colossal magnetoresistance (CMR) refers to dramatic changes in conductivity due to the presence of a magnetic field [112].

Therefore, it is natural to consider the occurrence of mixed glass phases as well. For instance, although not necessarily, it is widely known that positional disorder in amorphous states may give rise to spin glass [113, 114] and dipolar glass states [115]. In multiferroic systems, simultaneous vitrification of magnetic and dipolar degrees of freedom has also been reported in a magnetoelectric system [116]. Here we investigate the possibility of magnetic glass behavior induced by strain glass phase in ferromagnetic shape-memory alloys [61, 117, 118]. For this purpose, we extend the elastic model presented above to include magnetic degrees of freedom by means of the micromagnetic theory [119] and a magnetostructural coupling [120]. The total free energy F_T consists of three main contributions: $F_T = F_s + F_m + F_{m-s}$, where F_s corresponds to the pure elastic contribution explained in detail in the previous section. The term F_m includes magnetocrystalline anisotropy, exchange, magnetostatic and Zeeman energies:

$$F_m = D \int m_x^2(\mathbf{r}) m_y^2(\mathbf{r}) d^2r + J \int |\nabla \mathbf{m}(\mathbf{r})|^2 d^2r$$
$$-\mu_0 M_s \int \left(\frac{1}{2} \mathbf{H_d}(\mathbf{r}) + \mathbf{H}_{\text{ext}}(\mathbf{r}) \right) \cdot \mathbf{m}(\mathbf{r}) d^2r, \tag{9.5}$$

where $\mathbf{m} = (m_x, m_y, m_z)$ is the unit magnetization vector, D is the magnetocrystalline anisotropy constant, J is the exchange stiffness constant, M_s is the saturation magnetization, H_d is the demagnetizing field, and H_{ext} is the external magnetic field.

The magnetostructural energy F_{m-s} is taken to the lowest order allowed by symmetry:

$$F_{m-s} = B_1 \int [(m_x^2 + m_y^2)e_1 + (m_x^2 - m_y^2)e_2] d^2r + B_2 \int m_x m_y e_3 d^2r \tag{9.6}$$

where B_1 and B_2 are magnetostructural coupling constants. In the absence of thermal fluctuations, we assume that the Curie temperature is much higher than the ferroelastic transition temperature. In addition, for the sake of simplicity, the exchange stiffness constant J is taken to be positive and independent of position and distance between spins. Consequently, F_m is minimized when all the spins are parallel to each other. Then, the magnetic domains emerge from heterogeneous nucleation and coupling with the strain field from F_{m-s}. As F_{m-s} depends on the non-OP strains, the elastic compatibility constraint must be applied here as well. Despite the fact that the energy minimization here includes F_{m-s}, giving rise in this case to two additional Fourier kernels apart from that in Eq. (9.3), it does not give rise to new physics (see [118] for details). The magnetic dynamics is ruled by the Landau-Lifshitz-Gilbert equation [119].

At high temperatures, the magnetostructural coupling is absent as the strain field vanishes. Therefore, the spin field can arrange in a pure multidomain ferromagnetic configuration, with 180° domain walls arising from heterogeneous nucleation combined with the demagnetizing field that induces the magnetic field lines to form closed loops. Similarly, magnetic vortices may eventually emerge. Interestingly, on cooling below the ferroelastic transition, the magnetostructural term becomes relevant: The elastic twins impose a hierarchical arrangement of the magnetic domain walls: On the one hand, 90° walls correlated with the twin boundaries and, on the other hand, *magnetic stripes* with 180° walls inside the twins. These orientations are dictated by the interplay between the magnetostructural coupling that determines the magnetocrystalline anisotropy of each ferroelastic variant, and

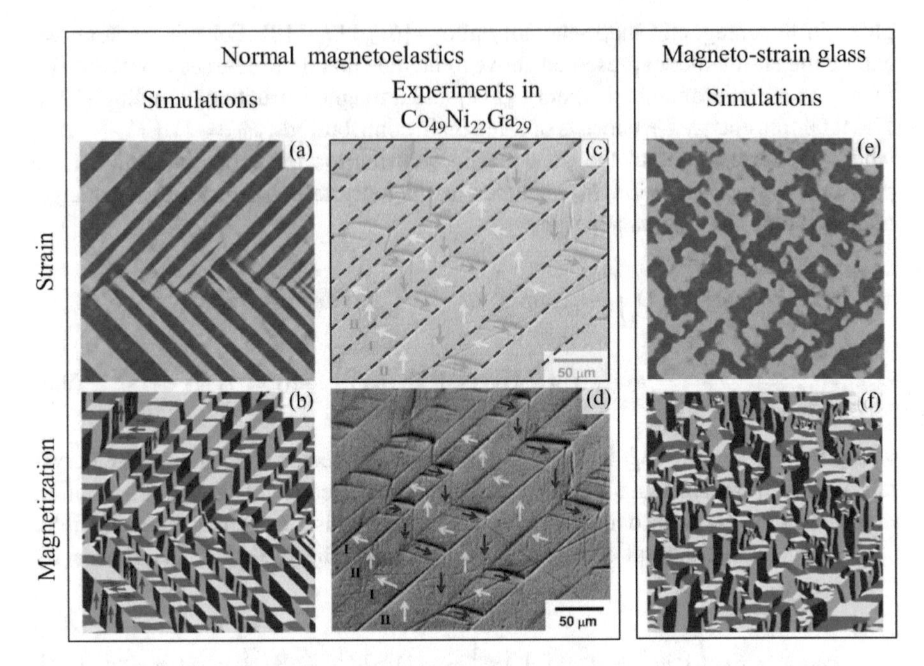

Fig. 9.11 Magnetostructures: strain (**a** and **e**) and magnetic (**b** and **f**) configurations obtained from the magnetoelastic model, for the normal magnetoelastic ferroic system (**a** and **b**) and strain-induced magnetostructural glass (**e** and **f**). Panel (**d**): experimental micromagnetic structure, taken from Ref. [106], whose ferroelastic domain boundaries are highlighted (red dashed lines) in panel (**c**). Arrows in magnetic configurations indicate the spin orientation within the magnetic stripes

the demagnetizing field. The exchange energy plays a role in determining the width of the magnetic stripes. Figures 9.11a and b show strain and magnetic configurations, respectively, in a normal magnetostructural multiferroic, which reproduce accurately the experimental image shown in Fig. 9.11d [106] (and other experimental observations [104, 105]). From the latter, we can deduce the domain walls, which have been highlighted with red dashed lines in Fig. 9.11c, for comparison with the simulated strain configuration.

Now we proceed as previously to induce a strain glass phase by increasing disorder and/or decreasing the elastic anisotropy. Notice that the pure magnetic terms are not modified, so that low anisotropy values influence the magnetization through the magnetostructural coupling only. The obtained strain and magnetic configurations are shown in Fig. 9.11e and f, respectively. In this case, the glassy strain nanodomains sweep the spin orientations along, with the subsequent loss of long-range order of the 180° magnetic domain walls. Therefore, despite 90° walls associated with magnetic stripes are still formed inside the ferroelastic nanodomains, magnetostructural correlations induce magnetic nanodomains associated with the clusters, assuming the slowing down of the strain dynamics and the consequent glassy state. Depending on the value of J and the transformed ferroelastic

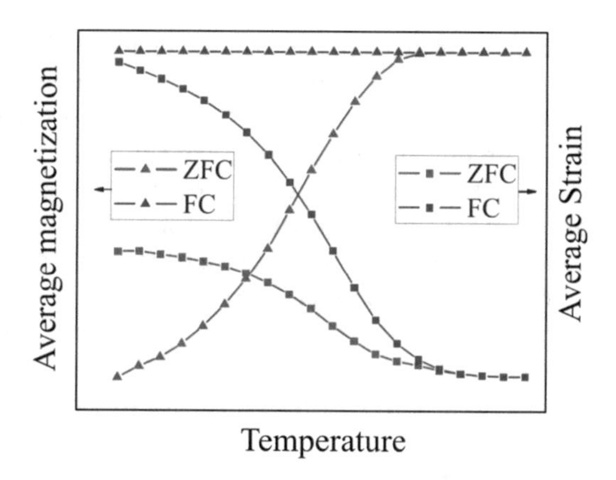

Fig. 9.12 Zero Magnetic Field Cooling (ZFC)/Magnetic Field Cooling (FC) protocols applied on the magnetostructural model. Strain (right) and magnetization (left) are displayed

fraction, the magnetic stripes may disappear, leading to a superspin glass-like configuration. Strain-induced magnetostructural glasses have been suggested in ferromagnetic shape-memory alloys such as Ni-Co-Mn-Sn [61].

To detect glassy features associated with the magnetostructural nanodomains, the system is subjected to Zero Magnetic Field Cooling/Magnetic Field Cooling (ZFC/FC) protocols while both the strain and magnetization are monitored. The resulting curves are shown in Fig. 9.12. The splitting between the two curves in both the magnetization and strain reveals glassy behavior in both ferroic variables. The decay of the magnetization at high temperatures will be expected above the Curie point, which is assumed to be well above the temperature range of simulations. It is worth mentioning that we have also performed ZSC/SC simulations (under application of stress instead of a magnetic field) and the splitting between ZSC and SC curves has been also obtained (not shown), which strengthens the existence of such clusterized magnetostructural glassy behavior.

9.5 Summary and Conclusions

Mesoscopic Ginzburg-Landau-based modelling has shown that anisotropy and disorder are the relevant parameters in the competition that rules the behavior of ferroelastics. Anisotropy-driven systems result in normal ferroelastics whereas disorder, above a critical threshold, breaks down long-range correlation and triggers the glassy behavior. It creates random strain fields that give rise to inhomogeneous nucleation of ferroelastic sites. Possible emerging incompatible strain interactions may result in the frustration of domain growth that slows down and finally freezes at the nanoscale, yielding the strain glass. This is consistent with numerous experimental observations showing glassy behavior beyond a critical off-stoichiometric composition, that introduces disorder basically as randomly distributed point defects.

The thermodynamic ferroelastic phase can be reached by application of external stress, which is a feature that in general differentiates ferroic cluster glasses from canonical structural glasses. This enables the occurrence, and eventually improvement, of functional properties typical from normal ferroelastics, namely shape-memory effect, superelasticity and elastocaloric effect, which renders them very useful in materials engineering as it widens the range of materials, temperatures, and performance in general that can be exploited in technological applications.

In summary, the characteristics of strain glass extend to cluster glasses the common broad framework shared by all ferroics. In turn, the magnetoelastic model predicts that multiferroic cross-variable response in magnetostructural systems could lead to magnetic glass mediated by strain.

Acknowledgements We acknowledge Prof. David Sherrington for fruitful discussions. This work was supported by CICyT (Spain) project MAT2013-40590-P, by DGU (Catalonia) project 2014SGR00581 and by the U.S. Department of Energy.

References

1. I. Gutzow, J. Schmelzer, *The Vitreous State. Thermodynamics, Structure, Rheology and Crystallization* (Springer, Berlin, 1995)
2. The difference between metastable relaxation and glassy dynamics may be so small that eventually cannot be distinguished. Therefore, glass has been described somewhere as metastable quasiequilibrium states. See, for instance: T. Loerting, V.V. Brazhkin, T. Morishita, in *Multiple Amorphous-Amorphous Transitions*, ed. by S. Rice. Advances in Chemical Physics, vol. 143 (Wiley, Hoboken, 2009), pp. 29–83
3. J.M.D. Coey, P.W. Readman, Nature **246**, 476–478 (1973)
4. C. Dekker, W. Eidelloth, R.H. Koch, Phys. Rev. Lett. **68**, 3347–3350 (1992)
5. D. Viehland, J.F. Li, S.J. Jang, L.E. Cross, M. Wuttig, Phys. Rev. B **46**, 8013–8017 (1992)
6. R. Fichtl et al., Phys. Rev. Lett. **94**, 027601 (2005)
7. P.G. Wolynes, V. Lubchenko, *Structural Glasses and Supercooled Liquids: Theory, Experiment, and Applications* (Wiley, New Jersey, 2012)
8. D.L. Stein, *Spin Glasses: Still Complex After All These Years?*, in Decoherence and Entropy in Complex Systems, ed. by T. Elze (Springer, Berlin, 2004)
9. S.H. Chen, H. Shi, B.M. Conger, J.C. Mastrangelo, T. Tsutsui, Adv. Mater. **8**, 998–1001 (1996)
10. S.H. Chen, D. Katsis, A.W. Schmid, J.C. Mastrangelo, T. Tsutsui, T.N. Blanton, Nat. Lett. **397**, 506–508 (1999)
11. R. Brand, P. Lunkenheimer, A. Loidl, J. Chem. Phys. **116**, 10386–10401 (2002)
12. C.T. Moynihan, A.J. Easteal, J. Wilder, J. Tucker, J. Phys. Chem. **78**, 2673–2677 (1974)
13. R. Moessner, A.P. Ramirez, Phys. Today **59**, 24–29 (2006)
14. M. Fujihala et al., Phys. Rev. B **85**, 012402 (2012)
15. M. Schmidt et al., Phys. A **438**, 416–423 (2015)
16. J.S. Gardner, M.J.P. Gingras, J.E. Greedan, Rev. Mod. Phys. **82**, 53–107 (2010)
17. R.F. Wang et al., Nat. Lett. **439**, 303–306 (2006)
18. F. Yen et al., Phys. B **403**, 1487–1489 (2008)
19. P.G. Debenedetti, F.H. Stillinger, Nature **410**, 259–267 (2001)
20. L.C. Pardo, A. Henao, A. Vispa, J. Non-Cryst. Sol. **407**, 220–227 (2015)
21. R.N. Bhowmik et al., Phys. Rev. B. **72**, 094405 (2005)

22. H.Y. Kwon et al., J. Magn. Magn. Mat. **324**, 2171–2176 (2012)
23. R.P. Erickson, D.L. Mills, Phys. Rev. B **43**, 11527(R) (1991)
24. W. Eerenstein, N.D. Mathur, J.F. Scott, Nature **442**, 759–765 (2006)
25. M. Porta, T. Castán, P. Lloveras, T. Lookman, A. Saxena, S.R. Shenoy, Phys. Rev. B **79**, 214117 (2009)
26. K. Battacharya, *Microstructure of martensite: Why it forms and how it gives rise to the shape-memory effect*. Oxford Series on Materials Modelling (Oxford University Press, Oxford, 2004)
27. S. Kaufmann et al., New J. Phys. **13**, 053029 (2011)
28. A.G. Khachaturyan, *Domain structure in martensitic transformation*, in *Proceedings of Advanced Materials'93*, ed. by K. Otsuka et al. (1994); published in Trans. Mat. Res. Soc. Jpn. **18B**, 799 (1994)
29. S. Kustov, D. Salas, E. Cesari, R. Santamarta, D. Mari, J. Van Humbeeck, Mat. Sci. Forum **738–739**, 274–275 (2013)
30. S. Kustov, D. Salas, E. Cesari, R. Santamarta, D. Mari, J. Van Humbeeck, Acta Mater. **73**, 275–286 (2014)
31. N. Shankaraiah, K.P.N. Murthy, T. Lookman, S.R. Shenoy, Phys. Rev. B **84**, 064119 (2011)
32. A. Millis, Solid State Commun. **126**, 3–8 (2003)
33. Y. Imry, M. Wortis, Phys. Rev. B **19**, 3580–3585 (1979)
34. Y. Nii et al., Phys. Rev. B **82**, 214104 (2010)
35. E.K.H. Salje, *Phase Transitions in Ferroelastic and Co-elastic Crystals* (Cambridge University Press, Cambridge, 1990)
36. Y. Murakami, H. Shibuya, D. Shindo, J. Microsc. **203**, 22–33 (2001)
37. S. Kartha, T. Castán, J.A. Krumhansl, J.P. Sethna, Phys. Rev. Lett. **67**, 3630–3633 (1991)
38. X. Ren et al., Philos. Mag. **90**, 141–157 (2010)
39. C. Paduani, A. Migliavacca, M.L. Sebben, J.D. Ardisson, M.I. Yoshida, S. Soriano, M. Kalisz, Solid State Commun. **141**, 145–149 (2007)
40. W. Ratcliff II et al., Phys. Rev. B **65**, 220406(R) (2002)
41. T. Chakrabarty, A.V. Mahajan, S. Kundu, J. Phys. Condens. Matter **26**, 405601 (2014)
42. J.L. Dormann, M. Nogues, J. Phys. Condens. Matter **2**, 1223–1237 (1990)
43. M. Nogues et al., J. Magn. Magn. Mater. **104–107**, 1641–1642 (1992)
44. J. Blasco, V. Cuartero, J. García, J.A. Rodríguez-Velamazán, J. Phys. Condens. Matter **24**, 076006 (2012)
45. S. Karmakar et al., Phys. Rev. B **74**, 104407 (2006)
46. C. Ang, Z. Yu, Z. Jing, Phys. Rev. B **61**, 957–961 (2000)
47. P.A. Sharma, S.B. Kim, T.Y. Koo, S. Guha, S.-W. Cheong, Phys. Rev. B **71**, 224416 (2005)
48. P. Toledano, D. Machon, Phys. Rev. B **71**, 024210 (2005)
49. S. Sarkar, X. Ren, K. Otsuka, Phys. Rev. Lett. **95**, 205702 (2005)
50. Y. Wang, X. Ren, K. Otsuka, Phys. Rev. Lett. **97**, 225703 (2006)
51. X. Ren, Y. Wang, K. Otsuka, P. Lloveras, T. Castán, M. Porta, A. Planes, A. Saxena, MRS Bull. **34**, 838–846 (2009)
52. Y. Zhou et al., Appl. Phys. Lett. **95**, 151906 (2009)
53. Y. Zhou et al., Acta Mater. **58**, 5433–5442 (2010)
54. J. Zhang et al., Phys. Rev. B **84**, 214201 (2011)
55. D.P. Wang et al., Europhys. Lett. **98**, 46004 (2012)
56. Y. Yao et al., Europhys. Lett. **100**, 17004 (2012)
57. Y. Wang et al., Appl. Phys. Lett. **102**, 141909 (2013)
58. Y. Wang et al., Sci. Rep. **4**, 3995 (2014)
59. Y. Zhou et al., Phys. Rev. Lett. **112**, 025701 (2014)
60. Y. Zhou et al., Phys. Status Solidi B **251**, 2027–2033 (2014)
61. P. Entel et al., Phys. Status Solidi B **251**, 2135–2148 (2014)
62. X. Ren et al., Mater. Sci. Eng. **312**, 196–206 (2001)
63. J. Zhang et al., Mater. Trans. JIM **40**, 385–388 (1999)
64. S. Muto et al., Acta Metall. Mater. **38**, 685–694 (1990)

65. E. Obradó et al., Phys. Rev. B **58**, 14245 (1998)
66. K. Enami et al., Scr. Metall. **10**, 879–884 (1976)
67. D. Ma et al., Phys. Status Solidi B **245**, 2642–2648 (2008)
68. K. Otsuka, C.M. Wayman, *Shape Memory Materials* (Cambridge University Press, Cambridge, 1990)
69. S. Miyazaki, K. Otsuka, S. Suzuki, Scr. Metall. **15**, 287–292 (1981)
70. Y. Wang, X. Ren, K. Otsuka, A. Saxena, Acta Mater. **56**, 2885–2896 (2008)
71. N. Nakanishi et al., Philos. Mag. **28**, 277–292 (1973)
72. T.-H. Nam et al., J. Mater. Sci. Lett. **21**, 1851–1853 (2002)
73. V.A. Chernenko et al., J. Appl. Phys. **93**, 2394–2399 (2003)
74. R. Kainuma et al., Nat. Lett. **439**, 957–960 (2006)
75. X. Ren, Phys. Status Solidi B **251**, 1982–1992 (2014)
76. J.M. Ball, R.D. James, Arch. Ration. Mech. Anal. **100**, 13–52 (1987)
77. Y. Wang, A.G. Khachaturyan, Acta Mater. **45**, 759–773 (1997)
78. A.G. Khachaturyan, *Theory of Structural Transformation in Solids* (Dover, New York, 2008)
79. O.U. Salman, A. Finel, R. Delville, D. Schryvers, J. Appl. Phys. **111**, 103517 (2012)
80. V.I. Levitas, D.L. Preston, Phys. Rev. B **66**, 134206 (2002); V.I. Levitas, D.L. Preston, Phys. Rev. B **66**, 134207 (2002)
81. S. Kartha , J.A. Krumhansl, J.P. Sethna, L.K. Wickham, Phys. Rev. B **52**, 803–822 (1995)
82. A. Saxena, T. Lookman, in Handbook of Materials Modeling, ed. by S. Yip, pp. 2143–2154 (Springer, Berlin, 2005)
83. P. Lloveras, T. Castán, M. Porta, A. Planes, A. Saxena, Phys. Rev. Lett. **100**, 165707 (2008)
84. P. Lloveras, T. Castán, M. Porta, A. Planes, A. Saxena, Phys. Rev. B **80**, 054107 (2009)
85. P. Lloveras, T. Castán, M. Porta, A. Planes, A. Saxena, Phys. Rev. B **81**, 214105 (2010)
86. S.R. Shenoy, T. Lookman, A. Saxena, A.R. Bishop, Phys. Rev. B **60**, R12537 (1999)
87. T. Castán, A. Planes, A. Saxena, Mat. Sci. Forum **738–739**, 155–159 (2013)
88. D.C. Mattis, Phys. Lett. **56A**, 421–422 (1976)
89. C. Lu et al., Sci. Rep. **4**, 4902 (2014)
90. S. Narayana Jammalamadaka, AIP Adv. **1**, 042151 (2011)
91. V.K. Pecharsky, K.A. Gschneidner Jr., C.B. Zimm, Adv. Cryog. Eng. Mater. **42**, 451–458 (1996)
92. S.M. Shapiro, J.Z. Larese, Y. Noda, S.C. Moss, L.E. Tanner, Phys. Rev. Lett. **57**, 3199–3202 (1986)
93. G. Arlt, D. Hennings, G. de With, J. Appl. Phys. **58**, 1619–1625 (1985)
94. L.S. Chumbley et al., IEEE Trans. Magn. **25**, 2337–2340 (1989)
95. T. Roy, T.E. Mitchell, Philos. Mag. A **63**, 225–232 (1991)
96. M.-S. Choi, T. Fukuda, T. Kakeshita, Scr. Mater. **53**, 869–873 (2005)
97. L. Zhang et al., Sci. Rep. **5**, 11477 (2015)
98. Y. Wang, X. Ren, K. Otsuka, A. Saxena, Phys. Rev. B **76**, 132201 (2007)
99. R.J. Hemley et al., Nature **334**, 52–54 (1988)
100. M. Paluch, K. Grzybowska, A. Grzybowski, J. Phys. Cond. Matt. **19**, 205117 (2007)
101. X. Moya, S. Karnarayan, N.D. Mathur, Nat. Mater. **13**, 439–450 (2014)
102. A.S. Mischenko, Q. Zhang, R.W. Whatmore, J.F. Scott, N.D. Mathur, Appl. Phys. Lett. **89**, 242912 (2006)
103. Z. Tang, Y. Wang, X. Liao, D. Wang, S. Yang, X. Song, J. Alloys Compd. **622**, 622–627 (2015)
104. Y. Murakami, D. Shindo, K. Oikawa, R. Kainuma, K. Ishida, Acta Mater. **50**, 2173–2184 (2002)
105. Y. Ge, O. Heczko, O. Soderberg, V.K. Lindroos, J. Appl. Phys. **96**, 2159–2163 (2004)
106. J.N. Armstrong et al., J. Appl. Phys. **103**, 023905 (2008)
107. A. Saxena et al., Phys. Rev. Lett. **92**, 197203 (2004)
108. E.K.H. Salje, M. Alexe, S. Kustov, M.C. Weber, J. Schiemer, G.F. Nataf, J. Kreisel, Sci. Rep. **6**, 27193 (2016)
109. E. Dagotto, Science **309**, 257–262 (2005)

110. R. James, M. Wuttig, Philos. Mag. A **77**, 1273–1299 (1998)
111. T. Fukuda et al., Mater. Trans. **45**, 188–192 (2004)
112. M. Uehara, S. Mori, C.H. Chen, S.-W. Cheong, Nature **399**, 560–563 (1999)
113. R.A. Pelcovits, E. Pytte, J. Rudnick, Phys. Rev. Lett. **40**, 476–479 (1978)
114. L. Krusin-Elbaum, A.P. Malozemoff, R.C. Taylor, Phys. Rev. B **27**, 562–565 (1983)
115. B.E. Vugmeister, M.D. Glinchuk, Rev. Mod. Phys. **62**, 993–1026 (1990)
116. A. Levstik et al., Appl. Phys. Lett. **91**, 012905 (2007)
117. M.H. Phan et al., Phys. Rev. B **81**, 094413 (2010)
118. P. Lloveras, G. Touchagues, T. Castán, T. Lookman, M. Porta, A. Saxena, A. Planes, Phys. Stat. Sol. B **251**, 2080–2087 (2014)
119. A. Aharoni, *Introduction to the Theory of Ferromagnetism* (Oxford University Press, New York, 1996)
120. C. Kittel, Rev. Mod. Phys. **21**, 541–583 (1949)

Chapter 10
Phase Field Model and Computer Simulation of Strain Glasses

Dong Wang, Xiaobing Ren, and Yunzhi Wang

Abstract Strain glass is a new structural state in ferroelastic materials, which offers unique transition behavior and properties. In this chapter, we introduce a phase field model of strain glass systems and study their transition behavior and the associated properties by computer simulations. Local stresses associated with randomly distributed defects, including point defects and extended defects (dislocations and concentration modulations), are found to play the most important role in the formation of strain glass, by suppressing autocatalysis in nucleation and confining the growth of martensitic domains. A broad distribution of defect strength leads to continued nucleation and growth of martensitic domains in a broad temperature or stress range and renders the otherwise sharp first-order martensitic transformation into a broadly smeared "diffuse" strain glass transition with slim hysteresis, nearly linear superelasticity, ultralow elastic modulus and Invar and Elinvar anomalies. New strategies for designing strain glass systems with large recoverable strain are discussed.

D. Wang (✉)
Center of Microstructure Science, Frontier Institute of Science and Technology, State Key Laboratory for Mechanical Behavior of Materials, Xi'an Jiaotong University, Xi'an, China
e-mail: wang_dong1223@mail.xjtu.edu.cn

X. Ren
Frontier Institute of Science and Technology and State Key Laboratory for Mechanical Behaviour of Materials, Xi'an Jiaotong University, Xi'an, China

Center for Functional Materials, National Institute for Materials Science, Tsukuba, Ibaraki, Japan

Y. Wang
Center of Microstructure Science, Frontier Institute of Science and Technology, State Key Laboratory for Mechanical Behavior of Materials, Xi'an Jiaotong University, Xi'an, China

Department of Materials Science and Engineering, The Ohio State University, Columbus, OH, USA

T. Lookman, X. Ren (eds.), *Frustrated Materials and Ferroic Glasses*, Springer Series in Materials Science 275, https://doi.org/10.1007/978-3-319-96914-5_10

10.1 Martensitic Transformation and Strain Glass Transition in Ferroelastic Systems

Martensitic phase transformation (MT) describes a diffusionless transition from high-symmetry parent phase (strain liquid) at high temperature to low-symmetry product phase (strain crystal) at low temperature [1]. Temperature/stress induced martensitic phase transformation is the physical origin of shape memory effect and superelasticity in shape memory, which produces wide applications in different fields [2–4]. However, conventional MT will form coarsen martensitic domains accompanying with large hysteresis, narrow temperature range. Recently, a new strain glass state (nanosized martensitic domains with randomly distribution) was reported in NiTi shape memory alloys with excess Ni doping, which has shown the possibility to design novel shape memory alloys with new properties [5–7]. Figure 10.1 shows a schematic drawing of three strain states in ferroelastic system in a temperature vs. point defect concentration phase diagram. High temperature shows austenite with high symmetry for all composition regions. With the decrease of temperature, high defect concentration and low defect concentration show different strain state at low temperature. Low defect concentration shows long-range ordered twinned martensite and high defect concentration shows short-range ordered nanoscale martensitic domains. Defects play important role in breaking the long-range order twinned martensite in strain glass systems. However, it is difficult to capture the physical origin of strain glass by experiments, e.g., how defects influence the martensitic phase transition and strain glass transition or whether all the martensitic systems could produce strain glass? Phase field model could be a powerful tool to help us study the microstructure evolution of MT and strain glass transition and the role of defects in strain glass transition.

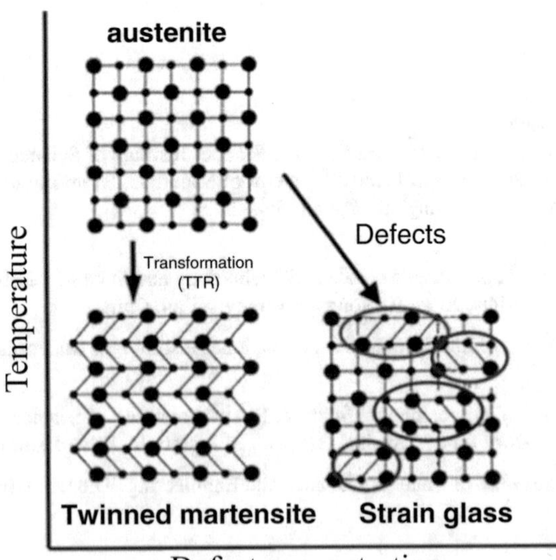

Fig. 10.1 Schematic drawing of the three strain states (austenite, twinned martensite, and strain glass) in ferroelastic systems

10.1.1 Phase Field Modeling of Martensitic Phase Transformation

Phase field approach can exhibit the microstructure evolution of solid-state phase transformations, which can be characterized by order parameters [8–13]. The order parameter for Ferroic phase transition is the local strain, polarization, and magnetization for ferroelastic [14, 15], ferroelectric [16–18], and ferromagnetic [19, 20] materials, respectively. Effective accommodation of elastic strain plays an important role in solid-state phase transformations, especially in ferroelastic materials. In a phenomenological description, the local free energy function is expressed as a polynomial of order parameters using a Landau expansion. All the terms of this expansion are required to respect to the symmetry operations of the high-temperature phase. The dependence of a phase transformation on strain is primarily determined by the coupling between the primary order parameter and strain. Martensitic systems include proper MT and improper MT; for proper MT, the primary order parameter is directly the strain, while for improper MT, the primary order parameter characterizes shuffle and strain are coupled to the primary order parameter [11, 21]. A phase field model describes domain patterns by spatial dependent order parameters (i.e., order parameter fields).

For martensitic systems, the total free energy of an inhomogeneous microstructure system described by nonconserved field variables $(\eta_1, \eta_2, \ldots)$ is given by:

$$F = \int \left[f\left(\eta_1, \eta_2, \ldots \eta_p\right) + \sum_{i=1}^{3} \sum_{j=1}^{3} \sum_{k=1}^{p} \beta_{ij} \nabla_i \eta_k \nabla_j \eta_k \right] d^3 r + \iint G\left(r - r^{'}\right) d^3 r d^3 r'$$

where f is the local free energy density as a function of field variables η_i, β_{ij} is the gradient energy coefficient. The second integral represents the contributions of long-range interactions, such as elastic interactions and electric dipole-dipole interaction, which also depend on the field variables.

For solid-state phase transformations, the local chemical free energy density f is expressed as a polynomial of field variables (i.e., the order parameters) by using a Landau-type expansion, which respects to the symmetry operations of the high-temperature phase. For example, for a martensitic phase transition from cubic to trigonal (R phase) in NiTi with four martensitic variants [22], the local free energy function is given by

$$f_{ch}\left(\eta_1, \eta_2, \ldots \eta_4\right) = f^0(c, T) + \frac{1}{2} A_2(c, T)\left(\eta_1^2 + \eta_2^2 + \eta_3^2 + \eta_4^2\right) - \frac{1}{4} A_4(c, T)\left(\eta_1^4 + \eta_2^4 + \eta_3^4 + \eta_4^4\right) + A_6(c, T)\left(\eta_1^2 + \eta_2^2 + \eta_3^2 + \eta_4^2\right)^3$$

where f^0 is the free energy of the parent phase, η_i are the structural order (SO) parameters that characterize the correspondence or deformation variants [23] of the low-symmetry martensitic phase, and A_2, A_4, A_6 are the expansion coefficients that are functions of temperature and composition. Any given microstructural state can be described by spatial distribution of these four order parameters (i.e., phase fields), e.g., $(\eta_1, \eta_2, \eta_3, \eta_4,) = (0, 0, 0, 0,)$ represents the austenite and $(\eta_1, \eta_2, \eta_3, \eta_4,) = (\eta_0, 0, 0, 0,), (0, \eta_0, 0, 0,), (0, 0, \eta_0, 0,), (0, 0, 0, \eta_0,)$represent four deformation variants of martensitic phases, respectively, where η_0 is the equilibrium value of the structural order parameter.

The gradient term f_{grad} describes the interfacial energy caused by structural inhomogeneities [24] such as interfaces between austenite and martensite and among different variants of the martensite

$$f_{\text{grad}} = \frac{1}{2}\sum_{i=1}^{3}\sum_{j=1}^{3}\sum_{p=1}^{4}\beta_{ij}^{\eta}(p)\nabla_i\eta_p(r)\nabla_j\eta_p(r) = \frac{1}{2}\sum\nolimits_{i=1}^{3}\sum\nolimits_{p=1}^{4}\beta_{ii}^{\eta}(p)\left(\nabla_{\eta_p}\right)^2$$

The elastic energy associated with long-range elastic interactions in ferroelastic system (caused by lattice mismatch between austenite and martensite and among different martensitic variants) is described by Khachaturyan's microelasticity theory [9]:

$$E_{\text{el}} = \frac{1}{2}C_{ijkl}\sum_{p=1}^{4}\sum_{q=1}^{4}\varepsilon_{ij}^{00}(p)\varepsilon_{kl}^{00}(q)\int \eta_p^2(\boldsymbol{r})\,\eta_q^2(\boldsymbol{r})\,d^3r$$
$$-\frac{1}{2}\sum_{p=1}^{4}\sum_{q=1}^{4}\int\frac{d^3k}{(2\pi)^3}B_{pq}\left(\frac{\vec{k}}{k}\right)\left\{\eta_p^2(\boldsymbol{r})\right\}_k\left\{\eta_q^2(\boldsymbol{r})\right\}_k^*$$

where C_{ijkl} is the elastic modulus tensor, $\varepsilon_{ij}^{00}(p)$ $(p = 1, 2, \ldots n)$ is the transformation strain tensor of the martensitic variant, and n is the number of martensitic variants, $\boldsymbol{k}$ is the wave vector defined in the reciprocal space, $\left\{\eta_p^2(\boldsymbol{r})\right\}_k$ is the Fourier transform of $\eta_p^2(\boldsymbol{r})$, $\left\{\eta_q^2(\boldsymbol{r})\right\}_k^*$ is the complex conjugate of $\left\{\eta_q^2(\boldsymbol{r})\right\}_k$, the kernel $B_{pq}\left(\vec{k}/k\right)$ is $B_{pq}\left(\vec{e}\right) = e_i\sigma_{ij}^0(p)\Omega(\vec{e})_{jk}\sigma_{kl}^0(q)e_l$, where $\vec{e} = \vec{k}/k$, $\sigma_{ij}^0 = C_{ijkl}\varepsilon_{ij}^{00}(p)$, $\Omega(\vec{e})_{ij}^{-1} = C_{ijkl}e_ke_l$.

Elastic energy caused by an external load can be described by:

$$E_{\text{load}} = -\sigma_{\text{load}}\varepsilon = -\int\sigma_{ij}^{\text{load}}\sum_{p=1}^{4}\varepsilon_{ij}^{00}(p)\eta_p^2 d^3r$$

where σ is the external load and ε is the average strain of the whole system.

The stochastic time-dependent Ginsburg-Landau equation is used for the time-evolution of the SO parameters:

$$\frac{\partial \eta_p(\boldsymbol{r},t)}{\partial t} = -M\frac{\delta F}{\delta \eta_p(\boldsymbol{r},t)} + \zeta_p(\boldsymbol{r},t),\ p = 1,2,\ldots n$$

where ζ is the noise term describing the thermal fluctuations, F is the total free energy, η is the structural order (SO) parameters, t is the time, M is the kinetic coefficient.

In numerical simulations [25], dimensionless parameters are often used, including reduced elastic strain energy $E^*_{\text{el}} = \frac{E_{\text{el}}}{\Delta f_{\text{scale}}}$ and elastic constants $C^*_{ij} = \frac{C_{ij}}{\Delta f_{\text{scale}}}$, reduced chemical free energy $f^*_{\text{ch}} = \frac{f_{\text{ch}}}{\Delta f_{\text{scale}}}$ and Landau expansion coefficients $A^*_i = \frac{A_i}{\Delta f_{\text{scale}}}$, reduced gradient energy $f^*_{\text{grad}} = \frac{f_{\text{grad}}}{\Delta f_{\text{scale}} l_0}$ and gradient energy coefficient $\beta^*_{ij} = \frac{\beta_{ij}}{\Delta f_{\text{scale}} l_0^2}$, reduced noise term $\zeta^* = \frac{\zeta}{\Delta f_{\text{scale}} M}$, and reduced length scale $r^* = \frac{r}{l_0}$ and time scale $t^* = \frac{tM\Delta f_{\text{scale}}}{M^*}$. Based on the relationship between velocity and driving force $v = M_{\text{in}}(-\Delta f)$ [26] of a dissipative process, and the relationship between physical interface mobility, M_{in}, and phase field mobility of the order parameter, M, we have $M = M_{\text{in}} \int_\delta \left(\frac{d\eta}{dx}\right)^2 dx = M_{\text{in}} \frac{\gamma_{\text{AM}}}{\beta}$, where δ (the integration limit) is the interface width in phase field simulations, γ_{AM} is the interfacial energy between austenite and martensite, β is the corresponding gradient energy coefficient.

The noise term in phase field simulations is assumed to be uncorrelated in space and time, it satisfy [25]:

$$\left\langle \zeta_p(\boldsymbol{r},t) \right\rangle = 0$$

$$\left\langle \zeta_p(\boldsymbol{r},t)\,\zeta_p(\boldsymbol{r}',t') \right\rangle = 2k_B T M \delta\left(t-t'\right)\left(\boldsymbol{r}-\boldsymbol{r}'\right)$$

In discrete form, we have:

$$\left\langle \zeta_p(n,m) \right\rangle = 0$$

$$\left\langle \zeta_p(n,m)\,\zeta_p\left(n',m'\right) \right\rangle = 2k_B T M \frac{\delta_{mm'}}{\Delta t}\frac{\delta_{nn'}}{l_0^d}$$

where n and m are the discrete spatial positions, Δt is the time step, l_0 is the grid size, d is the dimensionality of the space, and δ_{ij} is the Kronecker delta. The noise term can be emulated by a random number generator ρ_i with Gaussian distribution, and $\langle \rho_i \rangle = 0$ and $\langle \rho_i \rho_{i'} \rangle = \delta_{ii'}$, and we have $\zeta_p(n,m) = \sqrt{2k_B T M / \left(l_0^3 \Delta t\right)}\,\rho$.

According to above-described phase field simulations, temperature-dependent martensitic phase transition has been shown in Fig. 10.2. Figure 10.2 exhibits the volume fraction change upon cooling and heating for cubic to trigonal (R phase)

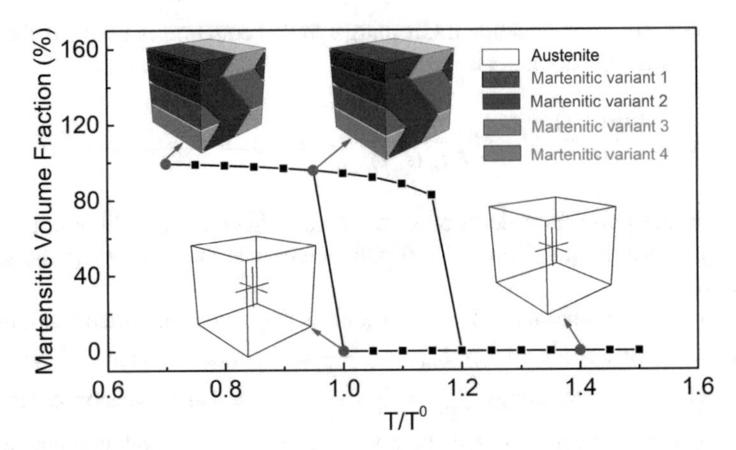

Fig. 10.2 Martensitic volume fraction vs. temperature and Microstructure evolution of martensitic phase transformation from Cubic to trigonal (R) by phase field simulations. White color describes the austenite and red, blue, green, and yellow color describes four martensitic variants respectively

martensitic transition in NiTi [22], which shows the formation of long-range ordered twinned martensite (inset picture) with large thermal hysteresis and sharp transition temperature.

10.1.2 Role of Point Defects

Point defects play an important role in changing transition temperature, transition sequence, and product phase for martensitic systems. It is believed that there exist three effects caused by point defects [27–31]: (1) global transition temperature effect (GTTE) caused by homogeneous composition change, which will influence the average phase transition temperature or the phase stability; (2) local transition temperature effect (LTTE) caused by inhomogeneous composition distribution, which will influence the local phase transition temperature or the local phase stability; (3) local field effect (LFE) caused by local stress/strain field associated with doped point defects, which will influence the local phase transition temperature and variants selections. Figure 10.3 shows the schematic pictures of the three effects caused by point defects. The GTTE or LTTE caused by point defects can be obtained by the expansion coefficient A_1 of Landau free energy in Eq. (2), $A_1 = A_1^0\left(T - T^0(c)\right)$, and $T^0(c) = T^{00} + bc$, where T^{00} is the phase transition temperature of defect-free materials (Fig. 10.3a), b is the strength of transition temperature effect associated with defect concentration. When $c = \overline{c}$, the equation describes the average composition and the GTTE (Fig. 10.3b), and if $c = c(\boldsymbol{r})$, the defect composition depends on position and describe the LTTE (Fig. 10.3c). In addition, doped point defects will produce lattice distortions $\varepsilon^{\mathrm{local}}(\boldsymbol{r})$ and generate excess energy $f_{\mathrm{LFE}}(\boldsymbol{r}) = C_{ijkl}\sum_{i,j,k,l}^{3}\varepsilon_{kl}^{\mathrm{local}}(\boldsymbol{r})\,\varepsilon_{ij}^{00}(\boldsymbol{r})$, which describe the LFE

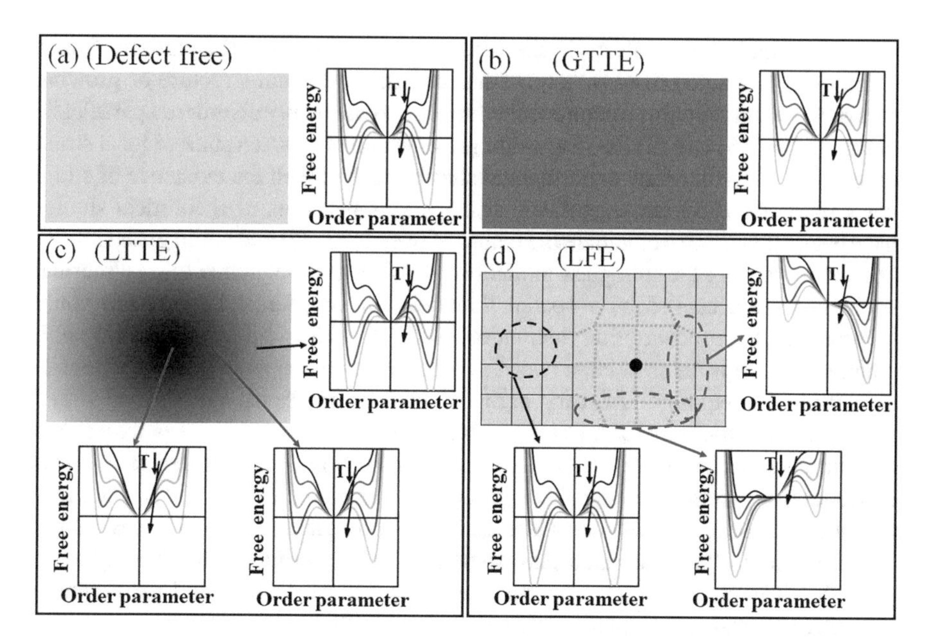

Fig. 10.3 Schematic pictures of effects caused by point defects. (**a**) The defect-free state with related Landau free energy. (**b**) GTTE with related Landau free energy. (**c**) LTTE with Landau free energy. Color in microstructure describes the different defect concentration. (**d**) LFE with related Landau free energy, black dot describes the defect position, and green lattice describes the lattice distortion caused by defects. Different colors in Landau free energy describe the free energy curves at different temperatures.

(Fig. 10.3d). Note that the schematic drawing of lattice distortion just shows local influence. The GTTE describes how the doped point defects alter the average thermodynamic stability of martensite and change the transition temperature of the whole system, and the LTTE describes such effect of spatial variation of the defect concentration (i.e., the composition inhomogeneity) and the Landau potential. The LFE associated with local lattice distortions caused by point defects creates spatial variation and symmetry-breaking of the Landau potential. Need to note that other extended defects such as precipitates, dislocations, and grain boundary may also produce composition change and lattice distortion, which also could influence the MT and strain glass transition.

10.2 Phase Field Simulation of Strain Glass Transition

Point defects such as dopants/impurities and vacancies have been known to play a key role in altering and controlling the transition behavior and designing novel properties of ferroelastic materials [32, 33]. In addition to the well-known paraelastic and ferroelastic states, it has been found that point defects can induce two other abnormal strain states: (1) the "precursory strain state" (or partially frozen strain order), characterized by a cross-hatched strain domain structure or growing nanosized strain domain structure embedded in a dynamically disordered paraelastic matrix [7, 34, 35] and (2) the new strain glass, which is a frozen state of local strain order [5, 6, 28, 36]. Many experimental studies have proven the existence of strain glass states for different ferroelastic systems [37–44]. Lots of theoretical studies have been attempted to elucidate the nature of tweed [27, 45] and strain glass [28, 30, 46] states by using the spin glass model [47, 48]. It was thought that these abnormal strain states are caused by a spatial fluctuation in M_S, i.e., LTTE, which could be produced by concentration fluctuation [18, 37]. At low defect concentration, it could produce spatial correlated tweed structure at high temperature and martensite at a lower temperature. At very high defect concentrations, the formed tweed structure could not transform to long-range ordered martensite and it is frozen into a strain glass state. However, experiments have shown that other defects without concentration fluctuation, such as high-density nano-precipitates and dislocations (as lattice distortion centers), can also generate strain glass [42, 49, 50], suggesting that local field effect may also play some role in the formation of strain glasses. To understanding the nature of strain glass, the role of defects, and the potential properties of strain glass, phase field simulations could be the best way.

Phase field simulations with considering two effects of point defects (LFE and GTTE) have been carried out [28], which reproduced a phase transition phase diagram (Fig. 10.4b) agreeing well with experimental observation (see Fig. 10.4a) [37]. According to the calculated phase diagram and related microstructure evolution in Fig. 10.5a, the strain glass transition can be understood easily. At high temperatures ($T > T_{nd}$), it is the austenite state which shows the existence of dynamic strain-domains in the system (i.e., strain liquid). When the temperature is lowered to $T < T_{nd}$, some dynamic strain-domains start to freeze (to long-range ordered martensite at $T < M_s$ or short-range ordered strain glass at $T < T_g$) and the system starts to lose ergodicity. For strain glass transition at high defect concentration, T_{nd} and T_g on defect concentration show opposite dependence, which can be attributed

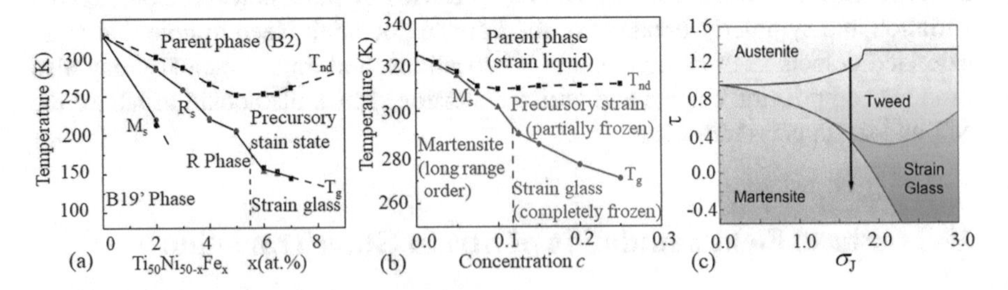

Fig. 10.4 (**a**) The experimental phase diagram of $Ti_{50}Ni_{50-x}Fe_x$ system [37]. R_s and M_s denote the martensitic transformation temperatures of R martensite and B19′ martensite, T_{nd} is the start temperature of static nanodomains, T_g is the strain glass transition temperature. (**b**) The calculated phase diagram according to phase field simulation [28]. (**c**) The calculated phase diagram in a temperature-disorder plane for spin model [51]

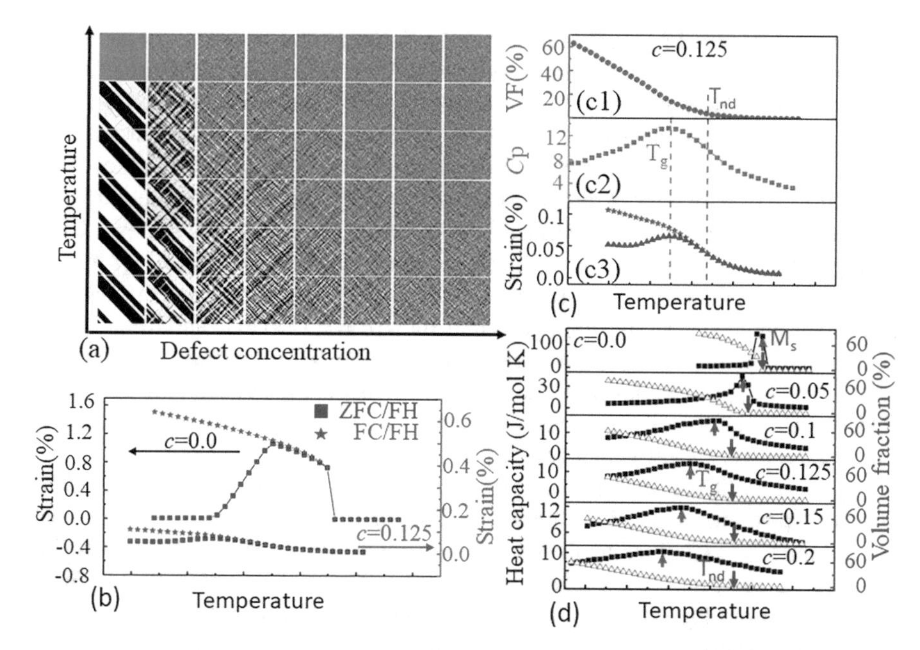

Fig. 10.5 (**a**) Microstructural evolution upon cooling for different defect concentration ($c = 0.0\sim0.2$). Gray color describes the parent phase, white and black color describe the two martensitic variants, respectively. (**b**) Comparison of zero-field cooling/field cooling curves for normal martensitic transformation ($c = 0.0$) and strain glass transition ($c = 0.125$). (**c1**) Volume fraction of nanosized martensite domains; (**c2**) heat capacity; (**c3**) ZFC/FC curves. (**d**) The heat capacity and martensite volume fraction curves at different defect concentrations. The arrows describe the peak temperatures (M_s or T_g) in heat capacity curves and T_{nd} in volume fraction curves

to the competition between the two effects: LFE and the GTTE. While the LFE is related to lattice distortion, which promotes the formation/freezing of local strain ordering (i.e., nanodomains) but prevents the formation of long-range ordered martensitic twins, the GTTE stabilizes the parent phase in our simulation (or Ti-Ni-Fe system [37]) and hence decreases M_s, T_g, and T_{nd}. The interplay of these two effects is the physical origin of the abnormal temperature dependence of T_{nd} and T_g as shown in Fig. 10.4b. Similar results can also be obtained by the spin model as shown in Fig. 10.4c [51].

Related microstructural evolution upon cooling from parent phase at different defect concentrations are shown in Fig. 10.5a. In the case of defect-free system ($c = 0.0$), the parent phase transformed into martensite with a typical long-range ordered twinned microstructure within a narrow temperature range. At low defect concentrations ($c = 0.025$–0.05), the system first generates some spatially correlated nanodomains of martensite and then transformed into the long-range ordered strain state (i.e., poly-twinned martensite) upon further cooling. When the defect concentration further increases ($c = 0.075$–0.2), the system will first form randomly distributed nanodomains and these domains will be frozen in nanoscale and did not transform into long-range ordered poly-twinned microstructure anymore upon cooling. These simulations are consistent with the experimental observation [5, 37, 39] that strain glass state is a static or frozen state of local lattice strains.

To understand the freezing process of strain glass transition at high defect concentration and capture the strain glass transition temperature, zero-field cooling/field cooling (ZFC/FC) calculation was carried out for the system as shown in Fig. 10.5b or Fig. 10.5(c3). Comparing with the sharp change of the ZFC/FC gap for normal martensitic transition ($c = 0.0$), the gradual increase of the gap between the ZFC and FC curves upon cooling for strain glass transition ($c = 0.125$) indicates a continuous breaking of the ergodicity of the strain glass system. This confirms the existence of the freezing process during the strain glass transition. Especially, the calculated ZFC/FC curves for strain glass are very similar to that obtained experimentally for strain glass [36], cluster-spin glass [52], and ferroelectric relaxor [53].

From the ZFC/FC curves, the glass transition temperature T_g is defined by the peak position in the ZFC curve [36, 54], and the strain glass freezing start temperature T_{nd} is defined by the branching point between ZFC and FC curves [37, 39]. Figure 10.5c shows the relationship among different properties (Fig. 10.5(c1) volume fraction, Fig. 10.5(c2) heat capacity and Fig. 10.5(c3) ZFC/FC), then the two glass transition characteristic temperatures T_g and T_{nd} have been labeled in the ZFC/FC curves. As the dash lines shown in Fig. 10.5c, we can easily find that glass freezing temperature is related to the maximum heat capacity value temperature (i.e., the peak position) and T_{nd} is related to the formation start temperature of martensitic domains. So the nanodomains occurrence temperatures and glass transition temperatures are determined by heat capacity and volume fraction curves in Fig. 10.5d. According to Fig. 10.5d, a complete phase diagram is established as shown in Fig. 10.4b including normal martensite, precursory strain state, and strain glass, which is in excellent agreement with the experimentally measured phase diagram for the $Ti_{50}Ni_{50-x}Fe_x$ system (Fig. 10.4a) [37, 39] and could help us to predict potential strain glass composition.

Besides the abovementioned defect concentration, defect strength and elastic anisotropy also influence the martensitic phase transition and strain glass transition. Lloveras et al. show that a decrease of anisotropy (strength of long-range interactions) may change the transition behavior and microstructural evolution as shown in Fig. 10.6a [30, 55]. Further study reports that the defect strength relative to the strength of the martensitic transformation plays an important role as shown in Fig. 10.6b [56]. This finding may shed light on why there is no report of the B19'/B19 strain glass in NiTi systems through point defect doping. B19'/B19 phase has large transformation strain ($\sim$10%) that is too large for point defects to confine the transformation when it starts. In our understanding, low strength of long-range interactions caused by martensitic phase and high strength/concentration of disorder caused by doped defects should be the origin of strain glass transition. Figure 10.7a shows the randomly spatial distribution of phase field voxels containing defects in the system and Fig. 10.7b–f shows the related spatial distribution of local Von Mises stress caused by doped defects with different defect strength. With the

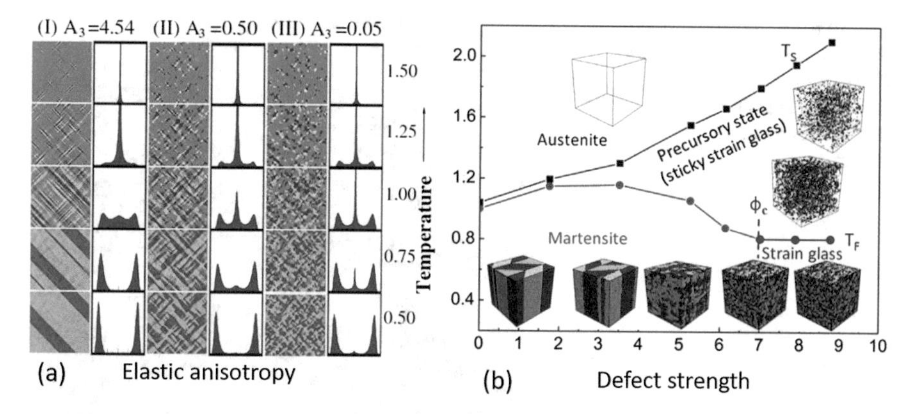

Fig. 10.6 (**a**) Elastic anisotropy's effect on martensitic phase transition with fixed defect concentration. (**b**) Defect strength's effect on martensitic phase transition with fixed defect concentration

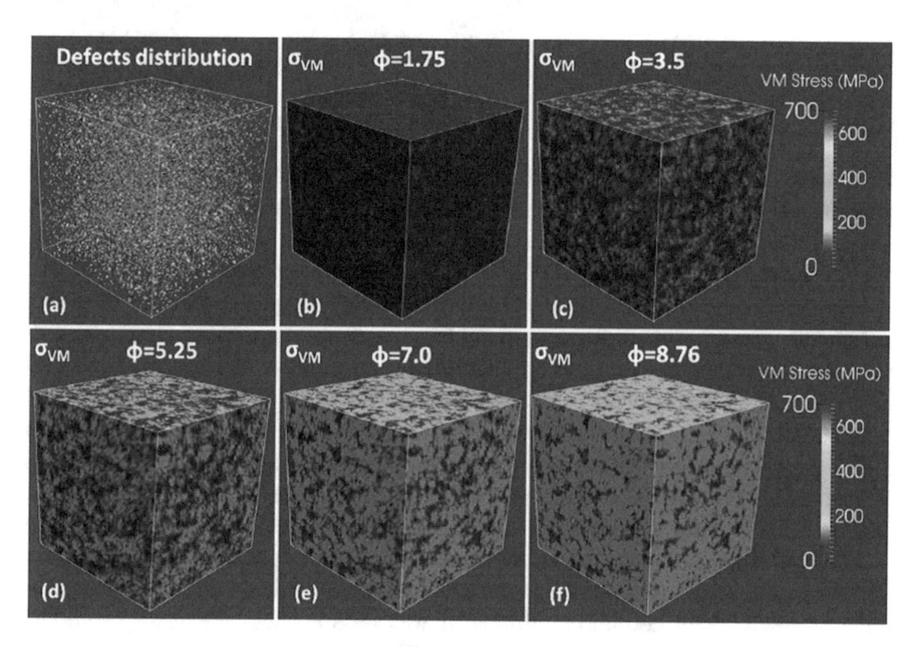

Fig. 10.7 (**a**) Spatial distribution of simulation grids containing point defects (white color). (**b–f**) Spatial distribution of local Von Mises stress σ_{VM} caused by point defects with different strength coefficient ϕ

increase of defect strength, the distribution of local stress caused by defects becomes more and more inhomogeneous/percolation and the volume of high-stress regions increase and these regions are separated by low-stress regions when defect strength coefficient is high. High-stress regions prefer the nucleation of certain M domains but limit its growth by the other local stress regions. The randomly distributed

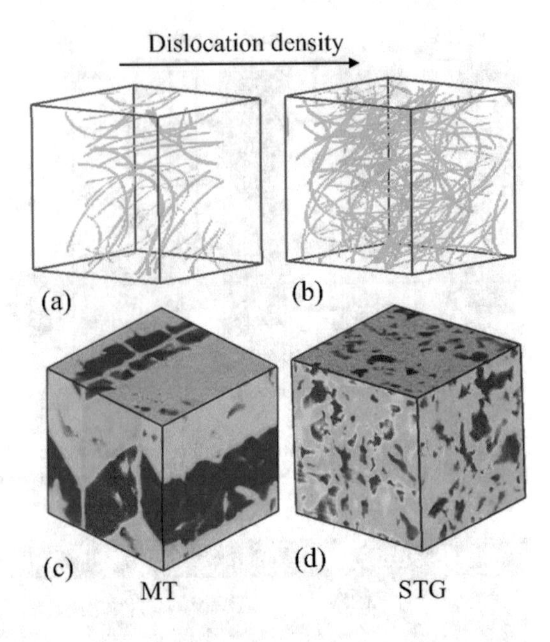

Fig. 10.8 (**a**, **b**) Pre-produced dislocation with different density. (**c**, **d**) Related martensitic microstructure at low temperature

local stress field and its percolation state influences the nucleation and growth of martensitic domains and should be the physical origin of strain glass transition and the existence of nanodomains.

For a given ferroelastic system, the elastic anisotropy is fixed, the way to produce strain glass is to adjust defect concentration and defect type (i.e., defect strength). Extended defects with high strength/density such as dislocations, grain boundaries, and precipitates could be a potential way to produce strain glass transition [56]. Figure 10.8 shows the dislocation density's effect on R martensitic phase transition; high-density dislocation could break the normal martensitic transition (MT) with long-range ordered martensitic twin structure and produce strain glass transition (STG) with short-range ordered martensitic nanodomain states as shown in Fig. 10.8. Similar experimental results and simulations have confirmed that the dislocation could influence martensitic transition and produce strain glass [42, 50].

LTTE could also influence the phase transition if we could produce a composition inhomogeneity with certain conditions. It was reported that precursory spinodal decomposition could be an efficient way [57, 58]. Figure 10.9a shows the simulation results of concentration evolution with spinodal decomposition in a single-crystal Ti2448 (with an average composition of Ti-15 at %Nb). Figure 10.9b, c shows the concentration wave along the body-diagonal of the computational cell and the statistical distribution of defect concentrations (Nb) with different aging time respectively. The Nb-lean and Nb-rich regions form a typical network structure due to spinodal decomposition. This concentration distribution leads to different stress-strain (SS) curves (in terms of the hysteresis and critical stress for the MT, and the pseudo-elastic behavior) for systems. Before aging (i.e., $t^* = 0$), the SS curve in Fig. 10.9(d1) shows a large stress hysteresis and obvious stress

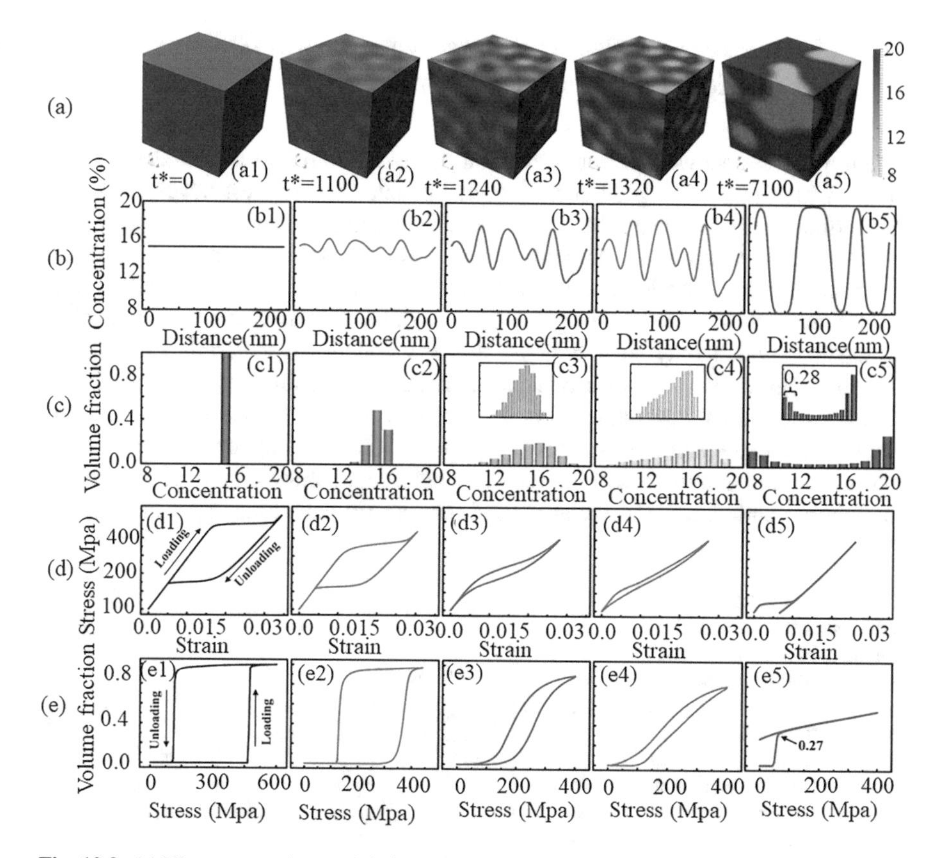

Fig. 10.9 (**a**) Nb concentration modulation induced by spinodal decomposition in the parent phase with different aging time. (**b1–b5**) One-dimensional Nb concentration profiles along the body-diagonal of the computation cells shown in (**a1–a5**). (**c1–c5**) The statistical distributions of Nb concentration in the computational cell of (**a1–a5**). (**d1–d5**) The stress-strain curves under uniaxial tension shown in (**a1–a5**). (**e1–e5**) The normalized volume fraction of martensite during cyclic loading shown in (**a1–a5**)

plateau (i.e., strongly nonlinear pseudo-elasticity) which are typical characteristics of conventional martensitic phase. However, the SS curves become slim ones with narrow hysteresis with the increase of aging time, as shown in Fig. 10.9(d2–d4). The volume fraction of martensite vs. stress curve shows a similar change from a square-like loop (sharp changes) (Fig. 10.9(e1)) to a slim and smooth loop (gradual changes) (Fig. 10.9(e3, e4), again demonstrating a transition from a sharp first-order MT to a higher-order like continuous transition. However, this simulation needs more experimental work to prove whether it is a strain glass transition.

10.3 Unique Properties Associated with Strain Glass Transition and Strain Glass State

Strain glass has shown different transition behavior from normal martensitic transition, which may solve previously reported puzzles and suggest novel properties. According to the microstructural evolution, the continuous nucleation and growth of nanoscaled martensitic domains from strain glass transition could produce continuous volume/strain change, or modulus change and slim hysteresis. Simulations have shown that strain glass could exhibit slim hysteresis upon cooling or loading, which may shed light on the new functional materials [22, 42, 44, 57, 59]. Figure 10.10a shows that the martensitic nanodomains volume fraction changes are almost reversible upon cooling and heating for strain glass system, showing a narrow thermal hysteresis. Figure 10.10b shows the enlarged images of microstructural evolution in Fig. 10.10a, which could show a gradual increase of the number and size for irregular shaped martensitic nanodomains in the systems

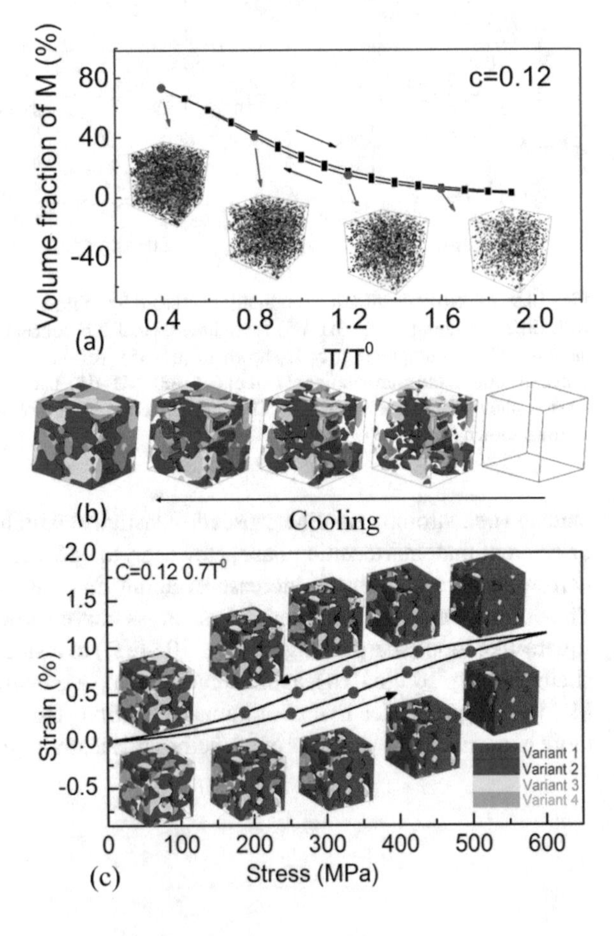

Fig. 10.10 (**a**) Calculated strain glass transition with volume fraction change and microstructural evolution upon cooling and heating. (**b**) Portions of the enlarged image of the microstructures shown in (**a**). (**c**) Calculated stress-strain curve for strain glass system and related corresponding microstructural evolution (to the red dots). Four colors represent the four variants of martensite

upon cooling, and the final domain size is limited by the randomly distributed point defects. Gradual nucleation caused by the continuous distribution of local stress field by point defects can be attributed to the physical origin of continuous transition characteristics of strain glass transition. High concentration of point defects or other defects also plays role in limiting the growth space of formed martensitic domains, which results in the existence of frozen nanoscaled martensitic domains. Different from the large hysteresis of stress-strain curves in the normal martensitic system, the stress-strain curve in strain glass system show superelasticity with narrow hysteresis without stress plateau and nearly zero remnant strain (see Fig. 10.10c). Insets of Fig. 10.10c show the microstructures evolution corresponding to the red dots in a stress-strain curve during loading and unloading. Upon loading, the systems evolve gradually from a state consisting of nanodomains of all martensitic variants having approximately equal volume fraction into nanodomains consisting dominantly of certain variant (red) that is favored by the load. Upon unloading, local fields associated with point defects tend to restore gradually the initial multi-variants state.

In addition, continuous strain glass transition could provide continuous volume change or strain change over a wide temperature range, which may help design tailorable Invar [38, 60] or Elinvar materials [61, 62]. Calculated temperature dependence of elastic modulus (including normal modulus hardening caused by anharmonic atomic vibration and softening caused by phase transition) with different defect concentrations are shown in Fig. 10.11. A normal modulus hardening with a constant thermoelastic coefficient is assumed for the model system. At low defect concentration (e.g., $X = 0.09$), the simulation result (open squares) shows a sharp decrease in the elastic modulus at the transition temperature and then a steady increase after the transition. A narrow and sharp peak is shown in the corresponding thermoelastic coefficient (Fig. 10.11b) (i.e., the derivative of the modulus with respect to temperature). With the increase of the defect concentration (e.g., $X = 0.30$), the sharp thermoelastic coefficient peak changes into a broadly smeared peak. When the defect concentration reaches a certain value, e.g., $X = 0.45$, the elastic modulus show almost invariant (open triangles in Fig. 10.11a) and the thermoelastic coefficient (Fig. 10.11b) is nearly zero over a wide temperature range ($\sim$100 K). Figure 10.11c shows the change in volume fraction of the M phase, and a gradual change of the volume fraction occurs at high defect concentration. These results suggest that impurity doping in ferroelastic systems is an effective way to adjust MT characteristics and tailor the thermoelastic coefficients.

10.4 Challenge and Opportunity

Strain glass not only exhibits important theoretical significance but also show wide applications. However, current strain glass system faces one key problem: small recoverable strain as shown in Fig. 10.12, i.e., local transition strain of nanoscaled martensitic domains is small. For example, there is no report of the existence of B19 or B19' (transformation strain $\sim$8%) strain glass in point defect doped NiTi

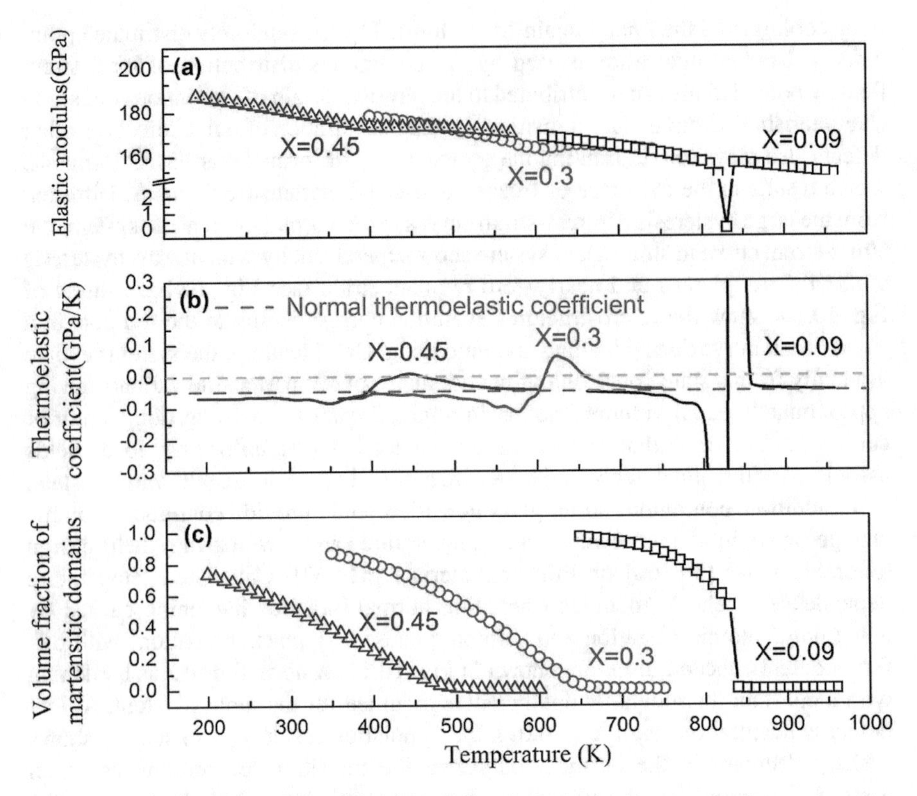

Fig. 10.11 Variations of (**a**) elastic modulus, (**b**) corresponding thermoelastic coefficient, and (**c**) volume fraction of martensitic domains as a function of temperature in doped ferroelastic systems have different point defect composition

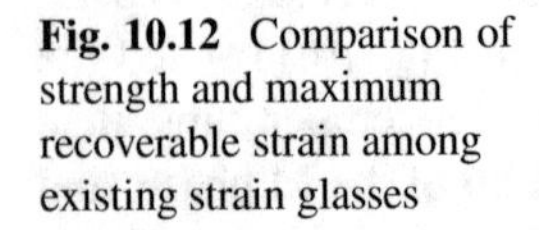

Fig. 10.12 Comparison of strength and maximum recoverable strain among existing strain glasses

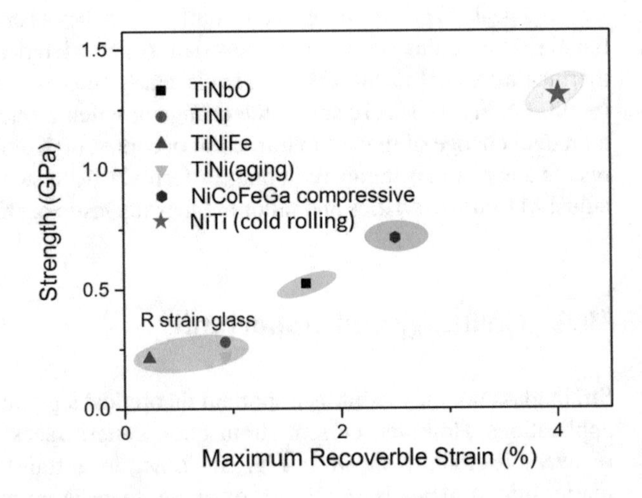

systems. Theoretical work has shown that both long-range interaction and defect strength/concentration play an important role. For certain system which includes martensitic phase with large transformation strain will have strong long-range interaction, the possible way to produce strain glass is to increase the defect strength and concentration. Based on the theoretical guidance, recent experimental work design materials with introducing the high strength dislocations by cold rolling and produce such B19' strain glass with large recoverable strain [42]. However, more work to help predict and design novel strain glass system with better properties is still a challenge. With the development of computer technology and method, simulations could be an efficient way to guide the new materials design with high speed and low cost in future.

References

1. K. Ōtsuka, C.M. Wayman, *Shape Memory Materials* (Cambridge University Press, Cambridge, 1998)
2. A.R. Pelton et al., Nitinol medical devices. Adv Mater Process **163**, 63–65 (2005)
3. T.W. Duerig, A.R. Pelton, An overview of superelastic stent design. Mater Sci Forum **394-3**, 1–8 (2001)
4. T.W. Duerig, *Engineering Aspects of Shape Memory Alloys* (Butterworth-Heinemann, Oxford, 1990)
5. S. Sarkar, X.B. Ren, K. Otsuka, Evidence for strain glass in the ferroelastic-martensitic system Ti50-xNi50 + x. Phys Rev Lett **95**, 205702 (2005). https://doi.org/10.1103/Physrevlett.95.205702
6. Y. Wang, X.B. Ren, K. Otsuka, Shape memory effect and superelasticity in a strain glass alloy. Phys Rev Lett **97**, 225703 (2006). https://doi.org/10.1103/Physrevlett.97.225703
7. X.B. Ren et al., Strain glass in ferroelastic systems: Premartensitic tweed versus strain glass. Philos Mag **90**, 141–157 (2010). https://doi.org/10.1080/14786430903074771
8. L.-Q. Chen, Phase-field models for microstructure evolution. Annual Review of Materials Research **32**, 113–140 (2002). https://doi.org/10.1146/annurev.matsci.32.112001.132041
9. A.G. Khachaturi͡an, *Theory of Structural Transformations in Solids* (Wiley, Hoboken, 1983)
10. D. Raabe, *Continuum Scale Simulation of Engineering Materials: Fundamentals, Microstructures, Process Applications* (Wiley-VCH, Hoboken, 2004), pp. 271–296
11. Y. Wang, J. Li, Phase field modeling of defects and deformation. Acta Materialia **58**, 1212–1235 (2010). https://doi.org/10.1016/j.actamat.2009.10.041
12. I. Steinbach, Phase-field model for microstructure evolution at the mesoscopic scale. Annual Review of Materials Research **43**, 89–107 (2013). https://doi.org/10.1146/annurev-matsci-071312-121703
13. K. Elder, H. Gould, J. Tobochnik, Langevin simulations of nonequilibrium phenomena. Computers in Physics **7**, 27–33 (1993). https://doi.org/10.1063/1.4823138
14. A. Artemev, Y. Jin, A.G. Khachaturyan, Three-dimensional phase field model of proper martensitic transformation. Acta Materialia **49**, 1165–1177 (2001). https://doi.org/10.1016/S1359-6454(01)00021-0
15. Y. Wang, A.G. Khachaturyan, Three-dimensional field model and computer modeling of martensitic transformations. Acta Materialia **45**, 759–773 (1997). https://doi.org/10.1016/S1359-6454(96)00180-2
16. Y.L. Li, S.Y. Hu, Z.K. Liu, L.Q. Chen, Phase-field model of domain structures in ferroelectric thin films. Applied Physics Letters **78**, 3878–3880 (2001). https://doi.org/10.1063/1.1377855

17. J. Wang, S.-Q. Shi, L.-Q. Chen, Y. Li, T.-Y. Zhang, Phase-field simulations of ferroelectric/ferroelastic polarization switching. Acta Materialia **52**, 749–764 (2004). https://doi.org/10.1016/j.actamat.2003.10.011
18. L.Q. Chen, Phase-field method of phase transitions/domain structures in ferroelectric thin films: A review. J Am Ceram Soc **91**, 1835–1844 (2008). https://doi.org/10.1111/j.1551-2916.2008.02413.x
19. L.J. Li, C.H. Lei, Y.C. Shu, J.Y. Li, Phase-field simulation of magnetoelastic couplings in ferromagnetic shape memory alloys. Acta Materialia **59**, 2648–2655 (2011). https://doi.org/10.1016/j.actamat.2011.01.001
20. Y.M. Jin, Domain microstructure evolution in magnetic shape memory alloys: Phase-field model and simulation. Acta Materialia **57**, 2488–2495 (2009). https://doi.org/10.1016/j.actamat.2009.02.003
21. Y.Z. Wang, A.G. Khachaturyan, Multi-scale phase field approach to martensitic transformations. Mat. Sci. Eng. A **438**, 55–63 (2006). https://doi.org/10.1016/j.msea.2006.04.123
22. D. Wang et al., Superelasticity of slim hysteresis over a wide temperature range by nanodomains of martensite. Acta Materialia **66**, 349–359 (2014). https://doi.org/10.1016/j.actamat.2013.11.022
23. Y.P. Gao, R.P. Shi, J.F. Nie, S.A. Dregia, Y.Z. Wang, Group theory description of transformation pathway degeneracy in structural phase transformations. Acta Materialia **109**, 353–363 (2016). https://doi.org/10.1016/j.actamat.2016.01.027
24. J.W. Cahn, J.E. Hilliard, Free energy of a nonuniform system. I. Interfacial free energy. The Journal of Chemical Physics **28**, 258–267 (1958). https://doi.org/10.1063/1.1744102
25. C. Shen, J.P. Simmons, Y. Wang, Effect of elastic interaction on nucleation: II. Implementation of strain energy of nucleus formation in the phase field method. Acta Materialia **55**, 1457–1466 (2007). https://doi.org/10.1016/j.actamat.2006.10.011
26. C.-J. Huang, D.J. Browne, S. McFadden, A phase-field simulation of austenite to ferrite transformation kinetics in low carbon steels. Acta Materialia **54**, 11–21 (2006). https://doi.org/10.1016/j.actamat.2005.08.033
27. S. Semenovskaya, A.G. Khachaturyan, Coherent structural transformations in random crystalline systems. Acta Materialia **45**, 4367–4384 (1997). https://doi.org/10.1016/S1359-6454(97)00071-2
28. D. Wang, Y.Z. Wang, Z. Zhang, X.B. Ren, Modeling abnormal strain states in ferroelastic systems: the role of point defects. Phys Rev Lett **105**, 205702 (2010). https://doi.org/10.1103/Physrevlett.105.205702
29. A.P. Levaniuk, A.S. Sigov, *Defects and Structural Phase Transitions* (Gordon and Breach Science Publishers, Philadelphia, 1988)
30. P. Lloveras, T. Castan, M. Porta, A. Planes, A. Saxena, Influence of elastic anisotropy on structural nanoscale textures. Phys Rev Lett **100**, 165707 (2008). https://doi.org/10.1103/Physrevlett.100.165707
31. D. Wang, X.B. Ren, Y.Z. Wang, Nanoscaled martensitic domains in ferroelastic systems: strain glass. Curr Nanosci **12**, 192–201 (2016). https://doi.org/10.2174/1573413711666150523001617
32. K. Otsuka, X. Ren, Physical metallurgy of Ti–Ni-based shape memory alloys. Progress in Materials Science **50**, 511–678 (2005). https://doi.org/10.1016/j.pmatsci.2004.10.001
33. E.K.H. Salje, *Phase Transitions in Ferroelastic and Co-Elastic Crystals: An Introduction for Mineralogists, Material Scientists, and Physicists* (Cambridge University Press, Cambridge, 1990)
34. S.M. Shapiro, J.Z. Larese, Y. Noda, S.C. Moss, L.E. Tanner, Neutron-scattering study of premartensitic behavior in Ni-Al alloys. Phys Rev Lett **57**, 3199–3202 (1986). https://doi.org/10.1103/PhysRevLett.57.3199
35. D. Shindo, Y. Murakami, T. Ohba, Understanding precursor phenomena for the R-phase transformation in Ti-Ni-based alloys. Mrs Bull **27**, 121–127 (2002). https://doi.org/10.1557/Mrs2002.48

36. Y. Wang, X. Ren, K. Otsuka, A. Saxena, Evidence for broken ergodicity in strain glass. Phys Rev B **76**, 132201 (2007). https://doi.org/10.1103/Physrevb.76.132201
37. D. Wang et al., Strain glass in Fe-doped Ti-Ni. Acta Materialia **58**, 6206–6215 (2010). https://doi.org/10.1016/j.actamat.2010.07.040
38. Y. Wang et al., Strain glass transition in a multifunctional beta-type Ti alloy. Sci Rep **4**, 3995 (2014). https://doi.org/10.1038/Srep03995
39. Z. Zhang et al., Phase diagram of Ti50-xNi50 + x: Crossover from martensite to strain glass. Phys Rev B **81**, 224102 (2010). https://doi.org/10.1103/Physrevb.81.224102
40. Y. Zhou et al., Strain glass in doped Ti50(Ni50 − xDx) (D = Co, Cr, Mn) alloys: Implication for the generality of strain glass in defect-containing ferroelastic systems. Acta Materialia **58**, 5433–5442 (2010). https://doi.org/10.1016/j.actamat.2010.06.019
41. Y.M. Zhou et al., High temperature strain glass in Ti-50(Pd50-xCrx) alloy and the associated shape memory effect and superelasticity. Applied Physics Letters **95**, 151906 (2009). https://doi.org/10.1063/1.3249580
42. Q. Liang et al., Novel B19\ensuremath{'} strain glass with large recoverable strain. Phys. Rev. Mater. **1**, 033608 (2017)
43. J. Liu et al., Strain glassy behavior and premartensitic transition in Au${}_{7}$Cu${}_{5}$Al${}_{4}$ alloy. Phys Rev B **84**, 140102 (2011)
44. Y. Nii, T.-h. Arima, H.Y. Kim, S. Miyazaki, Effect of randomness on ferroelastic transitions: Disorder-induced hysteresis loop rounding in Ti-Nb-O martensitic alloy. Phys Rev B **82**(214), 104 (2010)
45. S. Kartha, T. Castan, J.A. Krumhansl, J.P. Sethna, Spin-glass nature of tweed precursors in martensitic transformations. Phys Rev Lett **67**, 3630–3633 (1991). https://doi.org/10.1103/PhysRevLett.67.3630
46. R. Vasseur, T. Lookman, Effects of disorder in ferroelastics: A spin model for strain glass. Phys Rev B **81**, 094107 (2010). https://doi.org/10.1103/Physrevb.81.094107
47. D. Sherrington, S. Kirkpatrick, Solvable model of a spin-glass. Phys Rev Lett **35**, 1792–1796 (1975). https://doi.org/10.1103/PhysRevLett.35.1792
48. D. Sherrington, A simple spin glass perspective on martensitic shape-memory alloys. J. Phys. Condens. Matter **20**, 304213 (2008). https://doi.org/10.1088/0953-8984/20/30/304213
49. Y.C. Ji, X.D. Ding, T. Lookman, K. Otsuka, X.B. Ren, Heterogeneities and strain glass behavior: Role of nanoscale precipitates in low-temperature-aged Ti48.7Ni51.3 alloys. Phys Rev B **87**, 104110 (2013). https://doi.org/10.1103/Physrevb.87.104110
50. J. Zhang et al., Dislocation induced strain glass in Ti50Ni45Fe5 alloy. Acta Materialia **120**, 130–137 (2016). https://doi.org/10.1016/j.actamat.2016.08.015
51. R. Vasseur et al., Phase diagram of ferroelastic systems in the presence of disorder: Analytical model and experimental verification. Phys Rev B **86**, 184103 (2012). https://doi.org/10.1103/Physrevb.86.184103
52. N. Gayathri et al., Electrical transport, magnetism, and magnetoresistance in ferromagnetic oxides with mixed exchange interactions: A study of the La0.7Ca0.3Mn1-xCoxO3 system. Phys. Rev. B **56**, 1345–1353 (1997). https://doi.org/10.1103/PhysRevB.56.1345
53. D. Viehland, J.F. Li, S.J. Jang, L.E. Cross, M. Wuttig, Glassy Polarization Behavior Of Relaxor Ferroelectrics. Phys Rev B **46**, 8013–8017 (1992). https://doi.org/10.1103/PhysRevB.46.8013
54. J.A. Mydosh, *Spin Glasses: An Experimental Introduction* (Taylor & Francis, Oxfordshire, 1993)
55. P. Lloveras, T. Castán, M. Porta, A. Planes, A. Saxena, Thermodynamics of stress-induced ferroelastic transitions: Influence of anisotropy and disorder. Phys Rev B **81**, 214105 (2010)
56. D. Wang et al., Defect strength and strain glass state in ferroelastic systems. Journal of Alloys and Compounds **661**, 100–109 (2016). https://doi.org/10.1016/j.jallcom.2015.11.095
57. J. Zhu, Y. Gao, D. Wang, T.-Y. Zhang, Y. Wang, Taming martensitic transformation via concentration modulation at nanoscale. Acta Materialia **130**, 196–207 (2017). https://doi.org/10.1016/j.actamat.2017.03.042

58. D. Wang et al., Integrated computational materials engineering (ICME) approach to design of novel microstructures for Ti-alloys. JOM **66**, 1287–1298 (2014). https://doi.org/10.1007/s11837-014-1011-2
59. S.-J. Qin, J.-X. Shang, F.-H. Wang, Y. Chen, The role of strain glass state in the shape memory alloy Ni50 + xTi50 − x: Insight from an atomistic study. Materials & Design **120**, 238–254 (2017). https://doi.org/10.1016/j.matdes.2017.02.011
60. X.B. Ren, Strain glass and ferroic glass - Unusual properties from glassy nano-domains. Phys. Status Solidi B Basic Solid State Phys. **251**, 1982–1992 (2014). https://doi.org/10.1002/pssb.201451351
61. L. Zhang, D. Wang, X. Ren, Y. Wang, A new mechanism for low and temperature-independent elastic modulus. Sci Rep **5**, 11477 (2015). https://doi.org/10.1038/srep11477
62. J. Cui, X. Ren, Elinvar effect in co-doped TiNi strain glass alloys. Applied Physics Letters **105**, 061904 (2014). https://doi.org/10.1063/1.4893003

Index

T. Lookman, X. Ren (eds.), *Frustrated Materials and Ferroic Glasses*, Springer Series in Materials Science 275, https://doi.org/10.1007/978-3-319-96914-5

Printed in the United States
By Bookmasters